BENCHMARKS

FOR SCIENCE LITERACY

BENCHMARKS

FOR SCIENCE LITERACY

PROJECT 2061

AMERICAN ASSOCIATION FOR THE ADVANCEMENT OF SCIENCE

OXFORD UNIVERSITY PRESS

NEW YORK   OXFORD

1993

OXFORD UNIVERSITY PRESS

OXFORD   NEW YORK   TORONTO

DELHI   BOMBAY   CALCUTTA   MADRAS   KARACHI

KUALA LUMPUR   SINGAPORE   HONG KONG   TOKYO

NAIROBI   DAR ES SALAAM   CAPE TOWN

MELBOURNE   AUCKLAND   MADRID

AND ASSOCIATED COMPANIES IN

BERLIN   IBADAN

Library of Congress Cataloging-in-Publication Data
Benchmarks for science literacy.
p.   cm.
"Project 2061, American Association for the Advancement of Science."
Includes bibliographical references and index.
ISBN 0-19-508986-3
1. Science–Study and teaching–United States. 2. Mathematics–Study and teaching–United States.
3. Engineering–Study and teaching–United States. I. Project 2061 (American Association for the Advancement of Science)
Q183.3.A1B46 1993
507'.1'273–dc20
93-39107

8 9 7

Printed in the United States of America
on acid-free paper

Table of Contents

PROJECT 2061 FUNDERS

CARNEGIE CORPORATION OF NEW YORK

JOHN D. AND CATHERINE T. MACARTHUR FOUNDATION

ANDREW W. MELLON FOUNDATION

ROBERT N. NOYCE FOUNDATION

THE PEW CHARITABLE TRUSTS

INTERNATIONAL BUSINESS MACHINES CORPORATION

NATIONAL SCIENCE FOUNDATION

U.S. DEPARTMENT OF EDUCATION

CALIFORNIA STATE DEPARTMENT OF EDUCATION

GEORGIA DEPARTMENT OF EDUCATION

TEXAS EDUCATION AGENCY

WISCONSIN DEPARTMENT OF PUBLIC INSTRUCTION

The American Association for the Advancement of Science
wishes to express its gratitude to the major funders listed above and
to the many other businesses, school districts, and individuals who contributed
to Project 2061 and the work leading to this report.
The opinions, findings, conclusions,
and recommendations expressed in *Benchmarks for Science Literacy*
are those of Project 2061 and do not necessarily reflect
the views of these organizations.

PREFACE

Four years ago, 150 teachers and administrators in six school districts accepted a daunting challenge. They were asked to help shape the future of education in America, a future in which *all* students would become literate in science, mathematics, and technology by graduation from high school. That future is still some time away, to be sure, but this report—largely the result of the vision, commitment, and inventiveness of those in-the-trenches educators—gets us substantially closer to it.

Benchmarks for Science Literacy does this not by offering a standard curriculum to be adopted locally but by providing educators in every state and school district with a powerful tool to use in fashioning their own curricula. We believe that they will respond with zest and imagination to the opportunity to bring their own insights and experience to bear on the task of moving toward the grade-level standards for science literacy presented in this report from the American Association for the Advancement of Science.

This has been truly a grass-roots effort. As the Participants list shows, an unprecedented number of elementary-, middle-, and high-school teachers, school administrators, scientists, mathematicians, engineers, historians, and learning specialists participated in the development of *Benchmarks* and in its nationwide critique.

The Project 2061 staff brought vision and determination to the effort. They saw to it that the school-district teams were properly supported and that outstanding specialists were recruited to work with the teams and staff. Individual staff members made their own distinctive contributions to *Benchmarks* while working effectively with the teams and consultants to ensure that the best ideas emerging from the collaborative process were captured and faithfully expressed. Project 2061's advisory board, the National Council on Science and Technology Education, reshaped in 1990 to reflect the nature of the work ahead and led by Franklyn Jenifer, president of Howard University, provided precisely the inspiration and guidance that was needed to keep this ambitious, long-term project on track.

But none of this would have been possible without the generous support of private foundations, state and federal agencies, and business. On behalf of the National Council, AAAS, and Project 2061 staff, team members, and consultants, I would like to thank the organizations listed on the preceding page. We realize that in a nation that prizes quick results, it is not easy to support long-term, non-traditional efforts like this one, and hence our gratitude is all the more deeply felt.

F. JAMES RUTHERFORD
Director, Project 2061

NATIONAL COUNCIL ON SCIENCE AND TECHNOLOGY EDUCATION

CHAIRMAN   **Franklyn G. Jenifer** *President*, Howard University

MEMBERS   **Bill G. Aldridge** *Executive Director*, National Science Teachers Association

Raul Alvarado, Jr. *Senior Engineering Scientist*, McDonnell-Douglas Corporation

Francisco J. Ayala *Donald Bren Professor of Biological Sciences*, University of California, Irvine

William O. Baker *Retired Chairman of the Board*, AT&T Bell Telephone Laboratories

Catherine A. Belter *Chair, PTA Education Commission*, The National PTA

Diane J. Briars *Director*, Division of Mathematics, Pittsburgh Public Schools

Patricia L. Chavez *Statewide Executive Director*, New Mexico MESA, Inc.

Joan Duea *Professor of Education*, University of Northern Iowa

Stuart Feldman *Division Manager*, Computer Systems Research, Bellcore

Ernestine Friedl *James B. Duke Professor Emeritus*, Department of Cultural Anthropology, Duke University

Linda Froschauer *Teacher*, Weston Middle School, Connecticut

Mary Hatwood Futrell* *Senior Fellow and Associate Director*, Center for the Study of Education, The George Washington University

Patsy D. Garriott *Education Initiatives Representative*, Eastman Chemical Company

Robert Gauger *Chair*, Technology Department, Oak Park and River Forest High School, Illinois

Shirley A. Hill *Professor of Education and Mathematics*, University of Missouri-Kansas City

David Hornbeck* *Counsel*, Hogan & Hartson, Washington, DC

Gregory A. Jackson *Director*, Academic Computing, Massachusetts Institute of Technology

Cherry H. Jacobus *Former President*, Michigan State Board of Education

Donald W. Jones* *Deputy Assistant Secretary of Defense*, Military Manpower and Personnel Policy

Vera Katz* *Speaker of the House*, State of Oregon

David Kennedy *State Science Supervisor,* Washington

George J. Kourpias *President,* International Association of Machinists and Aerospace Workers

Ann Lynch* *President,* National PTA

Kenneth Manning *Professor of the History of Science,* Massachusetts Institute of Technology

Margaret L. A. MacVicar* *Dean for Undergraduate Studies,* Massachusetts Institute of Technology

Jose F. Mendez *President,* Ana G. Mendez Educational Foundation, Puerto Rico

Freda Nicholson *Executive Director,* Science Museums of Charlotte, Inc., North Carolina

James R. Oglesby* *Assistant to the Chancellor,* University of Missouri-Columbia

Gilbert S. Omenn *Dean,* School of Public Health and Community Medicine, University of Washington

Lee Etta Powell *Professor of Education,* The George Washington University

Vincent E. Reed *Vice President,* Communications, The Washington Post

Thomas Romberg *Director,* Education Research Center, University of Wisconsin

Mary Budd Rowe *Professor of Science Education,* Stanford University

F. James Rutherford *Chief Education Officer and Director,* Project 2061, American Association for the Advancement of Science

David Sanchez *Deputy Associate Director for Research and Education,* Los Alamos National Laboratory

Albert Shanker *President,* American Federation of Teachers

Gloria Takahashi *Teacher,* Science Department, La Habra High School, La Habra, California

Walter Waetjen *Chair,* Technology Education Advisory Council, International Technology Education Association

William Winter *Attorney-at-Law,* Watkins Ludlam & Stennis, Jackson, Mississippi

John Zola *Teacher,* Social Sciences, Fairview High School, Boulder, Colorado

* FORMER MEMBER, AFFILIATION AT THE TIME OF SERVICE

ABOUT BENCHMARKS

The terms and circumstances of human existence can be expected to change radically during the next human life span. Science, mathematics, and technology will be at the center of that change—causing it, shaping it, responding to it. Therefore, they will be essential to the education of today's children for tomorrow's world.

What should the substance and character of such education be?

The purpose of this report is to propose an answer to that question.

That was how *Science for All Americans (SFAA)*, the first Project 2061 publication, identified itself. Now, four years later, those words serve equally well to introduce *Benchmarks for Science Literacy*, a companion report. *SFAA* answers the question of what constitutes adult science literacy, recommending what all students should know and be able to do in science, mathematics, and technology by the time they graduate from high school. *Benchmarks* specifies how students should progress toward science literacy, recommending what they should know and be able to do by the time they reach certain grade levels. Together, the two publications can help guide reform in science, mathematics, and technology education.

ABOUT PROJECT 2061

Project 2061's benchmarks are statements of what *all* students should know or be able to do in science, mathematics, and technology by the end of grades 2, 5, 8, and 12. The grade demarcations suggest reasonable checkpoints for estimating student progress toward the science literacy goals outlined in *SFAA*. It is important to view the benchmarks in the context of the following Project 2061 premises concerning curriculum reform:

- Project 2061 promotes literacy in science, mathematics, and technology in order to help people live interesting, responsible, and productive lives. In a culture increasingly pervaded by science, mathematics, and technology, science literacy requires understandings and habits of mind that enable citizens to grasp what those enterprises are up to, to make some sense of how the natural and designed worlds work, to think critically and independently, to recognize and weigh alternative explanations of events and design trade-offs, and to deal sensibly with problems that involve evidence, numbers, patterns, logical arguments, and uncertainties.

- Curriculum reform should be shaped by our vision of the *lasting* knowledge and skills we want students to acquire by the time they become adults. This ought to include both a common core of learning—the focus of Project 2061—and learning that addresses the particular needs and interests of individual students.

- If we want students to learn science, mathematics, and technology well, we must radically reduce the sheer amount of material now being covered. The overstuffed curriculum places a premium on the

COMET HALLEY
Photographed from Las Campanas Observatory, Chile, March 1985.

ability to commit terms, algorithms, and generalizations to short-term memory and impedes the acquisition of understanding.

- Goals should be stated so as to reveal the intended character and sophistication of learning to be sought. Although goals for knowing and doing can be described separately, they should be learned together in many different contexts so that they can be used together in life outside of school.

- The common core of learning in science, mathematics, and technology should center on science literacy, not on an understanding of each of the separate disciplines. Moreover, the core studies should include connections among science, mathematics, and technology and between those areas and the arts and humanities and the vocational subjects.

- Common goals do not require uniform curricula, teaching methods, and materials. Project 2061 is developing tools to enable teachers to design learning experiences for students that take into account state and district requirements, student backgrounds and interests, teacher preferences, and the local environment.

- Reform must be comprehensive and long-term, if it is to be significant and lasting. It must center on *all* children, *all* grades, and *all* subjects. In addition, it must deal interactively with all aspects of the system—curriculum, teacher education, the organization of instruction, assessment, materials and technology, policy, and more. All of which takes time.

CHARACTERIZING BENCHMARKS

Benchmarks for Science Literacy is consistent with the above premises, but is sufficiently different in content, purpose, and style from other reform reports to require some clarification.

- ◆ *Benchmarks* **is a report from a cross-section of practicing educators.** In 1989, six school-district teams were formed in different parts of the nation to rethink the K-12 curriculum and outline alternative ways of achieving the literacy goals of *SFAA*. Each team, backed by consultants and Project 2061 staff, was made up of 25 teachers and administrators and cut across grade levels and subjects. Working together over four summers and three academic years, the teams developed a common set of benchmarks. Drafts of *Benchmarks* were critiqued in detail by hundreds of elementary-, middle-, and high-school teachers, as well as by administrators, scientists, mathematicians, engineers, historians, and experts on learning and curriculum design. Chapter 13: The Origin of Benchmarks, describes the process in greater detail.

- ◆ *Benchmarks* **is different from a curriculum, a curriculum framework, a curriculum design, or a plan for a curriculum.** It is a tool to be used by educators in designing a curricuium that makes sense to them and meets the standards for science literacy recommended in *SFAA*. Moreover, *Benchmarks* does not advocate any particular curriculum design. Far from pressing for one way of organizing instruction, Project 2061 pursues a reform strategy that will lead eventually to greater curriculum diversity than is common today.

- ◆ *Benchmarks* **is a compendium of specific science literacy goals that can be organized however one chooses.** As in most reference works, chapter order is unrelated to the relative importance of the benchmarks. Chapter 1 does not set the tone for all those that follow, nor does Chapter 12 culminate all that came before. Indeed, Project 2061 expects that benchmarks from the latter will appear in combination with those from various other chapters in most curriculum units that address science literacy goals. A version of *Benchmarks* on a computer disk will enable users to assemble benchmarks from various chapters into cogent sets.

- ◆ *Benchmarks* **specifies thresholds rather than average or advanced performance.** It describes levels of understanding and ability that *all* students are expected to reach on the way to becoming science-literate. A well-designed curriculum will provide students with the help and encouragement they need to meet those standards.

- ◆ *Benchmarks* **concentrates on the *common core* of learning that contributes to the science literacy of all students.** It does not spell out all of the science, mathematics, and technology goals that belong in the K-12 curriculum. Most students have interests, abilities, and ambitions that extend beyond the core studies, and some have learning difficulties that must be taken into account.

- ◆ *Benchmarks* **avoids technical language used for its own sake.** The number of technical terms that most adults must understand is relatively small. Accordingly, the 12th-grade benchmarks use only those technical terms that ought to be in the vocabulary of science-literate people. The language in the benchmarks for earlier grades is intended to signal the nature and sophistication of understandings to be sought. The Project 2061 analysis of these and other issues is summarized in Chapter 14: Issues and Language.

- ◆ *Benchmarks* **sheds only partial light on how to achieve the goals it recommends.** Deliberately. The means for realizing the ends listed in *Benchmarks* will be discussed in other Project 2061 materials. Although *Benchmarks* includes some commentary on aspects of instruction, that commentary is to clarify the meaning and intent of the benchmarks, not to present a systematic and detailed program of instruction.

- ◆ *Benchmarks* **is informed by research.** Research on students' understanding and learning bears significantly on the selection and grade placement of the benchmarks. Project 2061 surveyed the relevant research literature in the English language (and some in other languages) in search of solid findings on which to base benchmark decisions. The findings are discussed in Chapter 15: The Research Base.

- ◆ *Benchmarks* **is a developing product.** It will undergo periodic updates as more research on learning becomes available and as users of *Benchmarks* report their experiences. One of the important responsibilities of the Project 2061 school-district sites is to suggest revisions of *Benchmarks* based on their analysis of ongoing research and user recommendations.

◆ ***Benchmarks* is but one of a family of tools being designed by Project 2061.** To help educators bring about fundamental, lasting reform, *Benchmarks* and *SFAA* will be joined by other products. *Designs for Science Literacy* will describe Project 2061 models and curriculum blocks and will outline principles for configuring Project 2061 curricula. *Resources for Science Literacy* will be a continually updated database of outstanding learning and teaching materials suitable for curricula based on Project 2061 principles. *Blueprints for Reform* will recommend changes in the education system needed to make innovative K-12 curriculum reforms possible. A computerized curriculum-design and resource system is being developed to incorporate all of the Project 2061 products and link them interactively to each other and to educational resources. For more on this, see Chapter 16: Beyond Benchmarks.

◆ ***Benchmarks* is a companion for *SFAA*, not a substitute.** *SFAA* presents a vision of science literacy goals for all students to reach by the time they finish the 12th grade, and *Benchmarks* maps out the territory that students will have to traverse to get there. *SFAA* emphasizes cogency and connectedness. *Benchmarks* emphasizes analysis of the *SFAA* story into components and their sequence. In grades 9-12, where building coherence and connections becomes the main task, no list of components would be adequate to represent science literacy. (Indeed, not all of the detailed ideas in *SFAA* are represented in *Benchmarks*.) At the 9-12 level, therefore, reference to *SFAA* is more than ever necessary for a complete picture of science literacy, which the 9-12 *Benchmarks* only approximate. So, when working with *Benchmarks*, be sure to have a copy of *SFAA* at hand.

Using Benchmarks

Benchmarks was prepared as a tool to be used, along with *SFAA*, by everyone engaged in state or local efforts to transform learning in science, mathematics, and technology. The following suggestions for using *Benchmarks* came from Project 2061 team members, consultants, and staff, and from individuals who have seen prepublication draft versions of *Benchmarks:*

■ Study groups of teachers, administrators, school-board members, parents, interested citizens, and, whenever possible, scientists, engineers, and mathematicians can use *Benchmarks* to explore the concept of science literacy in relation to instruction in the early elementary, upper elementary, middle-, and high-school grades.

■ Cross-grade, cross-subject committees of teachers and curriculum specialists can use *Benchmarks* to gauge how well a K-12 curriculum or curriculum framework (state or local) addresses education for science literacy. Such an analysis can also lead to suggestions for making immediate and long-term curriculum and course improvement.

■ Developers of instructional materials can use *Benchmarks* to guide the creation of materials to support the work of teachers who are trying to foster science literacy for all students. Similarly, test writers can use *Benchmarks* to develop grade-level materials and techniques for assessing student progress toward science literacy.

■ Other reform efforts may find *Benchmarks* useful in supporting their work, just as Project 2061 has relied on so many of them for ideas and information. The federal programs that drew heavily on *SFAA*, such as the Statewide Systemic Initiatives (National Science Foundation), the Eisenhower Science and Mathematics Initiative (Department of Education), and the National Assessment of Educational Progress, have indicated that they intend also to use *Benchmarks*.

■ Universities and colleges that prepare elementary- and secondary-school teachers can use *Benchmarks* to supplement *SFAA*. Whereas *SFAA* explores the concept of science literacy in general, *Benchmarks* raises issues closer to the realities of curriculum and instruction.

■ Researchers can use *Benchmarks* to identify important topics for investigation. Such topics might include studies on the grade-level placement of benchmarks, the relationship between benchmarks and their precursors, effective ways to group benchmarks into instructional units, how to assess student progress toward science literacy, and how to evaluate learning materials and techniques used in support of the benchmarks. ■

Chapters 1 to 12 correspond to *SFAA*. Order of chapters is not related to their importance.

Sections, with few exceptions, correspond to *SFAA* sections.

Quotation from introductory section of this chapter in *SFAA*.

Chapter **5** THE LIVING ENVIRONMENT

A   DIVERSITY OF LIFE
B   HEREDITY
C   CELLS
D   INTERDEPENDENCE OF LIFE
E   FLOW OF MATTER AND ENERGY
F   EVOLUTION OF LIFE

People have long been curious about living things—how many different species there are, what they are like, where they live, how they relate to each other, and how they behave. Scientists seek to answer these questions and many more about the organisms that inhabit the earth. In particular, they try to develop the concepts, principles, and theories that enable people to understand the living environment better.

*Living organisms are made of the same components as all other matter, involve the same kinds of transformations of energy, and move using the same basic kinds of forces. Thus, all of the physical principles discussed in **Chapter 4: The Physical Setting** apply to life as well as to stars, raindrops, and television sets. But living organisms also have characteristics that can be understood best through the application of other principles.*

SCIENCE FOR ALL AMERICANS

What can be anywhere near as awe-inspiring as the vast array of living things that occupy every nook and cranny of the earth's surface, unless it is the array of extinct species that once occupied the planet? Biologists have already identified over a million living species, each with its own way of surviving, sometimes in the least likely places, each readily able to propagate itself in the next generation. Because only organisms with hard shells or skeletons are generally preserved, the fossil record does not preserve a good record of the even greater number of extinct species that have existed over the span of the earth's history.

Overall comments on the ideas to be learned and, in very general terms, the kinds of student experience that would foster learning.

99

5ʙ   Heredity

Building an observational base for heredity ought to be the first undertaking. Explanations can come later. The organisms children recognize are themselves, their classmates, and their pets. And that is the place to start studying heredity. However, it is important to be cautious about having children compare their own physical appearance to that of their siblings, parents, and grandparents. At the very least, the matter has to be handled with great delicacy so no one is embarrassed. Direct observations of generational similarities and differences of at least some plants and animals are essential.

Learning the genetic explanation for how traits are passed on from one generation to the next can begin in the middle years and carry into high school. The part played by DNA in the story should wait until students understand molecules. The interaction between heredity and environment in determining plant and animal behavior will be of interest to students. Examining specific cases can help them grasp the complex interactions of genetics and environment.

Section introductions comment on common difficulties in learning the ideas, on pacing over grade levels, and on clarification of the ideas themselves.

A summary of research findings relevant to how students think and learn about ideas in this section appear on the indicated page of Chapter 15, The Research Base.

Reasearch Notes

page 345

Chapter 4	D	Structure of Matter (DNA molecule)
6	B	Human Development (fertilization, cell differentiation)
	E	Physical Health (genetic disease)
7	A	Cultural Effects on Behavior (influence on behavior)
8	A	Agriculture (genetic manipulation)
	D	Communication (codes)
	E	Information Processing (programmed instructions)
	F	Health Technology (gene technologies)
9	D	Uncertainty (probability of gene combinations)
10	H	Explaining the Diversity of Life

Ideas in other chapters that relate to this section are found in the chapters or specific sections indicated in this box. Where the connection isn't obvious, a brief explanation is given in parentheses.

(Of course, there are also many connections to ideas in other sections of this same chapter — but these are too numerous and obvious to list here.)

Grade spans are only approximate. K-2, for example, might be labeled "primary" or "early elementary."

"Know" implies that students can explain ideas in their own words, relate the ideas to other benchmarks, and apply the ideas in novel contexts.

5B HEREDITY

Kindergarten through Grade 2

Teachers should lead students to make observations about how the offspring of familiar animals compare to one another and to their parents. Children know that animals reproduce their own kind—rabbits have rabbits (but you can usually tell one baby rabbit from another), cats have kittens that have different markings (but cats never have puppies), and so forth. This idea should be strengthened by a large number of examples, both plant and animal, that the children can draw on.

By the end of the 2nd grade, students should know that

▶ There is variation among individuals of one kind within a population.

▶ Offspring are very much, but not exactly, like their parents and like one another. ■

Grades 3 through 5

Students should move from describing individuals directly (she has blue eyes) to naming traits and classifying individuals with respect to those traits (eye color: blue). Students can be encouraged to keep lists of things that animals and plants get from their parents, things that they don't get, and things that the students are not sure about either way. This is also the time to start building the notion of a population whose members are alike in many ways but show some variation.

By the end of the 5th grade, students should know that

▶ Some likenesses between children and parents, such as eye color in human beings, or fruit or flower color in plants, are inherited. Other likenesses, such as people's table manners or carpentry skills, are learned.

▶ For offspring to resemble their parents, there must be a reliable way to transfer information from one generation to the next. ■

Grade-span comments, in order to clarify what "knowing" entails, sketch what students' experiences might likely include, and what difficulties students might have.

Growth toward these ideas can begin early in the grade span and need not be delayed until grade 5.

Benchmark statements, whenever possible, are cast in language that approximates the intended level of sophistication.

Ample margin space is provided for notes on connections, explanations, and learning experiences.

BENCHMARKS

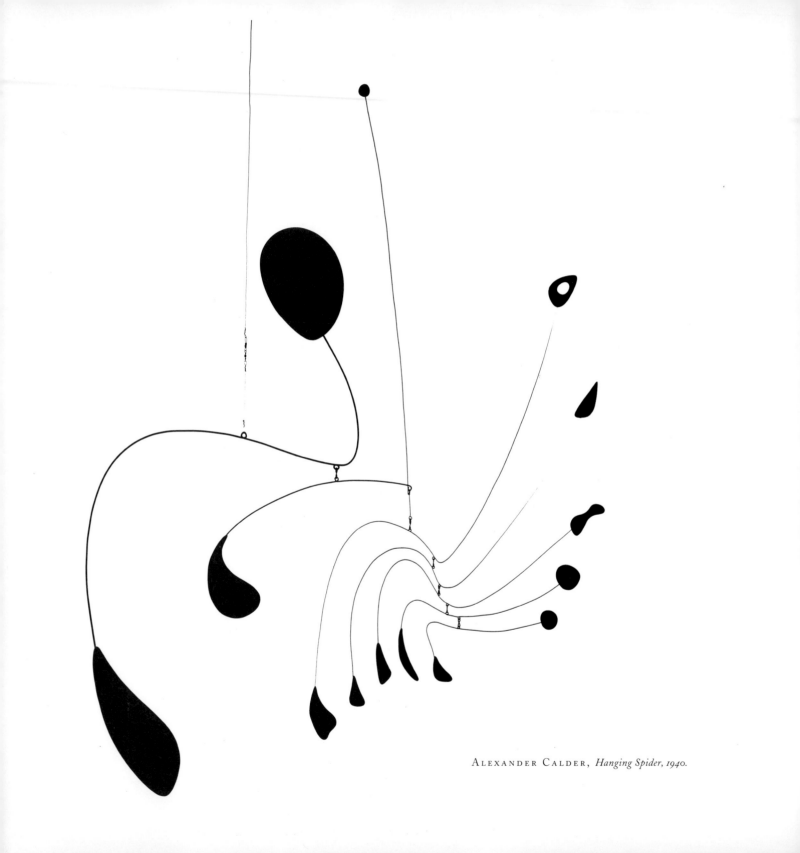

ALEXANDER CALDER, *Hanging Spider, 1940.*

Chapter 1 THE NATURE OF SCIENCE

A   THE SCIENTIFIC WORLD VIEW

B   SCIENTIFIC INQUIRY

C   THE SCIENTIFIC ENTERPRISE

Over the course of human history, people have developed many interconnected and validated ideas about the physical, biological, psychological, and social worlds. Those ideas have enabled successive generations to achieve an increasingly comprehensive and reliable understanding of the human species and its environment. The means used to develop these ideas are particular ways of observing, thinking, experimenting, and validating. These ways represent a fundamental aspect of the nature of science and reflect how science tends to differ from other modes of knowing.

It is the union of science, mathematics, and technology that forms the scientific endeavor and that makes it so successful. Although each of these human enterprises has a character and history of its own, each is dependent on and reinforces the others. Accordingly, the first three chapters of recommendations draw portraits of science, mathematics, and technology that emphasize their roles in the scientific endeavor and reveal some of the similarities and connections among them.

SCIENCE FOR ALL AMERICANS

The study of science as an intellectual and social endeavor—the application of human intelligence to figuring out how the world works—should have a prominent place in any curriculum that has science literacy as one of its aims. Consider the following:

When people know how scientists go about their work and reach scientific conclusions, and what the limitations of such conclusions are, they are more likely to react thoughtfully to scientific claims and less likely to reject them out of hand or accept them uncritically.

Once people gain a good sense of how science operates— along with a basic inventory of key science concepts as a basis for learning more later—they can follow the science adventure story as it plays out during their lifetimes.

The images that many people have of science and how it works are often distorted. The myths and stereotypes that young people have about science are not dispelled when science teaching focuses narrowly on the laws, concepts, and theories of science. Hence, the study of science as a way of knowing needs to be made explicit in the curriculum.

Acquiring scientific knowledge about how the world works does not necessarily lead to an understanding of how science itself works, and neither does knowledge of the philosophy and sociology of science alone lead to a scientific understanding of the world. The challenge for educators is to weave these different aspects of science together so that they reinforce one another.

For students in the early grades, the emphasis should overwhelmingly be on gaining experience with natural and social phenomena and on enjoying science. Abstractions of all kinds can gradually make their appearance as students mature and develop an ability to handle explanations that are complex and abstract. This phasing-in certainly applies to generalizations about the scientific world view, scientific inquiry, and the scientific enterprise.

That does not mean, however, that abstraction should be ignored altogether in the early grades. By gaining lots of experience *doing* science, becoming more sophisticated in conducting investigations, and explaining their findings, students will accumulate a set of concrete experiences on which they can draw to *reflect* on the process. At the same time, conclusions presented to students (in books and in class) about how scientists explain phenomena should gradually be augmented by information on how the science community arrived at those conclusions. Indeed, as students move through school, they should be encouraged to ask over and over, "How do we know that's true?"

History provides another avenue to the understanding of how science works, which is one of the chief reasons why a chapter on historical perspectives is included in both *Science for All Americans* and *Benchmarks*. Although that chapter emphasizes the great advances in science, it is equally important that students should come to realize that much of the growth of science and technology has resulted from the gradual accumulation of knowledge over many centuries.

This realization runs counter to widely held misconceptions about scientific progress. What has been called normal science, in contrast to scientific revolutions, is what goes on most of the time, engages most of the people, and yields most of the advances. While "breakthroughs" and "revolutions" attract people's attention more than step-by-step growth, focusing on those rare events exclusively will give students a distorted idea of science, in that both incremental growth and occasional radical shifts are part of the story of science.

For the same reason, not all of the historical emphasis should be placed on the lives of great scientists, those relatively few figures who, owing to genius and opportunity and good fortune, are best known. Students should learn that all sorts of people, indeed, people like themselves, have done and continue to do science.

To gain this understanding, students will need appropriate learning materials. Historical case studies, backed up by a solid collection of biographies and other reference works and films, will be essential. Also, science and history textbooks will need to be modified to include the history of science. Beginning with science, mathematics, and technology in the early Egyptian, Greek, Chinese, and Arabic cultures, these materials should extend to modern times and include information on the contributions of men and women from every part of the world. ■

1A   THE SCIENTIFIC WORLD VIEW

A scientific world view is not something that working scientists spend a lot of time discussing. They just do science. But underlying their work are several beliefs that are not always held by nonscientists. One is that by working together over time, people can in fact figure out how the world works. Another is that the universe is a unified system and knowledge gained from studying one part of it can often be applied to other parts. Still another is that knowledge is both stable and subject to change.

Little is gained by presenting these beliefs to students as dogma. For one thing, such beliefs are subtle. The first one cited above says only that scientists believe that the world *can* be understood, not that it ever *will* be so completely understood that science can shut down once and for all, the job done. Indeed, in finding answers to one set of questions about how the world works, scientists inevitably unearth new questions, so the quest will likely continue as long as human curiosity survives. Also, the human capacity for generating trustworthy knowledge about nature has limits. Scientific investigations often fail to find convincing answers to the questions they pursue. The claim that science will find answers always carries the implied disclaimers, "in many cases" and "in the very long run."

The belief that knowledge gained by studying one part of the universe can be applied to other parts is often confirmed but turns out to be true only part of the time. It happens, for example, that the behavior of a given organism is sometimes different when observed in a laboratory instead of its natural environment. Thus, a belief in the unity of the universe does not eliminate the need to show how far the findings in one situation can be extended.

The notion that scientific knowledge is always subject to modification can be difficult for students to grasp. It seems to oppose the certainty and truth popularly accorded to science, and runs counter to the yearning for certainty that is characteristic of most cultures, perhaps especially so among youth. Moreover, the picture of change in science is not simple. As new questions arise, new theories are proposed, new instruments are invented, and new techniques are developed. In response, new experiments are conducted, new specimens collected, new observations made, and new analyses performed. Some of the findings challenge existing theories, leading to their modification or to the invention, on very rare occasions, of entirely new theories—which, in turn, leads to new experiments, new observations...and so on.

But that ferment of change occurs mostly at the cutting edge of research. In fact, it is important not to overdo the "science always changes" theme, since the main body of scientific knowledge is very stable and grows by being corrected slowly and having its boundaries extended gradually. Scientists themselves accept the notion that scientific knowledge is always *open* to improvement and can never be declared absolutely certain.

RESEARCH NOTES
page 332

◁ ALSO SEE

5

Kindergarten through Grade 2

From their very first day in school, students should be actively engaged in learning to view the world scientifically. That means encouraging them to ask questions about nature and to seek answers, collect things, count and measure things, make qualitative observations, organize collections and observations, discuss findings, etc. Getting into the spirit of science and liking science are what count most. Awareness of the scientific world view can come later.

Anticipating an eventual understanding of the scientific world view, these early science experiences can be designed to bring out one aspect of the belief in the unity of nature: consistency. Students should sometimes repeat observations and investigations in the classroom, and then, when possible, do so again in the school yard and at home. For instance, students could be asked to compare what happens in different places when an egg is cooked, or how moving objects are affected when pushed or pulled, or what a seed looks like when it starts to grow. These activities should serve to stimulate curiosity and engage students in taking an interest in their environment and the workings of nature.

By the end of the 2nd grade, students should know that

▶ When a science investigation is done the way it was done before, we expect to get a very similar result.

▶ Science investigations generally work the same way in different places. ■

Grades 3 through 5

As children continue to investigate the world, the consistency premise can be strengthened by putting more emphasis on explaining inconsistency. When students observe differences in the way things behave or get different results in repeated investigations, they should suspect that something differs from trial to trial and try to find out what. Sometimes the difference results from methods, sometimes from the way the world is. The point is that different findings can lead to interesting new questions to be investigated.

This emphasis on scientific engagement calls for frequent hands-on activities. But that is not to say that students must, or even can, "discover" everything by direct experience. Stories about people making discoveries and inventions can be used to illustrate the kinds of convictions about the world and what can be learned from it that are shared by the varied people who do science.

By the end of the 5th grade, students should know that

▶ Results of similar scientific investigations seldom turn out exactly the same. Sometimes this is because of unexpected differences in the things being investigated, sometimes because of unrealized differences in the methods used or in the circumstances in which the investigation is carried out, and sometimes just because of uncertainties in observations. It is not always easy to tell which. ■

Most early adolescents have a more immediate interest in nature than in the philosophy of science. They should continue to be engaged in doing science and encouraged to reflect on the science they are engaged in, with the assumption that they will later acquire a more mature reflection on science as a world view.

Early adolescence, however, is not too early to begin to deal with the question of the durability of scientific knowledge, and particularly its susceptibility to change. Both incremental changes and more radical changes in scientific knowledge should be taken up. Radical changes in science sometimes result from the appearance of new information, and sometimes from the invention of better theories (for example, germ theory and geologic time, as discussed in Chapter 10: Historical Perspectives).

By the end of the 8th grade, students should know that

▶ **When similar investigations give different results, the scientific challenge is to judge whether the differences are trivial or significant, and it often takes further studies to decide. Even with similar results, scientists may wait until an investigation has been repeated many times before accepting the results as correct.**

▶ **Scientific knowledge is subject to modification as new information challenges prevailing theories and as a new theory leads to looking at old observations in a new way.**

▶ **Some scientific knowledge is very old and yet is still applicable today.**

▶ **Some matters cannot be examined usefully in a scientific way. Among them are matters that by their nature cannot be tested objectively and those that are essentially matters of morality. Science can sometimes be used to inform ethical decisions by identifying the likely consequences of particular actions but cannot be used to establish that some action is either moral or immoral. ■**

7

Grades 9 through 12

Aspects of the scientific world view can be illustrated in the upper grades both by the study of historical episodes in science and by reflecting on developments in current science. Case studies provide opportunities to examine such matters as the theoretical and practical limitations of science, the differences in the character of the knowledge the different sciences generate, and the tension between the certainty of accepted science and the breakthroughs that upset this certainty.

By the end of the 12th grade, students should know that

▶ **Scientists assume that the universe is a vast single system in which the basic rules are the same everywhere. The rules may range from very simple to extremely complex, but scientists operate on the belief that the rules can be discovered by careful, systematic study.**

▶ **From time to time, major shifts occur in the scientific view of how the world works. More often, however, the changes that take place in the body of scientific knowledge are small modifications of prior knowledge. Change and continuity are persistent features of science.**

▶ **No matter how well one theory fits observations, a new theory might fit them just as well or better, or might fit a wider range of observations. In science, the testing, revising, and occasional discarding of theories, new and old, never ends. This ongoing process leads to an increasingly better understanding of how things work in the world but not to absolute truth. Evidence for the value of this approach is given by the improving ability of scientists to offer reliable explanations and make accurate predictions. ■**

RESEARCH INSTITUTE

UNANSWERED QUESTIONS

UNQUESTIONED ANSWERS

Scientific inquiry is more complex than popular conceptions would have it. It is, for instance, a more subtle and demanding process than the naive idea of "making a great many careful observations and then organizing them." It is far more flexible than the rigid sequence of steps commonly depicted in textbooks as "the scientific method." It is much more than just "doing experiments," and it is not confined to laboratories. More imagination and inventiveness are involved in scientific inquiry than many people realize, yet sooner or later strict logic and empirical evidence must have their day. Individual investigators working alone sometimes make great discoveries, but the steady advancement of science depends on the enterprise as a whole. And so on.

If students themselves participate in scientific investigations that progressively approximate good science, then the picture they come away with will likely be reasonably accurate. But that will require recasting typical school laboratory work. The usual high-school science "experiment" is unlike the real thing: The question to be investigated is decided by the teacher, not the investigators; what apparatus to use, what data to collect, and how to organize the data are also decided by the teacher (or the lab manual); time is not made available for repetitions or, when things are not working out, for revising the experiment; the results are not presented to other investigators for criticism; and, to top it off, the correct answer is known ahead of time.

Of course, the student laboratory *can* be designed to help students learn about the nature of scientific inquiry. As a first step, it would help simply to reduce the number of experiments undertaken (making time available to probe questions more deeply) and eliminate many of their mechanical, recipe-following aspects. In making this change, however, it should be kept in mind that well-conceived school laboratory experiences serve other important purposes as well. For example, they provide opportunities for students to become familiar with the phenomena that the science concepts being studied try to account for.

Another, more ambitious step is to introduce some student investigations that more closely approximate sound science. Such investigations should become more ambitious and more sophisticated. Before graduating from high school, students working individually or in teams should design and carry out at least one major investigation. They should frame the question, design the approach, estimate the time and costs involved, calibrate the instruments, conduct trial runs, write a report, and finally, respond to criticism.

Such investigations, whether individual or group, might take weeks or months to conduct. They might happen in and out of school time and be broken up by periods when, for technical reasons, work cannot go forward. But the total time invested will probably be no more than the sum of all those weekly one-period labs that contribute little to student understanding of scientific inquiry.

Research Notes

page 332

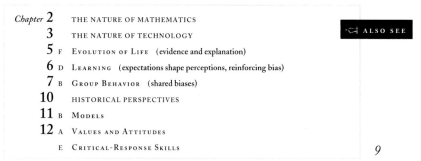

ALSO SEE

9

Students should be actively involved in exploring phenomena that interest them both in and out of class. These investigations should be fun and exciting, opening the door to even more things to explore. An important part of students' exploration is telling others what they see, what they think, and what it makes them wonder about. Children should have lots of time to talk about what they observe and to compare their observations with those of others. A premium should be placed on careful expression, a necessity in science, but students at this level should not be expected to come up with scientifically accurate explanations for their observations. Theory can wait.

By the end of the 2nd grade, students should know that

▶ **People can often learn about things around them by just observing those things carefully, but sometimes they can learn more by doing something to the things and noting what happens.**

▶ **Tools such as thermometers, magnifiers, rulers, or balances often give more information about things than can be obtained by just observing things without their help.**

▶ **Describing things as accurately as possible is important in science because it enables people to compare their observations with those of others.**

▶ **When people give different descriptions of the same thing, it is usually a good idea to make some fresh observations instead of just arguing about who is right.** ■

Children's strategies for finding out more and more about their surroundings improve as they gain experience in conducting simple investigations of their own and working in small groups. They should be encouraged to observe more and more carefully, measure things with increasing accuracy (where the nature of the investigations involves measurement), record data clearly in logs and journals, and communicate their results in charts and simple graphs as well as in prose. Time should be provided to let students run enough trials to be confident of their results. Investigations should often be followed up with presentations to the entire class to emphasize the importance of clear communication in science. Class discussions of the procedures and findings can provide the beginnings of scientific argument and debate.

Students' investigations at this level can be expected to bear on detecting the similarities and differences among the things they collect and examine. They should come to see that in trying to identify and explain likenesses and differences, they are doing what goes on in science all the time. What students may find most puzzling is when there are differences in the results they obtain in repeated investigations at different times or in different places, or when different groups of students get different results doing supposedly the same experiment. That, too, happens to scientists, sometimes because of the methods or materials used, but sometimes because the thing being studied actually varies.

Research studies suggest that there are some limits on what to expect at this level of student intellectual development. One limit is that the design of carefully controlled experiments is still beyond most students in the middle grades. Others are that such students confuse theory (explanation) with evidence for it and that they have difficulty making logical inferences. However, the studies say more about what students at this level do not learn in today's schools than about what they might possibly learn if instruction were more effective.

In any case, some children will be ready to offer explanations for why things happen the way they do. They should be encouraged to "check what you think against what you see." As explanations take on more and more importance, teachers must insist that students pay attention to the explanations of others and remain open to new ideas. This is an appropriate time to introduce the notion that in science it is legitimate to offer different explanations for the same set of observations, although this notion is apparently difficult for many youngsters to comprehend.

By the end of the 5th grade, students should know that

▶ **Scientific investigations may take many different forms, including observing what things are like or what is happening somewhere, collecting specimens for analysis, and doing experiments. Investigations can focus on physical, biological, and social questions.**

▶ **Results of scientific investigations are seldom exactly the same, but if the differences are large, it is important to try to figure out why. One reason for following directions carefully and for keeping records of one's work is to provide information on what might have caused the differences.**

▶ **Scientists' explanations about what happens in the world come partly from what they observe, partly from what they think. Sometimes scientists have different explanations for the same set of observations. That usually leads to their making more observations to resolve the differences.**

▶ **Scientists do not pay much attention to claims about how something they know about works unless the claims are backed up with evidence that can be confirmed and with a logical argument.** ■

Grades 6 through 8

At this level, students need to become more systematic and sophisticated in conducting their investigations, some of which may last for weeks or more. That means closing in on an understanding of what constitutes a good experiment. The concept of controlling variables is straightforward but achieving it in practice is difficult. Students can make some headway, however, by participating in enough experimental investigations (not to the exclusion, of course, of other kinds of investigations) and explicitly discussing how explanation relates to experimental design.

Student investigations ought to constitute a significant part—but only a part—of the total science experience. Systematic learning of science concepts must also have a place in the curriculum, for it is not possible for students to discover all the concepts they need to learn, or to observe all of the phenomena they need to encounter, solely through their own laboratory investigations. And even though the main purpose of student investigations is to help students learn how science works, it is important to back up such experience with selected readings. This level is a good time to introduce stories (true and fictional) of scientists making discoveries—not just world-famous scientists, but scientists of very different backgrounds, ages, cultures, places, and times.

By the end of the 8th grade, students should know that

▶ **Scientists differ greatly in what phenomena they study and how they go about their work. Although there is no fixed set of steps that all scientists follow, scientific investigations usually involve the collection of relevant evidence, the use of logical reasoning, and the application of imagination in devising hypotheses and explanations to make sense of the collected evidence.**

▶ **If more than one variable changes at the same time in an experiment, the outcome of the experiment may not be clearly attributable to any one of the variables. It may not always be possible to prevent outside variables from influencing the outcome of an investigation (or even to identify all of the variables), but collaboration among investigators can often lead to research designs that are able to deal with such situations.**

▶ **What people expect to observe often affects what they actually do observe. Strong beliefs about what should happen in particular circumstances can prevent them from detecting other results. Scientists know about this danger to objectivity and take steps to try and avoid it when designing investigations and examining data. One safeguard is to have different investigators conduct independent studies of the same questions.**

Students' ability to deal with abstractions and hypothetical cases improves in high school. Now the unfinished and tentative nature of science may make some sense to them. Students should not be allowed to conclude, however, that the mutability of science permits any belief about the world to be considered as good as any other belief. Theories compete for acceptance, but the only serious competitors are those theories that are backed by valid evidence and logical arguments.

The nature and importance of prediction in science can also be taken up at this level. Coverage of this topic should emphasize the use of statistics, probability, and modeling in making scientific predictions about complex phenomena often found in biological, meteorological, and social systems. Care also should be taken to dissociate the study of scientific prediction from the general public's notions about astrology and guessing the outcomes of sports events.

By the end of the 12th grade, students should know that

▶ **Investigations are conducted for different reasons, including to explore new phenomena, to check on previous results, to test how well a theory predicts, and to compare different theories.**

▶ **Hypotheses are widely used in science for choosing what data to pay attention to and what additional data to seek, and for guiding the interpretation of the data (both new and previously available).**

▶ **Sometimes, scientists can control conditions in order to obtain evidence. When that is not possible for practical or ethical reasons, they try to observe as wide a range of natural occurrences as possible to be able to discern patterns.**

▶ **There are different traditions in science about what is investigated and how, but they all have in common certain basic beliefs about the value of evidence, logic, and good arguments. And there is agreement that progress in all fields of science depends on intelligence, hard work, imagination, and even chance.**

▶ **Scientists in any one research group tend to see things alike, so even groups of scientists may have trouble being entirely objective about their methods and findings. For that reason, scientific teams are expected to seek out the possible sources of bias in the design of their investigations and in their data analysis. Checking each other's results and explanations helps, but that is no guarantee against bias.**

▶ **In the short run, new ideas that do not mesh well with mainstream ideas in science often encounter vigorous criticism. In the long run, theories are judged by how they fit with other theories, the range of observations they explain, how well they explain observations, and how effective they are in predicting new findings.**

▶ **New ideas in science are limited by the context in which they are conceived; are often rejected by the scientific establishment; sometimes spring from unexpected findings; and usually grow slowly, through contributions from many investigators. ■**

1c   THE SCIENTIFIC ENTERPRISE

Scientific activity is one of the main features of the contemporary world and distinguishes present times from earlier periods. As an endeavor for learning how the world works, it provides a living for a very large number of people. It is important for students to understand how science is organized because, as adults in a democracy, they will be in a position to influence what public support will be provided for basic and applied science. Students also need to be exposed to four other aspects of the scientific enterprise: its social structure, its discipline and institutional identification, its ethics, and the role of scientists in public affairs. These matters do not require explicit discussion in the early grades but should appear more and more frequently as students progress through school. By the time they graduate, students should feel comfortable talking in general terms about the nature of the scientific enterprise and should be able to understand discussions of science issues in the news.

RESEARCH NOTES

page 333

ALSO SEE

Chapter 2		THE NATURE OF MATHEMATICS
3		THE NATURE OF TECHNOLOGY
5	E	FLOW OF MATTER AND ENERGY (welfare of laboratory animals)
7	B	GROUP BEHAVIOR (shared ethics and biases)
	E	POLITICAL AND ECONOMIC SYSTEMS (government funding of science)
	G	GLOBAL INTERDEPENDENCE (international cooperation)
8	D	COMMUNICATION (sharing methods, ideas, findings)
	E	INFORMATION PROCESSING (computer storage and retrieval)
10		HISTORICAL PERSPECTIVES
11	A	SYSTEMS (science as a social system)
12	A	VALUES AND ATTITUDES

Kindergarten through Grade 2

Science should begin in kindergarten with students learning to work in small teams (rather than as isolated individuals) to ask and answer questions about their surroundings and to share their findings with classmates. Teachers and older students can help the groups learn how to share in deciding what to do, in collecting and organizing information, and in making presentations.

From the start, teachers should foster scientific values by recognizing instances of them in the work of individual students and student groups. For example, praise should be given for curiosity and creativity even when the investigations they lead to do not turn out as planned.

Given the value that science places on independent thought, it is important that students be assured that although they are part of a team, they are free to reach different conclusions from their classmates, and that when they do they should say so and say why. Because youngsters want to be liked, this notion that one can disagree with friends and still be friends is not easy to accept (and may not be true in the short run) and therefore has to be approached judiciously.

Student investigations usually involve collecting live animals to bring into the classroom for observation. Although most children want pet-like animals (goldfish, rabbits, etc.) to be treated carefully, not all do, and some children can be cruel. The use of animals in scientific research is a very complex issue, but long before students are ready to discuss it in any depth, they should have opportunities, in the context of science, to interact with living things in ways that promote respect. Teachers should all be familiar with

the National Science Teachers Association's guidelines for responsible use of animals in the classroom, published in the association's handbook.

The history of science and technology is mostly too advanced a subject for students in the earliest grades. But they are not too young to learn from their own collective experience that everyone can find some things out about nature, just as everyone can learn numbers, the alphabet, and how to read.

By the end of the 2nd grade, students should know that

▶ **Everybody can do science and invent things and ideas.**

▶ **In doing science, it is often helpful to work with a team and to share findings with others. All team members should reach their own individual conclusions, however, about what the findings mean.**

▶ **A lot can be learned about plants and animals by observing them closely, but care must be taken to know the needs of living things and how to provide for them in the classroom.** ■

Grades 3 through 5

As student research teams become more adept at doing science, more emphasis should be placed on how to communicate findings. As students learn to describe their procedures with enough detail to enable others to replicate them, make greater use of tables and graphs to summarize and interpret data, and submit their work to the criticism of others, they should understand that they are engaged in the scientific way of doing research.

Career information can be introduced to acquaint students with science as an occupation in which there is a wide variety of different kinds and levels of work. Films, books (science adventure, biographies), visits by scientists, and visits (if possible) to science centers and to university, industrial, and government laboratories provide multiple opportunities for students to become informed.

Teachers should emphasize the diversity to be found in the scientific community: different kinds of people (in terms of race, sex, age, nationality) pursuing different sciences and working in different places (from isolated field sites to labs to offices). Students can learn that some scientists and engineers use huge instruments (e.g., particle accelerators or telescopes), and others use only notebooks and pencils. And most of all, students can begin to realize that doing science involves more than "scientists," and that many different occupations are part of the scientific enterprise.

By the end of the 5th grade, students should know that

▶ **Science is an adventure that people everywhere can take part in, as they have for many centuries.**

▶ **Clear communication is an essential part of doing science. It enables scientists to inform others about their work, expose their ideas to criticism by other scientists, and stay informed about scientific discoveries around the world.**

▶ **Doing science involves many different kinds of work and engages men and women of all ages and backgrounds.** ■

Grades 6 through 8

Teachers should continue to seize opportunities for introducing information on science as a diverse line of work. Above all, children in early adolescence need to see science and science-related careers as a real option for themselves personally. That does not imply heavy, possibly premature recruiting, but means broadening student awareness of the possibilities and helping all students to keep themselves eligible for these possibilities. If such awareness develops in a proper context, then the knowledge gained will be valuable to all students when they become adult citizens, regardless of vocation.

By this level, student investigations should be more professional than could reasonably be expected in the elementary grades. For one thing, students must assess the risks associated with an investigation before being given permission to proceed. For another, students should now be using computers as scientists use them—namely to collect, store, and retrieve data, to help in data analysis, to prepare tables and graphs, and to write summary reports. If possible, students should have the opportunity to work on investigations in which they can use computers to communicate with students elsewhere who are working on the same problems.

By the end of the 8th grade, students should know that

▶ **Important contributions to the advancement of science, mathematics, and technology have been made by different kinds of people, in different cultures, at different times.**

▶ **Until recently, women and racial minorities, because of restrictions on their education and employment opportunities, were essentially left out of much of the formal work of the science establishment; the remarkable few who overcame those obstacles were even then likely to have their work disregarded by the science establishment.**

▶ **No matter who does science and mathematics or invents things, or when or where they do it, the knowledge and technology that result can eventually become available to everyone in the world.**

▶ **Scientists are employed by colleges and universities, business and industry, hospitals, and many government agencies. Their places of work include offices, classrooms, laboratories, farms, factories, and natural field settings ranging from space to the ocean floor.**

▶ **In research involving human subjects, the ethics of science require that potential subjects be fully informed about the risks and benefits associated with the research and of their right to refuse to participate. Science ethics also demand that scientists must not knowingly subject coworkers, students, the neighborhood, or the community to health or property risks without their prior knowledge and consent. Because animals cannot make informed choices, special care must be taken in using them in scientific research.**

continued

17

▶ Computers have become invaluable in science because they speed up and extend people's ability to collect, store, compile, and analyze data, prepare research reports, and share data and ideas with investigators all over the world.

▶ Accurate record-keeping, openness, and replication are essential for maintaining an investigator's credibility with other scientists and society. ■

At this level, science and history can support each other more elaborately. As students study science and mathematics, they should encounter some of the historical and cultural roots of the concepts they are learning. As they study the history of the different periods, cultures, and episodes, students should find that science, mathematics, and invention often played a central role.

Studies in history, government, and science can also come together to help students understand science as a social enterprise. Seminars based on actual case studies provide a way to approach issues of ethics in science and the role of scientists in social decision-making. There is no shortage of current issues, ranging from citizen resistance to potentially dangerous research in the community, to the use of human prisoners or animals in medical experiments, to charges of scientific fraud. Newspapers and magazines, the "news and comment" and "letters to the editor" sections of science journals, and congressional testimony all provide easily accessible documentary material. A seminar format can focus on informed discussion and debate rather than covering predetermined material to reach predetermined conclusions.

No matter how the curriculum is organized, it should provide students with opportunities to become aware of the great range of scientific disciplines that exist. There is no sense, however, in having students memorize definitions of anthropology, astrophysics, biochemistry, paleobacteriology, and the rest of the family. Individual students or small groups of students can study different disciplines in some detail—most scientific societies are happy to help out—and then share their findings with one another. The focus of such studies should be substantive (what are typical studies like in the discipline) and sociological (how is the field organized and who is in it), and they should probably involve, over an extended time, interviews, field trips, readings, data analysis, and, if possible, the conduct of small-scale experiments or field studies. Such activities will contribute to science literacy goals, and they should also help students realize how many different career possibilities exist in science.

By the end of the 12th grade, students should know that

▶ **The early Egyptian, Greek, Chinese, Hindu, and Arabic cultures are responsible for many scientific and mathematical ideas and technological inventions.**

▶ **Modern science is based on traditions of thought that came together in Europe about 500 years ago. People from all cultures now contribute to that tradition.**

▶ **Progress in science and invention depends heavily on what else is happening in society, and history often depends on scientific and technological developments.**

▶ **Science disciplines differ from one another in what is studied, techniques used, and outcomes sought, but they share a common purpose and philosophy, and all are part of the same scientific enterprise. Although each discipline provides a conceptual structure for organizing and pursuing knowledge, many problems are studied by scientists using information and skills from many disciplines. Disciplines do not have fixed boundaries, and it happens that new scientific disciplines are being formed where existing ones meet and that some subdisciplines spin off to become new disciplines in their own right.**

▶ **Current ethics in science hold that research involving human subjects may be conducted only with the informed consent of the subjects, even if this constraint limits some kinds of potentially important research or influences the results. When it comes to participation in research that could pose risks to society, most scientists believe that a decision to participate or not is a matter of personal ethics rather than professional ethics.**

▶ **Scientists can bring information, insights, and analytical skills to bear on matters of public concern. Acting in their areas of expertise, scientists can help people understand the likely causes of events and estimate their possible effects. Outside their areas of expertise, however, scientists should enjoy no special credibility. And where their own personal, institutional, or community interests are at stake, scientists as a group can be expected to be no less biased than other groups are about their perceived interests.**

continued

19

▶ The strongly held traditions of science, including its commitment to peer review and publication, serve to keep the vast majority of scientists well within the bounds of ethical professional behavior. Deliberate deceit is rare and likely to be exposed sooner or later by the scientific enterprise itself. When violations of these scientific ethical traditions are discovered, they are strongly condemned by the scientific community, and the violators then have difficulty regaining the respect of other scientists.

▶ Funding influences the direction of science by virtue of the decisions that are made on which research to support. Research funding comes from various federal government agencies, industry, and private foundations. ■

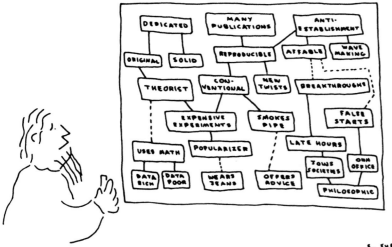

LINNAEUS ATTEMPTS TO CLASSIFY
SCIENTISTS, BUT DECIDES THAT
SPECIES WOULD BE EASIER.

C.D.DOXIADIAS, *model of boulevard network, central Paris, twentieth century.*

Chapter 2 THE NATURE OF MATHEMATICS

A   PATTERNS AND RELATIONSHIPS

B   MATHEMATICS, SCIENCE, AND TECHNOLOGY

C   MATHEMATICAL INQUIRY

Mathematics relies on logic and creativity, and it is pursued both for a variety of practical purposes and for its intrinsic interest. For some people, and not only professional mathematicians, the essence of mathematics lies in its beauty and its intellectual challenge. For others, including many scientists and engineers, the chief value of mathematics is how it applies to their own work. Because mathematics plays such a central role in modern culture, some basic understanding of the nature of mathematics is requisite for scientific literacy. To achieve this, students need to perceive mathematics as part of the scientific endeavor, comprehend the nature of mathematical thinking, and become familiar with key mathematical ideas and skills.

SCIENCE FOR ALL AMERICANS

No complex field of endeavor can be defined neatly in a few sentences or paragraphs. Not science or art or technology. Not mathematics. But the outsider can gradually develop a rich sense of the nature of any one of them by examining it from various perspectives and by doing some of the things that the insiders do. These same two kinds of experience—learning about a domain and practicing it—also lead to the development of skills that have wide utility in adult life, as well as to understanding.

As the science of patterns and relationships, mathematics shares many of the features of other sciences, as already described in Chapter 1: The Nature of Science. Particularly relevant similarities are the belief in underlying order, the ideals of honesty and openness in reporting research, the importance of criticism by colleagues in judging the value of new work, and the essential role played by imagination. Mathematics is also like science and technology in that it incorporates both finding answers to fundamental questions and solving practical problems.

The richness and ubiquity of mathematics underlie its treatment by Project 2061. One component of the project's approach is evident in both *Science for All Americans* and *Benchmarks for Science Literacy*. In both documents, Chapter 2 outlines what students should know about mathematics as a unique endeavor; Chapter 9 recommends what mathematical ideas students should acquire; Chapter 11 presents some powerful mathematical concepts, such as scale and models, that are widely useful as analytical tools; and Chapter 12 lists the mathematical skills students need, along with scientific and technological skills, to deal effectively with practical affairs in everyday life. Of course, the curriculum would not separate knowing and doing in that way.

Indeed, it is in curriculum design that the other component of the Project 2061 approach to mathematics will be found. As will be described fully in a forthcoming report, *Designs for Science Literacy*, nearly all of the building blocks of a Project 2061 curriculum will include mathematics. Students will encounter mathematics at every grade level, and learn mathematics in many different subject-matter and real-world contexts (some explicitly labeled as mathematics, some not). ∎

The universe is made up of galaxies, mountains, creatures, vehicles, and all manner of other things, each seemingly unique. Moreover, it is a chaotic affair in which those things intrude on one another in all sorts of ways, often violently but sometimes with great subtlety. But thanks to mathematics, people are able to think about the world of objects and happenings and to communicate those thoughts in ways that reveal unity and order.

The numbers, lines, angles, shapes, dimensions, averages, probabilities, ratios, operations, cycles, correlations, etc., that make up the world of mathematics enable people to make sense of a universe that otherwise might seem to be hopelessly complicated. Mathematical patterns and relationships have been developed and refined over the centuries, and the process is as vigorous and productive now as at any time in history. Perhaps that is because today mathematics is used in more fields of endeavor than ever before and has also become more essential in everyday life.

For purposes of general scientific literacy, it is important for students (1) to understand in what sense mathematics is the study of patterns and relationships, (2) to become familiar with some of those patterns and relationships, and (3) to learn to use them in daily life. The latter two of these general goals should be sought in parallel rather than sequentially. For the most part, learning mathematics in the abstract before seeking to use it has not proven to be effective. Thus, the curriculum should arrange instruction so that students encounter any given mathematical pattern or relationship in many different contexts before, during, and after its introduction in mathematics itself.

From time to time thereafter, in pursuit of the goal of understanding the nature of mathematics, students should have an opportunity in mathematics to reflect on the nature of patterns and relationships in a purely abstract way. Individual or class portfolios of examples of patterns and relationships collected over time could be used as the raw material for reflecting on how mathematics defines a pattern or relationship so that it transcends and is more powerful than individual instances of it.

The individual ideas created and used by mathematicians are traditionally aggregated, often for pedagogical reasons as well as for strictly conceptual ones, into families such as arithmetic, geometry, algebra, trigonometry, statistics, and calculus. Mathematicians look for patterns and relationships that link different ideas (themselves patterns and relationships) within such families and in between separate ones. Few achievements in mathematics are as satisfying as showing that what were previously thought of as two separate parts of mathematics are parallel or different examples of a single, more abstract formulation. If at all possible, all students should have the experience of discovering for themselves that an idea can be represented in different but analogous ways.

continued

RESEARCH NOTES
page 334

One line of research on how people learn emphasizes the helpfulness of making multiple representations of the same idea and translating from one to another. When a student can begin to represent a relationship in tables and in graphs and in symbols and in words, one can be confident that the student has really grasped its meaning. And, as the theory goes, the way students learn to make those representations and translations is to see them and practice them in contexts in which they care about what the answer is. Students engaged in this kind of activity will eventually get the idea of connectedness in mathematics—although they may need occasionally to look back on their own work and recognize the many connections they have made.

Kindergarten through Grade 2

In the first few grades, children think in very specific and concrete terms. They are little interested in grand categories such as mathematics, science, and technology, but they usually respond positively to the challenge of learning numbers and how to manipulate them, identifying shapes and simple patterns, collecting and describing collections, and building things. At some point, of course, they need to see that certain kinds of ideas and activities are mathematical, certain kinds scientific, and others technological. But when that labeling happens is less important than that, from the start, children study numbers and shapes and simple operations on them and do so in as many different contexts as possible.

By the end of the 2nd grade, students should know that

▶ Circles, squares, triangles, and other shapes can be found in things in nature and in things that people build.

▶ Patterns can be made by putting different shapes together or taking them apart.

▶ Things move, or can be made to move, along straight, curved, circular, back-and-forth, and jagged paths.

▶ Numbers can be used to count any collection of things. ■

Grades 3 through 5

If a basic strategy for learning about the nature of mathematics is to reflect on things already learned in various contexts, one possibility would be to have a list in the classroom, headed "What We Do In Mathematics," to which new items could be added month by month. From time to time, items could be grouped, or shown to be subsets of others, or shown to be similar to those in other lists called "What We Do in Science," and "What We Do in Language." If subject differences are to be deemphasized, an alternative would be to use "Numbers We Have Used," "Shapes We Have Used (or Made)," "Observations We Have Made," and so on.

The mathematics list might, for example, first include *count, measure, estimate,* and *see shapes in things,* then expand to include *add, subtract,* etc. Later, students could group together *add, subtract,* etc. under *do operations on numbers,* and group *make graphs, spread out data,* and *compare two groups of data* under *analyze data. Measure* would have to be included in the math and science lists, as would most of the *data* items, and would demonstrate linkages. Items such as *find patterns, describe relationships,* and *give reasons* would appear in the language list as well as the other lists. In this way, students would build their own inventory of mathematics and have a history of what they were learning, and newcomers could get an idea of what ideas (and language) would be expected of them.

By the end of the 5th grade, students should know that

▶ **Mathematics is the study of many kinds of patterns, including numbers and shapes and operations on them. Sometimes patterns are studied because they help to explain how the world works or how to solve practical problems, sometimes because they are interesting in themselves.**

▶ **Mathematical ideas can be represented concretely, graphically, and symbolically. ∎**

27

Grades 6 through 8

In earlier grades, students studied mathematical patterns and relationships in mathematics and, hopefully, in other classes as well. By now, they should have had considerable experience in making data tables, graphs, and geometric sketches and using them, along with symbols and clear English, to describe a wide variety of patterns and relationships. Thus students are now ready to concentrate more intensely than before on the creative aspects of mathematical problem-solving and to begin to develop a sense of how mathematicians go about their work.

Students start by beginning to reflect on what they do in mathematics. Groups of students should independently propose solutions to problems and compare their solutions with one another, defending and discussing differences. Groups should be encouraged to invent some of their own methods for making computations. One result could be recognition that more than one way works. But different groups might develop strong feelings about which method is best—thereby experiencing some of the heat that disagreements about mathematical abstractions have produced historically. Investigations of data sets should enable different groups of students to find different, perhaps even contradictory, relationships in them. Students can also begin to invent their own problems and see how they differ from those that other students find interesting.

To many students, the most "elegant" mathematics might seem to be the most complicated. Repeated reinforcement is necessary to establish that the simplest way of representing and connecting ideas is often what mathematicians value most. But a simple mathematical connection may have been found by very messy and prolonged study, which may include jumping back and forth from one part of the problem to another and sometimes running into dead ends.

By the end of the 8th grade, students should know that

▶ **Usually there is no one right way to solve a mathematical problem; different methods have different advantages and disadvantages.**

▶ **Logical connections can be found between different parts of mathematics.** ■

Grades 9 through 12

In addition to reflecting on personal problem-solving experience, case studies of how advances in mathematics have been made can be used to bring out some of the main features of how mathematics works and the kinds of patterns and relationships that have resulted from mathematical investigation. Students may occasionally discover mathematics for themselves, and although such discoveries are unlikely to be novel, a great deal should be made of them to encourage what may turn out to be a talent for mathematics.

Getting a grasp on the nature of today's mathematics, which is still engaged in eliciting new patterns and relationships, is likely to be a challenge for nearly all students. Modern theoretical mathematics may have to be suggested by the kind of practical problems it helps to solve—the coloring of maps, the optimizing of air routes, the recovery of detail from blurry images. If students believe that abstractions relevant to one practical situation are likely to be relevant to others as well, making the pedagogical transition from applied to abstract may not undermine the notion that the mathematician's interest is theoretical.

By the end of the 12th grade, students should know that

▶ **Mathematics is the study of any patterns or relationships, whereas natural science is concerned only with those patterns that are relevant to the observable world. Although mathematics began long ago in practical problems, it soon focused on abstractions from the material world, and then on even more abstract relationships among those abstractions.**

▶ **As in other sciences, simplicity is one of the highest values in mathematics. Some mathematicians try to identify the smallest set of rules from which many other propositions can be logically derived.**

▶ **Theories and applications in mathematical work influence each other. Sometimes a practical problem leads to the development of new mathematical theories; often mathematics developed for its own sake turns out to have practical applications.**

▶ **New mathematics continues to be invented, and connections between different parts of mathematics continue to be found.** ■

29

2B MATHEMATICS, SCIENCE, AND TECHNOLOGY

RESEARCH NOTES

page 334

Much of mathematics is done because of its intrinsic interest, without regard to its usefulness. Still, most mathematics does have applications, and much work in mathematics is stimulated by applied problems. Science and technology provide a large share of such applications and stimulants. In doing their work, scientists and engineers may attempt to do some useful mathematics themselves, or may call on mathematicians for help. The help may be to suggest some already-completed mathematics that will suffice or to develop some new mathematics to do the job. On the one hand, there have been some remarkable cases of finding new uses for centuries-old mathematics. On the other hand, the needs of natural science or technology have often led to the formulation of new mathematics.

Kindergarten through Grade 2

In the earliest grades, students make observations, collect and sort things, use tools, and build things. They are, for their level of development, doing science and using technology. In school practice, science and technology should contribute to understanding the value of mathematics, and mathematics should help in doing science and engineering. The usefulness of mathematics in science and technology will be clear to students if they experience it often.

▶ **No benchmarks for this level.** ■

Grades 3 through 5

The interaction should become more frequent and more sophisticated as students progress through the upper elementary and middle grades. Graphing, making tables, and making scale drawings should become commonplace in student inquiry and design projects, as should the use of geometric and mathematical concepts such as perpendicular, perimeter, volume, powers, roots, and negative numbers. Problems that are used to challenge students may take the form of contests and games, but at least some of the problems should stem directly from the science and technology being studied.

▶ **No benchmarks for this level.** ■

Grades 6 through 8

Science and technology are rich and especially important contexts in which to learn the value of mathematics and to develop mathematical problem-solving skills. But they are not the only ones. Art, music, social studies, history, physical education and sports, driver education, home economics, and other school subjects are appropriate places to *learn*, as well as use, mathematics.

By the end of the 8th grade, students should know that

▶ **Mathematics is helpful in almost every kind of human endeavor—from laying bricks to prescribing medicine or drawing a face. In particular, mathematics has contributed to progress in science and technology for thousands of years and still continues to do so. ■**

Grades 9 through 12

Students in this age range should be exposed to historical examples of how mathematics has contributed to the advancement of science and technology—and vice versa. The instances are so numerous that there is no trouble in finding some that are related to whatever mathematics is being studied. At some point, special attention should be paid to the use of mathematical models in both science and technology. Also, the curriculum needs to provide opportunities for students to examine explicitly the relationship of mathematics to science and technology.

By the end of the 12th grade, students should know that

▶ **Mathematical modeling aids in technological design by simulating how a proposed system would theoretically behave.**

▶ **Mathematics and science as enterprises share many values and features: belief in order, ideals of honesty and openness, the importance of criticism by colleagues, and the essential role played by imagination.**

▶ **Mathematics provides a precise language for science and technology—to describe objects and events, to characterize relationships between variables, and to argue logically.**

▶ **Developments in science or technology often stimulate innovations in mathematics by presenting new kinds of problems to be solved. In particular, the development of computer technology (which itself relies on mathematics) has generated new kinds of problems and methods of work in mathematics.**

▶ **Developments in mathematics often stimulate innovations in science and technology.** ∎

2_C MATHEMATICAL INQUIRY

Just what is it exactly that mathematicians *do* when they are doing mathematics? Most people have some sense, however inaccurate in detail, of what different occupations are because they encounter them personally or indirectly through books, movies, and television. They have little opportunity, however, to watch mathematicians at work or to have them explain what they do. Learning how to solve certain kinds of well-defined mathematics problems is important for students but does not automatically lead them to a broad understanding of how mathematical investigations are carried out.

Mathematics can be characterized as a cycle of investigation that is intended to lead to the development of valid mathematical ideas. That is the approach taken in *Science for All Americans* and in this section of *Benchmarks*. (Some of the same ground is covered in Chapter 11: Common Themes, in which the use of mathematical models is considered along with physical and conceptual models.)

RESEARCH NOTES

page 334

It is essential to keep in mind that mathematical discovery is no more the result of some rigid set of steps than is discovery in science. It is true that mathematical investigations sooner or later involve certain processes, but the order is not fixed and the emphasis placed on each process varies greatly. Each of the three parts of the cycle—representation, manipulation, and validation—should be studied in its own right as part of what constitutes learning mathematics. Students should have the chance to use the entire cycle in carrying out their own mathematical investigations. The purpose of this experience is to produce not professional mathematicians but adults who are familiar with mathematical inquiry.

Each part of the cycle poses some learning difficulties. The process of representing something by a symbol or expression is taken by many students to refer only to "real things." "Let A stand for the area of the floor in this room" is easier for young students to grasp than "Let Y equal the area of any rectangle." First, students have to be convinced that substituting abstract symbols for actual quantities is worth the effort. Then they need to work their way toward the realization that using symbols to represent abstractions, and abstractions of abstractions, also pays off in solving problems. Perhaps this means bringing students to see that in the world of mathematics, numbers, shapes, operations, symbols, and symbols that summarize sets of symbols are as "real" as blocks, cows, and dollars.

As to manipulation, there are two conditions that may seem contradictory to students. One is that there is always a set of rules that must be strictly adhered to; the other is that the rules can be made up. That is where the rigor and game-playing spirit of mathematics meet. Imagine some quantities, assign them properties, select some operations, represent everything by symbols, set a problem, and then, following the rules of logic that have been adopted, move the symbols around to see what solutions emerge.

But how good are the solutions? It depends—and that is what students may have trouble understanding. They are used to working mathematical problems in which the procedures are predetermined and "correct" answers are expected. But in real mathematical investigations, a good solution is one that results in new mathematical discoveries or that leads to practical outcomes in science, medicine, engineering, business, or elsewhere. Thus validation in mathematics is a matter of judgment, not authority. And where a solution is less than satisfactory, it may have as much to do with the sense of what is good enough or with how the problem was formulated as with how it was carried out.

CROCK RECHIN & WILDER

35

Kindergarten through Grade 2

Concrete objects should be employed routinely to help children discover and explain symbolic relationships. Students should come to see that numbers and shapes can be used to describe many things in the world around them. Eventually they should come to realize that just as letters and words make up a language in reading and writing, numbers and shapes make up a language in mathematics.

By the end of the 2nd grade, students should know that

▶ **Numbers and shapes can be used to tell about things.** ■

Grades 3 through 5

The routine use of concrete objects continues to be essential to help students connect real things and events with their abstract representations. The ability to picture and do things in their heads will be enhanced by frequent reference to real-world applications. Students should be encouraged to describe all sorts of things mathematically—in terms of numbers, shapes, and operations.

By the end of the 5th grade, students should know that

▶ **Numbers and shapes—and operations on them— help to describe and predict things about the world around us.**

▶ **In using mathematics, choices have to be made about what operations will give the best results. Results should always be judged by whether they make sense and are useful.** ■

Grades 6 through 8

Students should begin to assign letters as temporary names of objects—mathematical or not—for purposes of discussing these objects when no other name is known. Gradually the notion of a symbol standing in for a *particular* unknown can be extended to its standing for any of a *collection* of possible unknowns. Undoubtedly students will often have to return to concrete ideas as they learn new mathematics.

Students should examine the limitations of some mathematical models in describing and predicting events in the real world. (Disappointing results of mathematical modeling may be due to unpredictable variation in the real world, as well as to use of an inappropriate mathematical model.) Students should be encouraged to state their own criteria for what is a satisfactory result and to discuss their judgments in terms of their purposes.

The artificiality of problems should be minimized, so that there is not always a clear-cut right answer— and so that improvements and alternatives in the solution can be made through the mathematical cycle of trial, evaluation, and revision. A distinction should be drawn between mistakes (such as faulty multiplication) and reasonable choices that turn out to be unsuccessful (and can be reconsidered).

By the end of the 8th grade, students should know that

▶ **Mathematicians often represent things with abstract ideas, such as numbers or perfectly straight lines, and then work with those ideas alone. The "things" from which they abstract can be ideas themselves (for example, a proposition about "all equal-sided triangles" or "all odd numbers").**

▶ **When mathematicians use logical rules to work with representations of things, the results may or may not be valid for the things themselves. Using mathematics to solve a problem requires choosing what mathematics to use; probably making some simplifying assumptions, estimates, or approximations; doing computations; and then checking to see whether the answer makes sense. If an answer does not seem to make enough sense for its intended purpose, then any of these steps might have been inappropriate. ■**

Grades 9 through 12

So that students do not get the idea that there is always one best mathematical model for any science or technology problem, opportunities should be provided in which more than one mathematical description seems equally appropriate. The mathematical cycle of reasoning can first be considered explicitly by having students go back over how they solved problems before—and thereafter by recalling that to their attention whenever they approach new problems. The image of some mathematics as a "game" played with arbitrary rules should include the idea that the play is chosen with the goal that the results will be interesting and widely applicable. The rules of the game shouldn't be mutually contradictory—at least not within any intended applications.

By the end of the 12th grade, students should know that

▶ **Some work in mathematics is much like a game— mathematicians choose an interesting set of rules and then play according to those rules to see what can happen. The more interesting the results, the better. The only limit on the set of rules is that they should not contradict one another.**

▶ **Much of the work of mathematicians involves a modeling cycle, which consists of three steps: (1) using abstractions to represent things or ideas, (2) manipulating the abstractions according to some logical rules, and (3) checking how well the results match the original things or ideas. If the match is not considered good enough, a new round of abstraction and manipulation may begin. The actual thinking need not go through these processes in logical order but may shift from one to another in any order.** ■

39

FIG. 225

FIG. 226

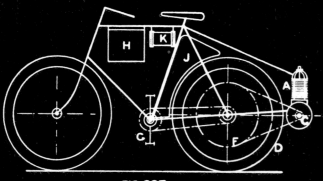

FIG. 227

FIG. 228

FIG. 229

FIG. 230

Chapter 3   THE NATURE OF TECHNOLOGY

A   TECHNOLOGY AND SCIENCE
B   DESIGN AND SYSTEMS
C   ISSUES IN TECHNOLOGY

As long as there have been people, there has been technology. Indeed, the techniques of shaping tools are taken as the chief evidence of the beginning of human culture. On the whole, technology has been a powerful force in the development of civilization, all the more so as its link with science has been forged. Technology—like language, ritual, values, commerce, and the arts—is an intrinsic part of a cultural system and it both shapes and reflects the system's values. In today's world, technology is a complex social enterprise that includes not only research, design, and crafts but also finance, manufacturing, management, labor, marketing, and maintenance.

In the broadest sense, technology extends our abilities to change the world: to cut, shape, or put together materials; to move things from one place to another; to reach farther with our hands, voices, and senses. We use technology to try to change the world to suit us better. The changes may relate to survival needs such as food, shelter, or defense, or they may relate to human aspirations such as knowledge, art, or control. But the results of changing the world are often complicated and unpredictable. They can include unexpected benefits, unexpected costs, and unexpected risks—any of which may fall on different social groups at different times. Anticipating the effects of technology is therefore as important as advancing its capabilities.

SCIENCE FOR ALL AMERICANS

In the United States, unlike in most developed countries in the world, technology as a subject has largely been ignored in the schools. It is not tied to graduation requirements, has no fixed place in elementary education, is absent altogether in the college preparatory curriculum, and does not constitute part of the content in science courses at any level.

However, that situation is now changing. There is growing awareness that technology works in everyday life to shape the character of civilization. Design projects are becoming more evident in the elementary grades, and the transformation of industrial arts and other subjects into technology education is gaining momentum. And the Science-Technology-Society (STS) emphasis in the curriculum is gaining adherents.

W.W. BEAUMONT
alternative designs for motorized bicycles, 1903.

The task ahead is to build technology education into the curriculum, as well as use technology to promote learning, so that all students become well informed about the nature, powers, and limitations of technology. As a human enterprise, technology has its own history and identity, quite apart from those of science and mathematics. In history, it preceded science and only gradually has come to draw on science—knowledge of how the natural world works—to help in controlling what happens in the world. In modern times, technology has become increasingly characterized by the interdependent relationships it has with science and mathematics. The benchmarks that follow suggest how students should develop their understanding of these relationships.

This chapter presents recommendations on what knowledge about the nature of technology is required for scientific literacy and emphasizes ways of thinking about technology that can contribute to using it wisely. Chapter 8: The Designed World presents principles relevant to some of the key technologies of today's world. Chapter 10: Historical Perspectives, includes a discussion of the Industrial Revolution. Chapter 12: Habits of Mind includes some skills relevant to participating in a technological world. ■

3A TECHNOLOGY AND SCIENCE

Technology is an overworked term. It once meant knowing how to do things—the practical arts or the study of the practical arts. But it has also come to mean innovations such as pencils, television, aspirin, microscopes, etc., that people use for specific purposes, and it refers to human activities such as agriculture or manufacturing and even to processes such as animal breeding or voting or war that change certain aspects of the world. Further, technology sometimes refers to the industrial and military institutions dedicated to producing and using inventions and know-how. In any of these senses, technology has economic, social, ethical, and aesthetic ramifications that depend on where it is used and on people's attitudes toward its use.

Sorting out these issues is likely to occur over many years as students engage in design and technology activities. First, they must use different tools to do different things in science and to solve practical problems. Through design and technology projects, students can engage in problem-solving related to a wide range of real-world contexts. By undertaking design projects, students can encounter technology issues even though they cannot define technology. They should have their attention called to the use of tools and instruments in science and the use of practical knowledge to solve problems before the underlying concepts are understood.

RESEARCH NOTES

page 334

ALSO SEE

Kindergarten through Grade 2

Young children are veteran technology users by the time they enter school. They ride in automobiles, use household utilities, operate wagons and bikes, use garden tools, help with the cooking, operate the television set, and so on. Children are also natural explorers and inventors, and they like to make things. School should give students many opportunities to examine the properties of materials, to use tools, and to design and build things. Activities should focus on problems and needs in and around the school that interest the children and that can be addressed feasibly and safely.

The task in these grades is to begin to channel the students' inventive energy and to increase their purposeful use of tools and—in the process—broaden their understanding of what constitutes a tool (a container, paper and pencil, camera, magnifier, etc.). Design and technology activities can be used to introduce students to measurement tools and techniques in a natural and meaningful manner. For example, five-year-olds have little trouble in designing and making things for their teddy bears built to an appropriate scale. Measurements should deal with magnitudes that are comprehensible to children of this age, which excludes, for example, the circumference of the earth or the diameter of a microbe.

By the end of the 2nd grade, students should know that

▶ **Tools are used to do things better or more easily and to do some things that could not otherwise be done at all. In technology, tools are used to observe, measure, and make things.**

▶ **When trying to build something or to get something to work better, it usually helps to follow directions if there are any or to ask someone who has done it before for suggestions.** ■

Grades 3 through 5

These years should build on the previous ones by increasing the sophistication of the design projects that students undertake. This approach entails students' increasing their repertoire of tools and techniques and improving their skills in measurement, calculation, and communication. Activities calling on the use of instruments such as microscopes, telescopes, cameras, and sound recorders to make observations and measurements are especially important for reinforcing the importance of the dependence of science on technology. Just as important, students should develop skill and confidence in using ordinary tools for personal, everyday purposes.

Students should begin now to write about technology, particularly about how technology helps people. Most of the complexities of the social consequences of the use of technology can wait, but students should begin to consider alternative ways of doing something and compare the advantages and disadvantages.

By the end of the 5th grade, students should know that

▶ **Throughout all of history, people everywhere have invented and used tools. Most tools of today are different from those of the past but many are modifications of very ancient tools.**

▶ **Technology enables scientists and others to observe things that are too small or too far away to be seen without them and to study the motion of objects that are moving very rapidly or are hardly moving at all.**

▶ **Measuring instruments can be used to gather accurate information for making scientific comparisons of objects and events and for designing and constructing things that will work properly.**

▶ **Technology extends the ability of people to change the world: to cut, shape, or put together materials; to move things from one place to another; and to reach farther with their hands, voices, senses, and minds. The changes may be for survival needs such as food, shelter, and defense, for communication and transportation, or to gain knowledge and express ideas.** ■

45

Grades 6 through 8

Students can now develop a broader view of technology and how it is both like and unlike science. They do not easily distinguish between science and technology, seeing both as trying to get things (including experiments) to happen the way one wants them to. There is no need to insist on definitions, but students' attention can be drawn to when they are clearly trying to find something out, clearly trying to make something happen, or doing some of each.

Furthermore, as students begin to think about their own possible occupations, they should be introduced to the range of careers that involve technology and science, including engineering, architecture, and industrial design. Through projects, readings, field trips, and interviews, students can begin to develop a sense of the great variety of occupations related to technology and to science, and what preparation they require.

By the end of the 8th grade, students should know that

▶ **In earlier times, the accumulated information and techniques of each generation of workers were taught on the job directly to the next generation of workers. Today, the knowledge base for technology can be found as well in libraries of print and electronic resources and is often taught in the classroom.**

▶ **Technology is essential to science for such purposes as access to outer space and other remote locations, sample collection and treatment, measurement, data collection and storage, computation, and communication of information.**

▶ **Engineers, architects, and others who engage in design and technology use scientific knowledge to solve practical problems. But they usually have to take human values and limitations into account as well. ■**

Grades 9 through 12

In addition to participating in major design projects to deepen their understanding of technology, students now should be helped to develop a richer sense of the relationships linking technology and science. That can come from reflection on the project experiences and from a study of the history of science and technology. Certain episodes in the history of science illustrate the importance of technology to science and the difficulty of clearly separating science and technology. The Industrial Revolution is especially important in this regard.

By the end of the 12th grade, students should know that

▶ **Technological problems often create a demand for new scientific knowledge, and new technologies make it possible for scientists to extend their research in new ways or to undertake entirely new lines of research. The very availability of new technology itself often sparks scientific advances.**

▶ **Mathematics, creativity, logic, and originality are all needed to improve technology.**

▶ **Technology usually affects society more directly than science because it solves practical problems and serves human needs (and may create new problems and needs). In contrast, science affects society mainly by stimulating and satisfying people's curiosity and occasionally by enlarging or challenging their views of what the world is like.** ■

3ʙ   Design and Systems

Engineering is the professional field most closely, or at least most deliberately, associated with technology. Engineers solve problems by applying scientific principles to practical ends. They design instruments, machines, structures, and systems to accomplish specified ends, and must do so while taking into account limitations imposed by time, money, law, morality, insufficient information, and more. In short, engineering has largely to do with the design of technological systems.

Perhaps the best way to become familiar with the nature of engineering and design is to do some. By participating in such activities, students should learn how to analyze situations and gather relevant information, define problems, generate and evaluate creative ideas, develop their ideas into tangible solutions, and assess and improve their solutions. To become good problem solvers, students need to develop drawing and modeling skills, along with the ability to record their analyses, suggestions, and results in clear language.

Gradually, as students participate in more sophisticated projects, they will encounter constraints and the need for making trade-offs. The concept of *trade-off* in technology—and more broadly in all social systems—is so important that teachers should put it into as many problem-solving contexts as possible. Students should be explicit in their own proposals about what is being traded off for what. They should learn to expect the same of others who propose technical, economic, or political solutions to problems.

Feedback should be another main concept learned in the study of technological systems. Students are likely to encounter it often in biology, physiology, politics, games, conversation, and even when operating tools and machines. Students should also learn that *technologies always have side effects* and that *all technological systems can fail.* These ideas can be introduced in simple form early and gradually become more prominent in the upper grades. Just as with *trade-off* and *feedback,* these new concepts should be encountered in a variety of contexts. Daily newspapers provide an inexhaustible supply of examples to analyze.

RESEARCH NOTES

page 334

Children should design and make things with simple tools and a variety of materials. They should identify a need or opportunity of interest to them, and then plan, design, make, evaluate, and modify the design with appropriate help. They might need help identifying problems that are both interesting to them and within their capabilities. After they gain experience working through one problem, they may find their next design project easier and feel more confident about trying it.

One design consideration to be introduced right away is *constraints*. Safety, time, cost, school policy, space, availability of materials, and other realities restrict student projects. Teachers can point out that adults also face constraints when they design things, and that the real challenge, for adults or children, is to devise solutions that give good results in spite of the restrictions. In the early grades, children may be inclined to go with their first design notion with little patience for testing or revision. Where possible, they should be encouraged to improve their ideas, but it is more important that they develop confidence in their ability to think up and carry out design projects. When their projects are complete, students can tell what they like about one another's designs.

By the end of the 2nd grade, students should know that

► **People may not be able to actually make or do everything that they can design.** ■

Students should become increasingly comfortable with developing designs and analyzing the product: "Does it work?" "Could I make it work better?" "Could I have used better materials?" The more experience students accrue, the less direct guidance they need. They should realize early that cooperative efforts and individual initiative are valuable in spotting and ironing out design glitches. They should begin to enjoy challenges that require them to clarify a problem, generate criteria for an acceptable solution, suggest possible solutions, try one out, and then make adjustments or start over with a newly proposed solution.

As students undertake more extensive design projects, emphasis should be placed on the notion that there usually is not one best design for a product or process, but a variety of alternatives and possibilities. One way to accomplish this goal is to have several groups design and execute solutions to the same problem and then discuss the advantages and disadvantages of each solution. Ideally, the problems should be "real" and engaging for the students.

By the end of the 5th grade, students should know that

► **There is no perfect design. Designs that are best in one respect (safety or ease of use, for example) may be inferior in other ways (cost or appearance). Usually some features must be sacrificed to get others. How such trade-offs are received depends upon which features are emphasized and which are down-played.**

continued

49

Grades **6** through **8**

▶ **Even a good design may fail. Sometimes steps can be taken ahead of time to reduce the likelihood of failure, but it cannot be entirely eliminated.**

▶ **The solution to one problem may create other problems.** ■

An idea to be developed in the middle grades is that complex systems require control mechanisms. The common thermostat for controlling room temperature is known to most students and can serve as a model for all control mechanisms. But students should explore how controls work in various kinds of systems— machines, athletic contests, politics, the human body, learning, etc. At some point, students should try to invent control mechanisms, which need not be mechanical or electrical, that they can actually put into operation.

The concept of side effects can be raised at this time, perhaps by using actual case studies of technologies (antibiotics, automobiles, spray cans, etc.) that turned out to have unexpected side effects. Students should also meet more interesting and challenging constraints as they work on design projects. Also, students should become familiar with many actual examples of how overdesign and redundancy are used to deal with uncertainty.

THE FAR SIDE   By GARY LARSON

Early experiments in transportation

By the end of the 8th grade, students should know that

▶ Design usually requires taking constraints into account. Some constraints, such as gravity or the properties of the materials to be used, are unavoidable. Other constraints, including economic, political, social, ethical, and aesthetic ones, limit choices.

▶ All technologies have effects other than those intended by the design, some of which may have been predictable and some not. In either case, these side effects may turn out to be unacceptable to some of the population and therefore lead to conflict between groups.

▶ Almost all control systems have inputs, outputs, and feedback. The essence of control is comparing information about what is happening to what people want to happen and then making appropriate adjustments. This procedure requires sensing information, processing it, and making changes. In almost all modern machines, microprocessors serve as centers of performance control.

▶ Systems fail because they have faulty or poorly matched parts, are used in ways that exceed what was intended by the design, or were poorly designed to begin with. The most common ways to prevent failure are pretesting parts and procedures, overdesign, and redundancy. ■

Grades 9 through 12

Adequate time should be spent fleshing out the concepts of resources (tools, materials, energy, information, people, capital, time), systems, control, and impacts introduced in earlier grades. Students should also move to higher levels of critical and creative thinking through progressively more demanding design and technology work. They need practice as individuals and as members of a group in developing and defining ideas using drawings and models.

New concepts to be introduced in high school include risk analysis and technology assessment. Students should become aware that designed systems are subject to failure but that the risk of failure can be reduced by a variety of means: overdesign, redundancy, fail-safe designs, more research ahead of time, more controls, etc. They should also come to recognize that these precautions add costs that may become prohibitive, so that few designs are ideal.

Because no number of precautions can reduce the risk of system failure to zero, comparing the estimated risks of a proposed technology to its alternatives is often necessary. The choice, usually, is not between a high-risk option and a risk-free one, but comes down to making a trade-off among actions, all of which involve some risk.

continued

51

Students should realize that analyzing risk entails looking at probabilities of events and at how bad the events would be if they were to happen. Through surveys and interviews, students can learn that comparing risks is difficult because people vary greatly in their perception of risk, which tends to be influenced by such matters as whether the risk is gradual or instantaneous (global warming versus plane crashes), how much control people think they have over the risk (cigarette smoking versus being struck by lightning), and how the risk is expressed (the number of people affected versus the proportion affected).

By the end of the 12th grade, students should know that

▶ **In designing a device or process, thought should be given to how it will be manufactured, operated, maintained, replaced, and disposed of and who will sell, operate, and take care of it. The costs associated with these functions may introduce yet more constraints on the design.**

▶ **The value of any given technology may be different for different groups of people and at different points in time.**

▶ **Complex systems have layers of controls. Some controls operate particular parts of the system and some control other controls. Even fully automatic systems require human control at some point.**

▶ **Risk analysis is used to minimize the likelihood of unwanted side effects of a new technology. The public perception of risk may depend, however, on psychological factors as well as scientific ones.**

▶ **The more parts and connections a system has, the more ways it can go wrong. Complex systems usually have components to detect, back up, bypass, or compensate for minor failures.**

▶ **To reduce the chance of system failure, performance testing is often conducted using small-scale models, computer simulations, analogous systems, or just the parts of the system thought to be least reliable. ■**

3c ISSUES IN TECHNOLOGY

More and more, citizens are called on to decide which technologies to develop, which to use, and how to use them. Part of being prepared for that responsibility is knowing about how technology works, including its alternatives, benefits, risks, and limitations. The long-term interests of society are best served when key issues concerning proposals to introduce or curtail technology are addressed before final decisions are made. Students should learn how to ask important questions about the immediate and long-range impacts that technological innovations and the elimination of existing technologies are likely to have. But intelligent adults disagree about wise use of technology. Schooling should help students learn how to think critically about technology issues, not what to think about them. Teachers can help students acquire informed attitudes on the various technologies and their social, cultural, economic, and ecological consequences. When teachers do express their personal views (to demonstrate that adults can have well-informed opinions), they should also acknowledge alternative views and fairly state the evidence, logic, and values that lead other people to have those views.

Understanding the potential impact of technology may be critical to civilization. Technology is not innately good, bad, or neutral. Typically, its effects are complex, hard to estimate accurately, and likely to have different values for different people at different times. Its effects depend upon human decisions about development and use. Human experience with technology, including the invention of processes and tools, shows that people have some control over their destiny. They can tackle problems by searching for better ways to do things, inventing solutions and taking risks.

Case studies of actual technologies provide an excellent way for students to discuss risk. There is a vast array of topics: the Aswan High Dam, the contraceptive pill, steam engines, pesticides, public-opinion polling, penicillin, standardized parts, refrigeration, nuclear power, fluoridated water, and hundreds more. Teachers and students can assemble case-study material or use commercially developed case studies. Good design projects and case studies can help students to develop insight into experience.

RESEARCH NOTES

page 335

ALSO SEE

Kindergarten through Grade 2

Design projects give students interesting opportunities to solve problems, use tools well, measure things carefully, make reasonable estimations, calculate accurately, and communicate clearly. And projects also let students ponder the effects their inventions might have. For example, if a group of the children in a class decides to build a large shallow tank to create an ocean habitat, the whole class should discuss what happens if the tank leaks, whether this project interferes with other projects or classroom activities, whether there are other ways to learn about ocean habitats, and so forth. More generally, young children can begin to learn about the effects that people have on their surroundings.

Students at this level are old enough to see that solving some problems may lead to other problems, but the social impact matters should not be pressed too hard now. That might overemphasize constraints and take much of the fun out of doing simple projects by requiring too much analysis.

By the end of the 2nd grade, students should know that

▶ **People, alone or in groups, are always inventing new ways to solve problems and get work done. The tools and ways of doing things that people have invented affect all aspects of life.**

▶ **When a group of people wants to build something or try something new, they should try to figure out ahead of time how it might affect other people.** ■

Grades 3 through 5

Students can become interested in comparing present technology with that of earlier times, as well as the technology in their everyday lives with that of other places in the world. They can imagine what life would be like without certain technology, as well as what new technology the future might hold. Reading about other civilizations or earlier times than their own will illustrate the central role that different technologies play. Students may get involved in current campaigns related to technology—saving energy, recycling materials, reducing litter, and the like. Waste disposal may be a particularly comprehensible and helpful topic in directing their attention to the side effects of technology.

By the end of the 5th grade, students should know that

▶ **Technology has been part of life on the earth since the advent of the human species. Like language, ritual, commerce, and the arts, technology is an intrinsic part of human culture, and it both shapes society and is shaped by it. The technology available to people greatly influences what their lives are like.**

▶ **Any invention is likely to lead to other inventions. Once an invention exists, people are likely to think up ways of using it that were never imagined at first.**

▶ **Transportation, communications, nutrition, sanitation, health care, entertainment, and other technologies give large numbers of people today the goods and services that once were luxuries enjoyed only by the wealthy. These benefits are not equally available to everyone.**

▶ Scientific laws, engineering principles, properties of materials, and construction techniques must be taken into account in designing engineering solutions to problems. Other factors, such as cost, safety, appearance, environmental impact, and what will happen if the solution fails also must be considered.

▶ Technologies often have drawbacks as well as benefits. A technology that helps some people or organisms may hurt others—either deliberately (as weapons can) or inadvertently (as pesticides can). When harm occurs or seems likely, choices have to be made or new solutions found.

▶ Because of their ability to invent tools and processes, people have an enormous effect on the lives of other living things. ■

To enrich their understanding of how technology has shaped how people live now, students should examine what life was like under different technological circumstances in the past. They should become aware that significant changes occurred in the lives of people when technology provided more and better food, control of sewage, heat and light for homes, and rapid transportation. Studying the past should engender respect for the inventions and constructions of earlier civilizations and cultures.

Both historical and literary approaches ought to be used to imagine what the future will bring and to reflect on people's somewhat limited ability to predict the future. Science fiction and novels set in future times suggest changes in human life that might occur because of yet-uninvented technology. Stories selected for this purpose should raise many different issues regarding the impact of technology, and students should probe beneath the plot to analyze those issues. Student groups can formulate and compare their own scenarios for some future time—say, when they are adults.

By the end of the 8th grade, students should know that

▶ The human ability to shape the future comes from a capacity for generating knowledge and developing new technologies—and for communicating ideas to others.

▶ Technology cannot always provide successful solutions for problems or fulfill every human need.

continued

▶ **Throughout history, people have carried out impressive technological feats, some of which would be hard to duplicate today even with modern tools. The purposes served by these achievements have sometimes been practical, sometimes ceremonial.**

▶ **Technology has strongly influenced the course of history and continues to do so. It is largely responsible for the great revolutions in agriculture, manufacturing, sanitation and medicine, warfare, transportation, information processing, and communications that have radically changed how people live.**

▶ **New technologies increase some risks and decrease others. Some of the same technologies that have improved the length and quality of life for many people have also brought new risks.**

▶ **Rarely are technology issues simple and one-sided. Relevant facts alone, even when known and available, usually do not settle matters entirely in favor of one side or another. That is because the contending groups may have different values and priorities. They may stand to gain or lose in different degrees, or may make very different predictions about what the future consequences of the proposed action will be.**

▶ **Societies influence what aspects of technology are developed and how these are used. People control technology (as well as science) and are responsible for its effects.** ■

As suggested earlier, the real-world work of students as supplemented by case studies probably provides the most effective way to examine issues related to how society responds to the promise or threat of technological change—whether by adopting new technologies or curtailing the use of existing ones. What must be avoided by teachers is turning the case studies into occasions for promoting a particular point of view. People tend to hold very strong opinions on the use of technologies, and not only of nuclear reactors and genetic engineering. The teacher's job is not to provide students with the "right" answers about technology but to see to it that students know what questions to ask.

Students can also add detail to their awareness of the effects of the human presence on life. For instance, they should be able to cite several examples of how the introduction of foreign species has changed an ecosystem. Out of this should come an awareness that people can make some decisions about what life on earth will survive and a sense of responsibility about exercising power. Students also should learn that people cannot shape every aspect of life to their own liking.

For example, most Americans recognize that technology has provided new goods and services, but not that industrialization of agriculture, by eliminating the need for children to work in the fields, made it possible for them to attend school, thereby increasing the general educational level of the population. These kinds of social impacts should be studied as well as those that affect human health and the environment.

By the end of the 12th grade, students should know that

▶ Social and economic forces strongly influence which technologies will be developed and used. Which will prevail is affected by many factors, such as personal values, consumer acceptance, patent laws, the availability of risk capital, the federal budget, local and national regulations, media attention, economic competition, and tax incentives.

▶ Technological knowledge is not always as freely shared as scientific knowledge unrelated to technology. Some scientists and engineers are comfortable working in situations in which some secrecy is required, but others prefer not to do so. It is generally regarded as a matter of individual choice and ethics, not one of professional ethics.

▶ In deciding on proposals to introduce new technologies or to curtail existing ones, some key questions arise concerning alternatives, risks, costs, and benefits. What alternative ways are there to achieve the same ends, and how do the alternatives compare to the plan being put forward? Who benefits and who suffers? What are the financial and social costs, do they change over time, and who bears them? What are the risks associated with using (or not using) the new technology, how serious are they, and who is in jeopardy? What human, material, and energy resources will be needed to build, install, operate, maintain, and replace the new technology, and where will they come from? How will the new technology and its waste products be disposed of and at what costs?

▶ The human species has a major impact on other species in many ways: reducing the amount of the earth's surface available to those other species, interfering with their food sources, changing the temperature and chemical composition of their habitats, introducing foreign species into their ecosystems, and altering organisms directly through selective breeding and genetic engineering.

▶ Human inventiveness has brought new risks as well as improvements to human existence. ∎

THOMAS HART BENTON, *Trail Riders, 1964-65.*

Chapter 4 THE PHYSICAL SETTING

Humans have never lost interest in trying to find out how the universe is put together, how it works, and where they fit in the cosmic scheme of things. The development of our understanding of the architecture of the universe is surely not complete, but we have made great progress. Given a universe that is made up of distances too vast to reach and of particles too small to see and too numerous to count, it is a tribute to human intelligence that we have made as much progress as we have in accounting for how things fit together. All humans should participate in the pleasure of coming to know their universe better.

SCIENCE FOR ALL AMERICANS

One of the grand success stories of science is the unification of the physical universe. It turns out that all natural objects, events, and processes are connected to each other in such a way that only a relatively few concepts are needed to make sense of them.

In a way, this fact complicates efforts to delineate what students should know about the makeup and structure of the universe. Any one arrangement of topics inevitably neglects many cross-connections among topics. In the arrangement used here (and also in *Science for All Americans*), benchmarks dealing with gravity, electromagnetism, and scale appear in several different sections. For example, 4A: The Universe, 4B: The Earth, 4F: Motion, and 4G: Forces of Nature are intimately linked by ideas of gravitational attraction and immense scales of distance, mass, and time. And 4D: Structure of Matter, 4E: Energy Transformations, and 4G: Forces of Nature are linked by ideas of electromagnetism and minute scales of distance, mass, and energy. Benchmarks for any section are connected to others and should be read in the context of the others.

The physical universe is a subject in which many ideas make high demands on students' comprehension and imagination. Students in elementary school can only begin to form notions of stars and matter. The drastically different scales of astronomical and atomic phenomena can be learned only over many years. But it is important that all students develop a sense of the context of place, time, and physical interactions in which their lives occur. Students in the early years are especially curious about how the world works.

Consequently, there is a dilemma about when to introduce ideas into the curriculum. On the one hand, rushing to teach elementary students about atoms or galaxies is not likely to be productive. Most students will only learn to recite facts about them, with little comprehension. On the other hand, discussion and images about such imponderables are common in the popular media, and avoiding them seems unreasonable. The curriculum can focus on experiences and ideas that are accessible to children—for example, how different other planets are from the earth, or the different kinds of materials found in nature. And it can build in precursors to eventual understanding, such as observable motions in the sky and observable changes in materials. ■

Calvin and Hobbes   by Bill Watterson

4_A   THE UNIVERSE

In earlier times, people everywhere were much more aware of the stars and were familiar with them in ways that few people today are. Back then, people knew the patterns of stars in the night sky, the regularity of the motions of the stars, and how those motions related to the seasons. They used their knowledge to plan the planting of crops and to navigate boats. The constellations, along with the sun, the moon, and the "wanderers"—the planets—have always figured in the efforts of people to explain themselves and their world through stories, myths, religions, and philosophies.

For all of that, and for the sheer wonder the stars provoke on a clear, moonless night far from city lights—awe that has inspired the expressive powers of poets, musicians, and artists—science is not needed. Why, then, insist that everyone become familiar with the heavens as portrayed by science? Consider that in cities the night sky is no longer a familiar part of a person's neighborhood. Many people today live in circumstances that deprive them of the chance to see the sky often enough to become personally familiar with it. Fortunately, telescopes, photography, computers, and space probes make up the difference by revealing more of the cosmos in greater detail than ever before. Thus, science education can bring back the sky—not the same sky, but one that is richer and more varied than people's eyes alone had ever led them to imagine.

Finding our place in the cosmic scheme of things and how we got here is a task for the ages—past, present, and future. The scientific effort to understand the universe is part of that enduring human imperative, and its successes are a tribute to human curiosity, resourcefulness, intelligence, and doggedness. If being educated means having an informed sense of time and place, then it is essential for a person to be familiar with the scientific aspects of the universe and know something of its origin and structure.

In thinking about what students should learn about the heavens, at least three aspects of the current scientific view ought to be taken into account: (1) the composition of the cosmos and its scale of space and time; (2) the principles on which the universe seems to operate; and (3) how the modern view of the universe emerged. The benchmarks in this section deal primarily with composition and scale; principles are dealt with in subsequent sections of the chapter, and some rudiments of the history of the scientific picture appear in Chapter 10: Historical Perspectives.

RESEARCH NOTES
page 335

◁ **ALSO SEE**

Kindergarten through Grade 2

During these years, learning about objects in the sky should be entirely observational and qualitative, for the children are far from ready to understand the magnitudes involved or to make sense out of explanations. The priority is to get the students noticing and describing what the sky looks like to them at different times. They should, for example, observe how the moon appears to change its shape. But it is too soon to name all the moon's phases and much too soon to explain them.

By the end of the 2nd grade, students should know that

▶ There are more stars in the sky than anyone can easily count, but they are not scattered evenly, and they are not all the same in brightness or color.

▶ The sun can be seen only in the daytime, but the moon can be seen sometimes at night and sometimes during the day. The sun, moon, and stars all appear to move slowly across the sky.

▶ The moon looks a little different every day, but looks the same again about every four weeks. ■

Grades 3 through 5

Students should begin to develop an inventory of the variety of things in the universe. Planets can be shown to be different from stars in two essential ways—their appearance and their motion. When a modest telescope or pair of binoculars is used instead of the naked eyes, stars only look brighter—and more of them can be seen. The brighter planets, however, clearly are disks. (Not very large disks except in good-sized telescopes, but impressive enough after seeing a lot of stars.) The fixed patterns of stars should be made more explicit, although learning the constellation names is not important in itself. When students know that the star patterns stay the same as they move across the sky (and gradually shift with the seasons), they can then observe that the planets change their position against the pattern of stars.

Once students have looked directly at the stars, moon, and planets, use can be made of photographs of planets and their moons and of various collections of stars to point out their variety of size, appearance, and motion. No particular educational value comes from memorizing their names or counting them, although some students will enjoy doing so. Nor should students invest much time in trying to get the scale of distances firmly in mind. As to numbers of stars in the universe, few children will have much of an idea of what a billion is; thousands are enough of a challenge. (At this stage, a billion means more than a person could ever count one-at-a-time in an entire lifetime.)

Students' grasp of many of the ideas of the composition and magnitude of the universe has to grow slowly over time. Moreover, in spite of its common depiction, the sun-centered system seriously conflicts with common intuition. Students may need compelling reasons to really abandon their earth-centered views. Unfortunately, some of the best reasons are subtle and make sense only at a fairly high level of sophistication.

Some ideas about light and sight are prerequisite to understanding astronomical phenomena. Children should learn early that a large light source at a great distance looks like a small light source that is much closer. This phenomenon should be observed directly (and, if possible, photographically) outside at night. How things are seen by their reflected light is a difficult concept for children at this age, but is probably necessary for them to learn before phases of the moon will make sense.

By the end of the 5th grade, students should know that

▶ **The patterns of stars in the sky stay the same, although they appear to move across the sky nightly, and different stars can be seen in different seasons.**

▶ **Telescopes magnify the appearance of some distant objects in the sky, including the moon and the planets. The number of stars that can be seen through telescopes is dramatically greater than can be seen by the unaided eye.**

▶ **Planets change their positions against the background of stars.**

▶ **The earth is one of several planets that orbit the sun, and the moon orbits around the earth.**

▶ **Stars are like the sun, some being smaller and some larger, but so far away that they look like points of light.** ■

Grades 6 through 8

Students should add more detail to their picture of the universe, pay increasing attention to matters of scale, and back up their understanding with activities using a variety of astronomical tools. Student access to star finders, telescopes, computer simulations of planetary orbits, or a planetarium can be useful at this level. Figuring out and constructing models of size and distance—for example, of the planets within the solar system—is probably the most effective activity. Models with three dimensions are preferable to pictures and diagrams. Everyone should experience trying to fashion a physical model of the solar system in which the same scale is used for the sizes of the objects and the distances between them (as distinct from most illustrations, in which distances are underrepresented by a factor of 10 or more).

Some experiences with how apparent positions of objects differ from different points of observation will make plausible the estimation of distances to the moon and sun. Finding distances with scale drawings will help students to understand how the distances to the moon and sun were estimated and why the stars must be very much farther away. (The dependence of apparent size on distance can be used to pose the historically important puzzle that star patterns do *not* appear any larger from one season to the next, even though the earth swings a hundred million miles closer to them.)

Using light years to express astronomical distances is not as straightforward as it seems. (Many adults think of light years as a measure of time.) Beginning with analogs such as "automobile hours" may help.

By the end of the 8th grade, students should know that

▶ The sun is a medium-sized star located near the edge of a disk-shaped galaxy of stars, part of which can be seen as a glowing band of light that spans the sky on a very clear night. The universe contains many billions of galaxies, and each galaxy contains many billions of stars. To the naked eye, even the closest of these galaxies is no more than a dim, fuzzy spot.

▶ The sun is many thousands of times closer to the earth than any other star. Light from the sun takes a few minutes to reach the earth, but light from the next nearest star takes a few years to arrive. The trip to that star would take the fastest rocket thousands of years. Some distant galaxies are so far away that their light takes several billion years to reach the earth. People on earth, therefore, see them as they were that long ago in the past.

▶ Nine planets of very different size, composition, and surface features move around the sun in nearly circular orbits. Some planets have a great variety of moons and even flat rings of rock and ice particles orbiting around them. Some of these planets and moons show evidence of geologic activity. The earth is orbited by one moon, many artificial satellites, and debris.

▶ Large numbers of chunks of rock orbit the sun. Some of those that the earth meets in its yearly orbit around the sun glow and disintegrate from friction as they plunge through the atmosphere— and sometimes impact the ground. Other chunks of rocks mixed with ice have long, off-center orbits that carry them close to the sun, where the sun's radiation (of light and particles) boils off frozen material from their surfaces and pushes it into a long, illuminated tail. ■

This is the time for all of the pieces to come together. Concepts from physics and chemistry, insights from history, mathematical ways of thinking, and ideas about the role of technology in exploring the universe all contribute to a grasp of the character of the cosmos. In particular, the role of gravity in forming and maintaining planets, stars, and the solar system should become clear. The scale of billions will make better sense, and the speed of light can be used to express relative distances conveniently.

By the end of the 12th grade, students should know that

▶ **The stars differ from each other in size, temperature, and age, but they appear to be made up of the same elements that are found on the earth and to behave according to the same physical principles. Unlike the sun, most stars are in systems of two or more stars orbiting around one another.**

▶ **On the basis of scientific evidence, the universe is estimated to be over ten billion years old. The current theory is that its entire contents expanded explosively from a hot, dense, chaotic mass. Stars condensed by gravity out of clouds of molecules of the lightest elements until nuclear fusion of the light elements into heavier ones began to occur. Fusion released great amounts of energy over millions of years. Eventually, some stars exploded, producing clouds of heavy elements from which other stars and planets could later condense. The process of star formation and destruction continues.**

▶ **Increasingly sophisticated technology is used to learn about the universe. Visual, radio, and x-ray telescopes collect information from across the entire spectrum of electromagnetic waves; computers handle an avalanche of data and increasingly complicated computations to interpret them; space probes send back data and materials from the remote parts of the solar system; and accelerators give subatomic particles energies that simulate conditions in the stars and in the early history of the universe before stars formed.**

▶ **Mathematical models and computer simulations are used in studying evidence from many sources in order to form a scientific account of the universe.** ■

65

4ʙ   THE EARTH

An integrated picture of the earth has to develop over many years, with some concepts being visited over and over again in new contexts and greater detail. Some aspects can be learned in science, others in geography; some parts can be purely descriptive, others must draw on physical principles. The benchmarks in this section complement those of the previous section that locate the earth in the cosmos and those of the following section that focus on the surface of the earth. This arrangement does not imply any particular order of teaching. Often, teaching near-at-hand phenomena before teaching the far-distant ones makes sense; on the other hand, sometimes the near-to-far progression that makes sense

cognitively may not correspond to what interests children.

Perhaps the most important reason for students to study the earth repeatedly is that they take years to acquire the knowledge that they need to complete the picture. The full picture requires the introduction of such concepts as temperature, the water cycle, gravitation, states of matter, chemical concentration, and energy transfer. Understanding of these concepts grows slowly as children mature and encounter them in different contexts.

The benchmarks here call for students to be able to explain two phenomena—the seasons and the phases of the moon—that are usually not learned well. Most adults are unable to give even approximately correct explanations for them. Most students are *told* by teachers what causes the seasons and the phases of the moon, and they read about them without understanding. Moon phases are difficult because of students' unfamiliarity with the geometry of light and "seeing." To help figure out the geometry, students can act out the sun-earth-moon relationships and make physical models. In trying to understand the seasons, students have difficulties regarding geometry and solar radiation. Students need direct experience with light and surfaces—shadows, reflection, and warming effects at different angles.

RESEARCH NOTES
page 335

ALSO SEE ☞

Kindergarten through Grade 2

There are many ways to acquaint children with earth-related phenomena that they will only come to understand later as being cyclic. For instance, students can start to keep daily records of temperature (hot, cold, pleasant) and precipitation (none, some, lots), and plot them by week, month, and years. It is enough for students to spot the pattern of ups and downs, without getting deeply into the nature of climate. They should become familiar with the freezing of water and melting of ice (with no change in weight), the disappearance of wetness, and the appearance of water on cold surfaces. Evaporation and condensation will mean nothing different from disappearance and appearance, perhaps for several years, until students begin to understand that the evaporated water is still present in the form of invisibly small molecules.

By the end of the 2nd grade, students should know that

▶ **Some events in nature have a repeating pattern. The weather changes some from day to day, but things such as temperature and rain (or snow) tend to be high, low, or medium in the same months every year.**

▶ **Water can be a liquid or a solid and can go back and forth from one form to the other. If water is turned into ice and then the ice is allowed to melt, the amount of water is the same as it was before freezing.**

▶ **Water left in an open container disappears, but water in a closed container does not disappear. ■**

Grades 3 through 5

During this period, students can begin to learn some of the surface features of the earth and also the earth's relation to the sun, moon, and other planets. Films, computer simulations, a planetarium, and telescopic observations will help, but it is essential that all students, sometimes working together in small groups, make physical models and explain what the models show. At the same time, students can begin learning about scale (counting, comparative distances, volumes, times, etc.) in interesting, readily understood activities and readings. However, scale factors larger than thousands, and even the idea of ratios, may be difficult before early adolescence.

An important point to be made along the way is that one cannot determine how the solar system is put together just by looking at it. Diagrams show what the system would look like if people could see it from far away, a feat that cannot be accomplished. Telescopes and other instruments do provide information, but a model is really needed to make sense out of the information. (The realization that people are not able to see, from the outside, how the solar system is constructed will help students understand the basis for the Copernican Revolution when the topic arises later.)

In making diagrams to show, say, the relative sizes of the planets and the distances of the planets from the sun, students may try to combine them using a single scale—and quickly become frustrated. Perhaps this can lead to a discussion of the general limits of graphic methods (including photographs) for showing reality. In any case, at this stage a rough picture of the organization of the solar system is enough.

Water offers another important set of experiences for students at this level. Students can conduct investigations that go beyond the observations made in the earlier grades to learn the connection between liquid and solid forms, but recognizing that water can also be a gas, while much more difficult, is still probably accessible. Perhaps the main thrust there is to try to figure out where water in an open container goes. This is neither self-evident nor easy to detect. But the water cycle is of such profound importance to life on earth that students should certainly have experiences that will in time contribute to their understanding of evaporation, condensation, and the conservation of matter.

By the end of the 5th grade, students should know that

▶ **Things on or near the earth are pulled toward it by the earth's gravity.**

▶ **Like all planets and stars, the earth is approximately spherical in shape. The rotation of the earth on its axis every 24 hours produces the night-and-day cycle. To people on earth, this turning of the planet makes it seem as though the sun, moon, planets, and stars are orbiting the earth once a day.**

▶ **When liquid water disappears, it turns into a gas (vapor) in the air and can reappear as a liquid when cooled, or as a solid if cooled below the freezing point of water. Clouds and fog are made of tiny droplets of water.**

▶ **Air is a substance that surrounds us, takes up space, and whose movement we feel as wind.** ■

Grades 6 through 8

Students can now consolidate their prior knowledge of the earth (as a planet) by adding more details (especially about climate), getting a firmer grasp of the geometry involved in explaining the seasons and phases of the moon, improving their ability to handle scale, and shifting their frame of reference away from the earth when needed. An inevitable paradox of the large scales involved is that an ocean that is difficult to imagine being 7 miles deep also can be considered a "relatively thin" layer on the earth's surface. Students should exercise their understanding of the paradox, perhaps by debating provocative questions such as "Is the ocean amazingly deep or amazingly shallow?"

Gravity, earlier thought of as acting toward the ground, can by now be thought of as acting toward the center of the spherical earth and reaching indefinitely into space. It is also time for students to begin to look at the planet's role in sustaining life—a complex subject that involves many different issues and benchmarks. In this section, the emphasis is on water and air as essential resources.

The cause of the seasons is a subtle combination of global and orbital geometry and of the effects of radiation at different angles. Students can learn part of the story at this grade level, but a complete picture cannot be expected until later.

By the end of the 8th grade, students should know that

▶ **We live on a relatively small planet, the third from the sun in the only system of planets definitely known to exist (although other, similar systems may be discovered in the universe).**

▶ **The earth is mostly rock. Three-fourths of its surface is covered by a relatively thin layer of** water (some of it frozen), and the entire planet is surrounded by a relatively thin blanket of air. It is the only body in the solar system that appears able to support life. The other planets have compositions and conditions very different from the earth's.

▶ Everything on or anywhere near the earth is pulled toward the earth's center by gravitational force.

▶ Because the earth turns daily on an axis that is tilted relative to the plane of the earth's yearly orbit around the sun, sunlight falls more intensely on different parts of the earth during the year. The difference in heating of the earth's surface produces the planet's seasons and weather patterns.

▶ The moon's orbit around the earth once in about 28 days changes what part of the moon is lighted by the sun and how much of that part can be seen from the earth—the phases of the moon.

▶ Climates have sometimes changed abruptly in the past as a result of changes in the earth's crust, such as volcanic eruptions or impacts of huge rocks from space. Even relatively small changes in atmospheric or ocean content can have widespread effects on climate if the change lasts long enough.

▶ The cycling of water in and out of the atmosphere plays an important role in determining climatic patterns. Water evaporates from the surface of the earth, rises and cools, condenses into rain or

snow, and falls again to the surface. The water falling on land collects in rivers and lakes, soil, and porous layers of rock, and much of it flows back into the ocean.

▶ Fresh water, limited in supply, is essential for life and also for most industrial processes. Rivers, lakes, and groundwater can be depleted or polluted, becoming unavailable or unsuitable for life.

▶ Heat energy carried by ocean currents has a strong influence on climate around the world.

▶ Some minerals are very rare and some exist in great quantities, but—for practical purposes— the ability to recover them is just as important as their abundance. As minerals are depleted, obtaining them becomes more difficult. Recycling and the development of substitutes can reduce the rate of depletion but may also be costly.

▶ The benefits of the earth's resources—such as fresh water, air, soil, and trees—can be reduced by using them wastefully or by deliberately or inadvertently destroying them. The atmosphere and the oceans have a limited capacity to absorb wastes and recycle materials naturally. Cleaning up polluted air, water, or soil or restoring depleted soil, forests, or fishing grounds can be very difficult and costly. ■

69

Grades 9 through 12

Two important strands of understanding can now be pulled together to enrich students' views of the physical setting. One strand connects such physical concepts and principles as energy, gravitation, conservation, and radiation to the descriptive picture that students have built in their minds about the operation of the planets. The other strand consists of the Copernican Revolution, which illustrates the place of technology, mathematics, experimentation, and theory in scientific breakthroughs. In the context of thinking about how the solar system is put together, this historical event unites physics and astronomy, involves colorful personalities, and raises deep philosophical and political issues.

By the end of the 12th grade, students should know that

▶ Life is adapted to conditions on the earth, including the force of gravity that enables the planet to retain an adequate atmosphere, and an intensity of radiation from the sun that allows water to cycle between liquid and vapor.

▶ Weather (in the short run) and climate (in the long run) involve the transfer of energy in and out of the atmosphere. Solar radiation heats the land masses, oceans, and air. Transfer of heat energy at the boundaries between the atmosphere, the land masses, and the oceans results in layers of different temperatures and densities in both the ocean and atmosphere. The action of gravitational force on regions of different densities causes them to rise or fall—and such circulation, influenced by the rotation of the earth, produces winds and ocean currents. ■

4c Processes that Shape the Earth

Students should learn what causes earthquakes, volcanos, and floods and how those events shape the surface of the earth. Students, however, may show more interest in the phenomena than in the role the phenomena play in sculpting the earth. So teachers should start with students' immediate interests and work toward the science. Students may find it harder to take seriously the less-obvious, less-dramatic, long-term effects of erosion by wind and water, annual deposits of sediment, the creep of continents, and the rise of mountains. Students' recognition of those effects will depend on an improving sense of long time periods and familiarity with the effect of multiplying tiny fractions by very large numbers (in this case, slow rates by long times).

Students can start in the early grades with the ways in which organisms, themselves included, modify their surroundings. As people have used earth resources, they have altered some earth systems. Students can gradually come to recognize how human behavior affects the earth's capacity to sustain life. Questions of environmental policy should be pursued when students become interested in them, usually in the middle grades or later, but care should be taken not to bypass science for advocacy. Critical thinking based on scientific concepts and understanding is the primary goal for science education.

Research Notes
page 336

◁ ALSO SEE

Kindergarten through Grade 2

Teaching geological facts about how the face of the earth changes serves little purpose in these early years. Students should start becoming familiar with all aspects of their immediate surroundings, including what things change and what seems to cause change. Perhaps "changing things" can be a category in a class portfolio of things students observe and read about. At some point, students can start thinking up and trying out safe and helpful ways to change parts of their environment.

By the end of the 2nd grade, students should know that

▶ **Chunks of rocks come in many sizes and shapes, from boulders to grains of sand and even smaller.**

▶ **Change is something that happens to many things.**

▶ **Animals and plants sometimes cause changes in their surroundings.** ■

Grades 3 through 5

In these years, students should accumulate more information about the physical environment, becoming familiar with the details of geological features, observing and mapping locations of hills, valleys, rivers, etc., but without elaborate classification. Students should also become adept at using magnifiers to inspect a variety of rocks and soils. The point is not to classify rigorously but to notice the variety of components.

Students should now observe elementary processes of the rock cycle—erosion, transport, and deposit. Water and sand boxes and rock tumblers can provide them with some first-hand examples. Later, they can connect the features to the processes and follow explanations of how the features came to be and still are changing. Students can build devices for demonstrating how wind and water shape the land and how forces on materials can make wrinkles, folds, and faults. Films of volcanic magma and ash ejection dramatize another source of buildup.

By the end of the 5th grade, students should know that

▶ **Waves, wind, water, and ice shape and reshape the earth's land surface by eroding rock and soil in some areas and depositing them in other areas, sometimes in seasonal layers.**

▶ **Rock is composed of different combinations of minerals. Smaller rocks come from the breakage and weathering of bedrock and larger rocks. Soil is made partly from weathered rock, partly from plant remains—and also contains many living organisms.** ■

Grades 6 through 8

At this level, students are able to complete most of their understanding of the main features of the physical and biological factors that shape the face of the earth. This understanding will still be descriptive because the theory of plate tectonics will not be encountered formally until high school. Of course, students should see as great a variety of landforms and soils as possible.

It is especially important that students come to understand how sedimentary rock is formed periodically, embedding plant and animal remains and leaving a record of the sequence in which the plants and animals appeared and disappeared. Besides the relative age of the rock layers, the absolute age of those remains is central to the argument that there has been enough time for evolution of species. The process of sedimentation is understandable and observable. But imagining the span of geologic time will be difficult for students.

By the end of the 8th grade, students should know that

▶ **The interior of the earth is hot. Heat flow and movement of material within the earth cause earthquakes and volcanic eruptions and create mountains and ocean basins. Gas and dust from large volcanoes can change the atmosphere.**

▶ **Some changes in the earth's surface are abrupt (such as earthquakes and volcanic eruptions) while other changes happen very slowly (such as uplift and wearing down of mountains). The earth's surface is shaped in part by the motion of water and wind over very long times, which act to level mountain ranges.**

▶ **Sediments of sand and smaller particles (sometimes containing the remains of organisms) are gradually buried and are cemented together by dissolved minerals to form solid rock again.**

▶ **Sedimentary rock buried deep enough may be reformed by pressure and heat, perhaps melting and recrystallizing into different kinds of rock. These re-formed rock layers may be forced up again to become land surface and even mountains. Subsequently, this new rock too will erode. Rock bears evidence of the minerals, temperatures, and forces that created it.**

▶ **Thousands of layers of sedimentary rock confirm the long history of the changing surface of the earth and the changing life forms whose remains are found in successive layers. The youngest layers are not always found on top, because of folding, breaking, and uplift of layers.**

▶ **Although weathered rock is the basic component of soil, the composition and texture of soil and its fertility and resistance to erosion are greatly influenced by plant roots and debris, bacteria, fungi, worms, insects, rodents, and other organisms.**

▶ **Human activities, such as reducing the amount of forest cover, increasing the amount and variety of chemicals released into the atmosphere, and intensive farming, have changed the earth's land, oceans, and atmosphere. Some of these changes have decreased the capacity of the environment to support some life forms. ■**

73

Grades 9 through 12

The thrust of study should now turn to modern explanations for the pheomena the students have learned descriptively. Knowledge of radioactivity helps them understand how rocks can be dated, which helps them appreciate the scale of geologic time.

By the end of the 12th grade, students should know that

▶ **Plants alter the earth's atmosphere by removing carbon dioxide from it, using the carbon to make sugars and releasing oxygen. This process is responsible for the oxygen content of the air.**

▶ **The formation, weathering, sedimentation, and reformation of rock constitute a continuing "rock cycle" in which the total amount of material stays the same as its forms change.**

▶ **The slow movement of material within the earth results from heat flowing out from the deep interior and the action of gravitational forces on regions of different density.**

▶ **The solid crust of the earth—including both the continents and the ocean basins—consists of separate plates that ride on a denser, hot, gradually deformable layer of the earth. The crust sections move very slowly, pressing against one another in some places, pulling apart in other places. Ocean-floor plates may slide under continental plates, sinking deep into the earth. The surface layers of these plates may fold, forming mountain ranges.**

▶ **Earthquakes often occur along the boundaries between colliding plates, and molten rock from below creates pressure that is released by volcanic eruptions, helping to build up mountains. Under the ocean basins, molten rock may well up between separating plates to create new ocean floor. Volcanic activity along the ocean floor may form undersea mountains, which can thrust above the ocean's surface to become islands.** ■

This section may have the most implications for students' eventual understanding of the picture that science paints of how the world works. And it may offer great challenges too. Atomic theory powerfully explains many phenomena, but it demands imagination and the joining of several lines of evidence. Students must know about the properties of materials and their combinations, changes of state, effects of temperature, behavior of large collections of pieces, the construction of items from parts, and even about the desirability of nice, simple explanations. All of these elements should be introduced in middle school so the unifying idea of atoms can begin by the end of the 8th grade.

The scientific understanding of atoms and molecules requires combining two closely related ideas: All substances are composed of invisible particles, and all substances are made up of a limited number of basic ingredients, or "elements." These two merge into the idea that combining the particles of the basic ingredients differently leads to millions of materials with different properties.

Students often get the idea that atoms somehow just fill matter up rather than the correct idea that the atoms *are* the matter. Middle-school students also have trouble with the idea that atoms are in continual motion. Coming to terms with these concepts is necessary for students to make sense of atomic theory and its explanatory power.

The strategy here is to describe the complexity of atoms gradually, using evidence and explanations from several connected story lines. Students first learn the notion that atoms make up objects, not merely occupy space inside them; then they are introduced to crystal arrays and molecules. With this understanding, they can imagine how molecules and crystals lead to visible, tangible matter. Only then should the study of the internal structure of atoms be taken up.

Bringing atomic and molecular theory into the earlier grades is a great temptation, but most students are not ready to understand atomic theory before adolescence. The theory is certainly essential to much of modern scientific explanation, but moving atomic/molecular theory forward to the earlier grades should be resisted. The tiny size and huge number of atoms in even a sand grain are vastly beyond even adult experience. Having students memorize the names of invisible things and their parts gets things backward and wastes time. Concrete perceptions must come before abstract explanations. Students need to become familiar with the physical and chemical properties of many different kinds of materials through firsthand experience before they can be expected to consider theories that explain them.

There seems to be no tidy and consistent way to relate the terms atom, molecule, ion, polymer, and crystal. A facility in discussing these terms will grow slowly over time. Students should also not rush into discussions of nuclear theory. The abstractions are too formidable. The emptiness of the atom and its electrical balance, isotopes, decay, and radiation challenge the human mind. The preparations for these concepts should be developed carefully over several years so they can converge in high school.

RESEARCH NOTES
page 336

◄ ALSO SEE

Kindergarten through Grade 2

Students should examine and use a wide variety of objects, categorizing them according to their various observable properties. They should subject materials to such treatments as mixing, heating, freezing, cutting, wetting, dissolving, bending, and exposing to light to see how they change. Even though it is too early to expect precise reports or even consistent results from the students, they should be encouraged to describe what they did and how materials responded.

Students should also get a lot of experience in constructing things from a few kinds of small parts ("Tinkertoys" and "Legos"), then taking them apart and rearranging them. They should begin to consider how the properties of objects may differ from properties of the materials they are made of. And they should begin to inspect things with a magnifying glass to discover features not visible without it.

By the end of the 2nd grade, students should know that

▶ **Objects can be described in terms of the materials they are made of (clay, cloth, paper, etc.) and their physical properties (color, size, shape, weight, texture, flexibility, etc.).**

▶ **Things can be done to materials to change some of their properties, but not all materials respond the same way to what is done to them.** ■

Grades 3 through 5

The study of materials should continue and become more systematic and quantitative. Students should design and build objects that require different properties of materials. They should write clear descriptions of their designs and experiments, present their findings whenever possible in tables and graphs (designed by the *students*, not the teacher), and enter their data and results in a computer database.

Objects and materials can be described by more sophisticated properties—conduction of heat and electricity, buoyancy, response to magnets, solubility, and transparency. Students should measure, estimate, and calculate sizes, capacities, and weights. If young children can't feel the weight of something, they may believe it to have no weight at all. Many experiences of weighing (if possible on increasingly sensitive balances)—including weighing piles of small things and dividing to find the weight of each—will help. It is not obvious to elementary students that wholes weigh the same as the sum of their parts. That idea is preliminary to, but far short of, the conservation principle to be learned later that weight doesn't change in spite of striking changes in other properties as long as all the parts (including invisible gases) are accounted for.

With magnifiers, students should inspect substances composed of large collections of particles, such as salt and talcum powder, to discover the unexpected details at smaller scales. They should also observe and describe the behavior of large collections of pieces—powders, marbles, sugar cubes, or wooden blocks (which can, for example, be "poured" out of a container) and consider that the collections may have new properties that the pieces do not.

By the end of the 5th grade, students should know that

▶ **Heating and cooling cause changes in the properties of materials. Many kinds of changes occur faster under hotter conditions.**

▶ **No matter how parts of an object are assembled, the weight of the whole object made is always the same as the sum of the parts; and when a thing is broken into parts, the parts have the same total weight as the original thing.**

▶ **Materials may be composed of parts that are too small to be seen without magnification.**

▶ **When a new material is made by combining two or more materials, it has properties that are different from the original materials. For that reason, a lot of different materials can be made from a small number of basic kinds of materials.** ■

The structure of matter is difficult for this grade span. Historically, much of the evidence and reasoning used in developing atomic/molecular theory was complicated and abstract. In traditional curricula too, very difficult ideas have been offered to children before most of them had any chance of understanding. The law of definite proportions in chemical combinations, so obvious when atoms (and proportions) are well understood, is not likely to be helpful at this level. The behavior of gases—such as their compressibility and their expansion with temperature—may be investigated for qualitative explanation; but the mathematics of quantitative gas laws is likely to be more confusing than helpful to most students. When students first begin to understand atoms, they cannot confidently make the distinction between atoms and molecules or make distinctions that depend upon it—among elements, mixtures, and compounds, or between "chemical" and "physical" changes. An understanding of how things happen on the atomic level—making and breaking bonds—is more important than memorizing the official definitions (which are not so clear in modern chemistry anyway). Definitions can, of course, be memorized with no understanding at all.

Going into details of the structure of the atom is unnecessary at this level, and holding back makes sense. By the end of the 8th grade, students should have sufficient grasp of the general idea that a wide variety of phenomena can be explained by alternative arrangements of vast numbers of invisibly tiny, moving parts. Possible differences in atoms of the same element should be avoided at this stage. Historically, the identical nature of atoms of the same element was an assumption of atomic theory for a very long time.

When isotopes are introduced later, to explain subsequent observations, they can be a surprise and a lesson in the nature of progress in science. The alternative—teaching atoms' variety at the same time as the notion of their identity—seems likely to be prohibitively confusing to most students.

To that end, students should become familiar with characteristics of different states of matter—now including gases—and transitions between them. Most important, students should see a great many examples of reactions between substances that produce new substances very different from the reactants. Then they can begin to absorb the rudiments of atomic/molecular theory, being helped to see that the value of the notion of atoms lies in the explanations it provides for a wide variety of behavior of matter. Each new aspect of the theory should be developed as an explanation for some observed phenomenon and grasped fairly well before going on to the next.

By the end of the 8th grade, students should know that

▶ All matter is made up of atoms, which are far too small to see directly through a microscope. The atoms of any element are alike but are different from atoms of other elements. Atoms may stick together in well-defined molecules or may be packed together in large arrays. Different arrangements of atoms into groups compose all substances.

▶ Equal volumes of different substances usually have different weights.

▶ Atoms and molecules are perpetually in motion. Increased temperature means greater average energy of motion, so most substances expand when heated. In solids, the atoms are closely locked in position and can only vibrate. In liquids, the atoms or molecules have higher energy, are more loosely connected, and can slide past one another; some molecules may get enough energy to escape into a gas. In gases, the atoms or molecules have still more energy and are free of one another except during occasional collisions.

▶ The temperature and acidity of a solution influence reaction rates. Many substances dissolve in water, which may greatly facilitate reactions between them.

▶ Scientific ideas about elements were borrowed from some Greek philosophers of 2,000 years earlier, who believed that everything was made from four basic substances: air, earth, fire, and water. It was the combinations of these "elements" in different proportions that gave other substances their observable properties. The Greeks were wrong about those four, but now over 100 different elements have been identified, some rare and some plentiful, out of which everything is made. Because most elements tend to combine with others, few elements are found in their pure form.

▶ There are groups of elements that have similar properties, including highly reactive metals, less-reactive metals, highly reactive nonmetals (such as chlorine, fluorine, and oxygen), and some almost completely nonreactive gases (such as helium and neon). An especially important kind of reaction between substances involves combination of oxygen with something else—as in burning or rusting. Some elements don't fit into any of the categories; among them are carbon and hydrogen, essential elements of living matter.

▶ No matter how substances within a closed system interact with one another, or how they combine or break apart, the total weight of the system remains the same. The idea of atoms explains the conservation of matter: If the number of atoms stays the same no matter how they are rearranged, then their total mass stays the same. ∎

Understanding the general architecture of the atom and the roles played by the main constituents of the atom in determining the properties of materials now becomes relevant. Having learned earlier that all the atoms of an element are identical and are different from those of all other elements, students now come up against the idea that, on the contrary, atoms of the same element can differ in important ways. This revelation is an opportunity as well as a complication—scientific knowledge grows by modifications, sometimes radical, of previous theories. Sometimes advances have been made by neglecting small inconsistencies, and then further advances have been made later by attending closely to those inconsistencies.

Students may at first take isotopes to be something in addition to atoms or as only the unusual, unstable nuclides. The most important features of isotopes (with respect to general scientific literacy) are their nearly identical chemical behavior and their different nuclear stabilities. Insisting on the rigorous use of isotope and nuclide is probably not worthwhile, and the latter term can be ignored.

The idea of half-life requires that students understand ratios and the multiplication of fractions, and be somewhat comfortable with probability. Games with manipulative or computer simulations should help them in getting the idea of how a constant proportional rate of decay is consistent with declining measures that only gradually approach zero. The mathematics of inferring backwards from measurements to age is not appropriate for most students. They need only know that such calculations are possible.

79

By the end of the 12th grade, students should know that

▶ Atoms are made of a positive nucleus surrounded by negative electrons. An atom's electron configuration, particularly the outermost electrons, determines how the atom can interact with other atoms. Atoms form bonds to other atoms by transferring or sharing electrons.

▶ The nucleus, a tiny fraction of the volume of an atom, is composed of protons and neutrons, each almost two thousand times heavier than an electron. The number of positive protons in the nucleus determines what an atom's electron configuration can be and so defines the element. In a neutral atom, the number of electrons equals the number of protons. But an atom may acquire an unbalanced charge by gaining or losing electrons.

▶ Neutrons have a mass that is nearly identical to that of protons, but neutrons have no electric charge. Although neutrons have little effect on how an atom interacts with others, they do affect the mass and stability of the nucleus. Isotopes of the same element have the same number of protons (and therefore of electrons) but differ in the number of neutrons.

▶ The nucleus of radioactive isotopes is unstable and spontaneously decays, emitting particles and/or wavelike radiation. It cannot be predicted exactly when, if ever, an unstable nucleus will decay, but a large group of identical nuclei decay at a predictable rate. This predictability of decay rate allows radioactivity to be used for estimating the age of materials that contain radioactive substances.

▶ Scientists continue to investigate atoms and have discovered even smaller constituents of which electrons, neutrons, and protons are made.

▶ When elements are listed in order by the masses of their atoms, the same sequence of properties appears over and over again in the list.

▶ Atoms often join with one another in various combinations in distinct molecules or in repeating three-dimensional crystal patterns. An enormous variety of biological, chemical, and physical phenomena can be explained by changes in the arrangement and motion of atoms and molecules.

▶ The configuration of atoms in a molecule determines the molecule's properties. Shapes are particularly important in how large molecules interact with others.

▶ The rate of reactions among atoms and molecules depends on how often they encounter one another, which is affected by the concentration, pressure, and temperature of the reacting materials. Some atoms and molecules are highly effective in encouraging the interaction of others. ■

4ᴇ   Energy Transformations

Energy is a mysterious concept, even though its various forms can be precisely defined and measured. At the simplest level, children can think of energy as something needed to make things go, run, or happen. But they have difficulty distinguishing energy needs from other needs—plants need water to live and grow; cars need water, oil, and tires; people need sleep, etc. People in general are likely to think of energy as a substance, with flow and conservation analogous to that of matter. That is not correct, but for most people it can be an acceptable analogy. Although learning about energy does not make it much less mysterious, it is worth trying to understand because a wide variety of scientific explanations are difficult to follow without some knowledge of the concept of energy.

Energy is a major exception to the principle that students should understand ideas before being given labels for them. Children benefit from talking about energy before they are able to define it. Ideas about energy that students encounter outside of school—for example, getting "quick energy" from a candy bar or turning off a light so as not to "waste energy"— may be imprecise but are reasonably consistent with ideas about energy that we want students to learn.

Three energy-related ideas may be more important than the idea of energy itself. One is energy transformation. All physical events involve transferring energy or changing one form of energy into another— radiant to electrical, chemical to mechanical, and so on. A second idea is the conservation of energy. Whenever energy is reduced in one place, it is increased somewhere else by exactly the same amount. A third idea is that whenever there is a transformation of energy, some of it is likely to go into heat, which spreads around and is therefore not available for use.

Heat energy itself is a surprisingly difficult idea for students, who thoroughly confound it with the idea of temperature. A great deal of work is required for students to make the distinction successfully, and the heat/temperature distinction may join mass/weight, speed/acceleration, and power/energy distinctions as topics that, for purposes of literacy, are not worth the extraordinary time required to learn them. Because dissipated heat energy is at a lower temperature, some students' confusion about heat and temperature leads them to infer that the amount of energy has been reduced. On the other hand, some students' idea that dissipated heat energy has been "exhausted" or "expended" may be tolerably close to the truth.

Similarly, units and formulas for kinetic and potential energy are more difficult than they are worth for the semiquantitative understanding that we seek here. But the notion of potential energy is still useful for some situations in which motion might occur (for example, gravitational energy in water behind a dam, mechanical energy in a cocked mousetrap, or chemical energy in a flashlight battery or sugar molecule).

Research Notes
page 337

◁ **ALSO SEE**

Work, in the specialized sense used in physics, is often considered a useful, even necessary, concept for dealing with ideas of energy. These benchmarks propose to do without a technical definition of work for purposes of basic literacy, because it is so greatly confused with the common English-language meaning of the word. The calculation of work as force times distance is not essential to understanding many important ideas about energy. Running makes you tired; rubbing your hands together makes them warmer; coming out of water makes you feel cool.

Older students can grasp these ideas in a general way, but even they should not be expected to understand them deeply. For young students, it may be enough at first to convince them that energy is needed to get physical things to happen and that they should get in the habit of wondering where the energy came from. Then, as they study physical, chemical, and biological systems, many opportunities arise for them to see the many different forms energy takes and to find out how useful the energy concepts are.

Teachers have to decide what constitutes a sufficient understanding of energy and its transformations and conservation. As the benchmarks below indicate, in harmony with *Science for All Americans*, qualitative approximations are more important and should have priority. Much time can be invested in having students memorize definitions—for heat, temperature, system, transformation, entropy, and the like—with little to show for it in the way of understanding.

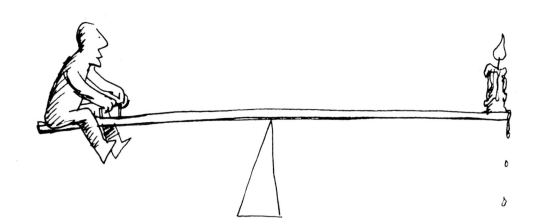

No effort to introduce energy as a scientific idea ought to be organized in these first years. If children use the term *energy* to indicate how much pep they have, that is perfectly all right, in that the meaning is clear and no technical mischief has been done. By the end of the 2nd grade, students should be familiar with a variety of ways of making things go and should consider "What makes it go?" to be an interesting question to ask. Once they learn that batteries wear down and cars run out of gasoline, turning off unneeded appliances can be said to "save on batteries" and "save on gas." The idea that is accessible at this age is that keeping anything going uses up some resource. (Little is gained by having children answer, "Energy.")

By the end of the 2nd grade, students should know that

▶ **The sun warms the land, air, and water.** ■

Investing much time and effort in developing formal energy concepts can wait. The importance of energy, after all, is that it is a *useful* idea. It helps make sense out of a very large number of things that go on in the physical and biological and engineering worlds. But until students have reached a certain point in their understanding of bits and pieces of the world, they gain little by having such a tool. It is a matter of timing.

The one aspect of the energy story in which students of this age can make some headway is heat, which is produced almost everywhere. In their science and technology activities during these years, students should be alerted to look for things and processes that give off heat—lights, radios, television sets, the sun, sawing wood, polishing surfaces, bending things, running motors, people, animals, etc.—and then for those that seem not to give off heat. Also, the time is appropriate to explore how heat spreads from one place to another and what can be done to contain it or shield things from it.

Students' ideas of heat have many wrinkles. In some situations, *cold* is thought to be transferred rather than heat. Some materials may be thought to be intrinsically warm (blankets) or cold (metals). Objects that keep things warm—such as a sweater or mittens—may be thought to be sources of heat. Only a continuing mix of experiment and discussion is likely to dispel these ideas.

Students need not come out of this grade span understanding heat or its difference from temperature. In this spirit, there is little to be gained by having youngsters refer to heat as *heat energy*. More important, students should become familiar with the warming of objects that start out cooler than their environment, and vice versa. Computer labware probes and graphic displays that detect small changes in temperature and plot them can be used by students to examine many instances of heat exchange. Because many students think of cold as a substance that spreads like heat, there may be some advantage in translating descriptions of transfer of cold into terms of transfer of heat.

By the end of the 5th grade, students should know that

▶ **Things that give off light often also give off heat. Heat is produced by mechanical and electrical machines, and any time one thing rubs against something else.**

▶ **When warmer things are put with cooler ones, the warm ones lose heat and the cool ones gain it until they are all at the same temperature. A warmer object can warm a cooler one by contact or at a distance.**

▶ **Some materials conduct heat much better than others. Poor conductors can reduce heat loss.** ■

At this level, students should be introduced to energy primarily through energy transformations. Students should trace where energy comes from (and goes next) in examples that involve several different forms of energy along the way: heat, light, motion of objects, chemical, and elastically distorted materials. To change something's speed, to bend or stretch things, to heat or cool them, to push things together or tear them apart all require transfers (and some transformations) of energy.

At this early stage, there may be some confusion in students' minds between energy and *energy sources*. Focusing on energy transformations may get around this somewhat. Food, gasoline, and batteries obviously get used up. But the energy they contain does not disappear; it is changed into other forms of energy.

The most primitive idea is that the energy needed for an event must come from somewhere. That should trigger children's interest in asking, for any situation, where the energy comes from and (later) asking where it goes. Where it comes from is usually much more evident than where it goes, because some usually diffuses away as radiation and random molecular motion.

A slightly more sophisticated proposition is the semiquantitative one that whenever *some* energy seems to show up in one place, *some* will be found to disappear from another. Eventually, the energy idea can become quantitative: If we can keep track of how much energy of each kind increases and decreases, we find that whenever the energy in one place decreases, the energy in other places increases by *just the same amount*. This energy-cannot-be-created-or-destroyed way of stating conservation fully may be more intuitive than the abstraction of a constant energy total within an isolated system. The quantitative (equal amounts) idea should probably wait until high school.

Convection is not so much an independent means of heat transfer as it is an aid to transfer of heat by conduction and radiation. Convection currents appear spontaneously when density differences caused by heating (conduction and radiation) are acted on by a gravitational field. (Though not in space stations, unless they are rotating.) But these subtleties are not appropriate for most 8th graders.

By the end of the 8th grade, students should know that

▶ **Energy cannot be created or destroyed, but only changed from one form into another.**

▶ **Most of what goes on in the universe—from exploding stars and biological growth to the operation of machines and the motion of people—involves some form of energy being transformed into another. Energy in the form of heat is almost always one of the products of an energy transformation.**

▶ **Heat can be transferred through materials by the collisions of atoms or across space by radiation. If the material is fluid, currents will be set up in it that aid the transfer of heat.**

▶ **Energy appears in different forms. Heat energy is in the disorderly motion of molecules; chemical energy is in the arrangement of atoms; mechanical energy is in moving bodies or in elastically distorted shapes; gravitational energy is in the separation of mutually attracting masses.** ■

Grades 9 through 12

The concepts acquired in the earlier grades should now be extended to nuclear realms and living organisms. Revisiting energy concepts in new contexts provides opportunities to improve student understanding of the basic concepts and to see just how powerful they are.

Two other major ideas merit introduction during these years, but without resort to mathematics. One of these is that the total amount of energy available for useful transformation is almost always decreasing; the other is that energy changes on the atomic scale occur only in discrete jumps. The first of those is not too difficult or implausible for students because they can experience in many ways a wide variety of actions that give off heat. The emphasis should probably be on the practical consequences of the loss of useful energy through heat dissipation.

On the other hand, the notion that energy changes in atoms can occur in only fixed amounts with no intermediate values is strange to begin with and hard to demonstrate. Some evidence should be presented for this scientific belief but not in great detail. The easiest phenomenon to show, which is also a major reason for including quantum jumps in literacy, is the discrete colors of light emitted by separate atoms, as in sodium-vapor or mercury-vapor lights. Another major reason for having students encounter the quantum idea is to illustrate the point that in science it is sometimes useful to invent ideas that run counter to intuition and prior experience.

continued

An important application of the atom/energy relationship to bring to the attention of students is that the distinctive light energies emitted or absorbed by different atoms enable them to be identified on earth, in our sun, and even on the other side of the universe. This fact is a prime example of the "rules are the same everywhere" principle.

By the end of the 12th grade, students should know that

▶ **Whenever the amount of energy in one place or form diminishes, the amount in other places or forms increases by the same amount.**

▶ **Heat energy in a material consists of the disordered motions of its atoms or molecules. In any interactions of atoms or molecules, the statistical odds are that they will end up with less order than they began—that is, with the heat energy spread out more evenly. With huge numbers of atoms and molecules, the greater disorder is almost certain.**

▶ **Transformations of energy usually produce some energy in the form of heat, which spreads around by radiation or conduction into cooler places. Although just as much total energy remains, its being spread out more evenly means less can be done with it.**

▶ **Different energy levels are associated with different configurations of atoms and molecules. Some changes of configuration require an input of energy whereas others release energy.**

▶ **When energy of an isolated atom or molecule changes, it does so in a definite jump from one value to another, with no possible values in between. The change in energy occurs when radiation is absorbed or emitted, so the radiation also has distinct energy values. As a result, the light emitted or absorbed by separate atoms or molecules (as in a gas) can be used to identify what the substance is.**

▶ **Energy is released whenever the nuclei of very heavy atoms, such as uranium or plutonium, split into middleweight ones, or when very light nuclei, such as those of hydrogen and helium, combine into heavier ones. The energy released in each nuclear reaction is very much greater than the energy given off in each chemical reaction.** ■

4ꜰ Motion

Nothing in the universe is at rest. Motion is as essential to understanding the physical world as matter and energy are. Following the organization of *Science for All Americans*, the benchmarks for motion constitute a wide range of topics, from the movement of objects to vibrations and the behavior of waves. Rotary motion, as interesting as it is, poses much greater difficulties for students and is not included in the benchmarks.

The benchmarks for understanding the motion of objects and repeating patterns of motion do not demand the use of equations. For purposes of science literacy, a qualitative understanding is sufficient. Equations may clarify relationships for the most mathematically apt students, but for many students they are difficult and may obscure the ideas rather than clarify them. For example, almost all students can grasp that the effect of a force on an object's motion will be greater if the force is greater and will be less if the object has more mass—but learning $a=F/m$ (which to many teachers seems like the same thing) is apparently much harder.

Newton's laws of motion are simple to state, and sometimes teachers mistake the ability of students to recite the three laws correctly as evidence that they understand them. The fact that it took such a long time, historically, to codify the laws of motion suggests that they are not self-evident truths, no matter how obvious they may seem to us *once we understand them well*. Much research in recent years has documented that students typically have trouble relating formal ideas of motion and force to their personal view of how the world works. These are three of the obstacles:

1. A basic problem is the ancient perception that sustained motion requires sustained force. The contrary notion that it takes force to *change* an object's motion, that something in motion will move in a straight line forever without slowing down unless a force acts on it, runs counter to what we can see happening with our eyes.

2. Limitations in describing motion may keep students from learning about the effect of forces. Students of all ages tend to think in terms of motion or no motion. So the first task may be to help students divide the category of motion into steady motion, speeding up, and slowing down. For example, falling objects should be described as falling faster and faster rather than just falling down. As indicated earlier, the basic idea expressed in Newton's second law of motion is not difficult to grasp, but vocabulary may get in the way if students have to struggle over the meaning of force and acceleration. Both terms have many meanings in common language that confound their specialized use in science.

3. Like inertia, the action-equals-reaction principle is counterintuitive. To say that a book presses down on the table is sensible enough, but then to say that the table pushes back up with exactly the same force (which disappears the instant you pick up the book) seems false on the face of it.

continued

RESEARCH NOTES

page 338

What is to be done? Students should have lots of experiences to shape their intuition about motion and forces long before encountering laws. Especially helpful are experimentation and discussion of what happens as surfaces become more elastic or more free of friction.

Vibrations treated only descriptively bring no special problems, other than the occasional confusion caused by the word *speed* being used in English for both frequency and velocity. Does a guitar string move quickly (back and forth a thousand times a second) or slowly (only 15 miles or so per hour)? Similarly, is the earth's rotation slow (once a day) or fast (1,000 miles per hour at the equator)? In the overall story of motion, vibrations serve in good part to introduce the ideas of frequency and amplitude. Because there are so many examples of vibrating systems that students can experience directly, they easily see vibration as a common way for some things to move and see frequency as a measure of that motion.

Waves, on the other hand, present a greater challenge. Wave motion is familiar to children through their experience with water. Surface waves on water provide the standard image of what waves are, and ropes and springs can also be used to show some of the properties of waves. Without formal schooling, young people learn that many other kinds of waves exist: radio waves, x rays, radar, microwaves, sound waves, ultraviolet radiation, and more. But they still might not know what these things are, how they relate to one another, what they have to do with motion, or in what sense such waves are waves.

From the outset, students should view, describe, and discuss all kinds of moving things—themselves, insects, birds, trees, doors, rain, fans, swings, volleyballs, wagons, stars, etc.—keeping notes, drawing pictures to suggest their motion, and raising questions: Do they move in a straight line? Is their motion fast or slow? How can you tell? How many ways does a growing plant move? The questions count more than the answers, at this stage. And students should gain varied experiences in getting things to move or not to move and in changing the direction or speed of things that are already in motion.

Presumably students will start "making music" from the first day in school, and this provides an opportunity to introduce vibrations as a phenomenon rather than a theory. With the drums, bells, stringed and other instruments they use, including their own voices, they can feel the vibrations on the instruments as they hear the sounds. These experiences are important for their own sake and at this point do not need elaboration.

By the end of the 2nd grade, students should know that

▶ **Things move in many different ways, such as straight, zigzag, round and round, back and forth, and fast and slow.**

▶ **The way to change how something is moving is to give it a push or a pull.**

▶ **Things that make sound vibrate.** ■

Students should continue describing motion. And they can be more experimental and more quantitative as their measurement skills sharpen. Determining the speed of fast things and slow things can present a challenge that students will readily respond to. They also can work out for themselves some of the general relationships between force and change of motion and internalize the notion of force as a push or pull of one thing on another—whether rubber bands, magnets, or explosions.

Students should also increase their inventory of examples of periodic motion and perhaps devise ways of measuring different rates of vibration. And students should use prisms to see that white light produces a whole "rainbow" of colors. (The idea that white light is "made up of" different colors is difficult and should be postponed to later grades.) There is nothing to be gained at this stage, however, from linking light to wave motion.

By the end of the 5th grade, students should know that

▶ **Changes in speed or direction of motion are caused by forces. The greater the force is, the greater the change in motion will be. The more massive an object is, the less effect a given force will have.**

▶ **How fast things move differs greatly. Some things are so slow that their journey takes a long time; others move too fast for people to even see them.** ■

89

Grades 6 through 8

The force/motion relationship can be developed more fully now and the difficult idea of inertia be given attention. Students have no trouble believing that an object at rest stays that way unless acted on by a force; they see it every day. The difficult notion is that an object in motion will continue to move unabated unless acted on by a force. Telling students to disregard their eyes will not do the trick—the things around them *do* appear to slow down of their own accord unless constantly pushed or pulled. The more experiences the students can have in seeing the effect of reducing friction, the easier it may be to get them to imagine the friction-equals-zero case.

Students can now learn some of the properties of waves by using water tables, ropes, and springs, and quite separately they can learn about the electromagnetic spectrum, including the assertion that it consists of wavelike radiations. Wave length should be the property receiving the most attention but only minimal calculation.

By the end of the 8th grade, students should know that

▶ **Light from the sun is made up of a mixture of many different colors of light, even though to the eye the light looks almost white. Other things that give off or reflect light have a different mix of colors.**

▶ **Something can be "seen" when light waves emitted or reflected by it enter the eye—just as something can be "heard" when sound waves from it enter the ear.**

▶ **An unbalanced force acting on an object changes its speed or direction of motion, or both. If the force acts toward a single center, the object's path may curve into an orbit around the center.**

▶ **Vibrations in materials set up wavelike disturbances that spread away from the source. Sound and earthquake waves are examples. These and other waves move at different speeds in different materials.**

▶ **Human eyes respond to only a narrow range of wavelengths of electromagnetic radiation— visible light. Differences of wavelength within that range are perceived as differences in color.** ■

Grades 9 through 12

At this level, students learn about relative motion, the action/reaction principle, wave behavior, the interaction of waves with matter, the Doppler effect now used in weather observations, and the red shift of distant galaxies. Relative motion is fun—students find it interesting to figure out their speeds in different reference frames, and many activities and films illustrate this principle. Learning this concept is important for its own sake and for the part it plays in the changing reference frames of the Copernican Revolution, and in simple relativity.

This level is also a time to show the power of mathematics. Once students are fully convinced that change in motion is proportional to the force applied, then mathematical logic requires that when F = 0, there be no change in motion. (So Newton's first law is just a special case of his second.) Students can move from a qualitative understanding of the force/motion relationship (more force changes motion more; more mass is harder to change) to one that is more quantitative (the change in motion is directly proportional to the amount of force and inversely proportional to the mass). Experimentally, they can learn that the change in motion of an object is proportional to the applied force and inversely proportional to the mass—a step beyond knowing that change in motion goes up with increasing force and down with increasing mass.

Students should come to understand qualitatively that (1) doubling the force on an object of a given mass doubles the effect the force has, tripling triples the effect, and so on; and (2) that whatever effect a given force has on an object, it will have half the effect on an object having twice the mass, a third on one having triple the mass, and so on. This need not entail having students solving lots of numerical problems.

The qualitative principle also applies to waves. Even as simple a relationship as speed = wavelength x frequency poses difficulties for many students. A sufficient minimum is that students develop semiquantitative notions about waves—for example, higher frequencies have shorter wavelengths and those with longer wavelengths tend to spread out more around obstacles.

The effect of wavelength on how waves interact with matter can be developed through intrinsically interesting phenomena—such as the blueness of the sky and redness of sunsets resulting from light of short wavelengths being scattered most by the atmosphere, or the color of grass resulting from its absorbing light of both shorter and longer wavelengths while reflecting the intermediate green. Electromagnetic waves with different wavelengths have different effects on the human body. Some pass through the body with little effect, some tan or injure the skin, and some are absorbed in different amounts by internal organs (sometimes injuring cells).

By the end of the 12th grade, students should know that

▶ **The change in motion of an object is proportional to the applied force and inversely proportional to the mass.**

▶ **All motion is relative to whatever frame of reference is chosen, for there is no motionless frame from which to judge all motion.**

continued

91

▶ Accelerating electric charges produce electromagnetic waves around them. A great variety of radiations are electromagnetic waves: radio waves, microwaves, radiant heat, visible light, ultraviolet radiation, x rays, and gamma rays. These wavelengths vary from radio waves, the longest, to gamma rays, the shortest. In empty space, all electromagnetic waves move at the same speed—the "speed of light."

▶ Whenever one thing exerts a force on another, an equal amount of force is exerted back on it.

▶ The observed wavelength of a wave depends upon the relative motion of the source and the observer. If either is moving toward the other, the observed wavelength is shorter; if either is moving away, the wavelength is longer. Because the light seen from almost all distant galaxies has longer wavelengths than comparable light here on earth, astronomers believe that the whole universe is expanding.

▶ Waves can superpose on one another, bend around corners, reflect off surfaces, be absorbed by materials they enter, and change direction when entering a new material. All these effects vary with wavelength. The energy of waves (like any form of energy) can be changed into other forms of energy. ■

For a good many school years, force may be treated as the originator of motion, and an explanation of force itself may be postponed. But the force between a bat and a ball has an entirely different origin than that between the earth and the moon. In helping students broaden their understanding of the fundamental forces of nature, the emphasis should be on gravitational and electromagnetic forces.

The general idea of universal gravitation and how weak it is compared to other kinds of forces is sufficient. Working out numerical problems adds little and is very likely to leave many students behind. The math is not hard but the units are baffling. A paradoxical idea for students is how weak gravity is compared to electric and magnetic forces. Gravity becomes appreciable only when very large accumulations of matter figure, such as that of a student and the entire earth. To students, gravitational forces seem strong compared to the trivial electric forces on dry hair charged by combing. But they can be led to see quite the opposite: The whole earth is required to pull a hair down by gravity, while only a small amount of charge is needed to force it up electrically against gravity.

Electric and magnetic forces and the relationship between them ought also to be treated qualitatively. Fields can be introduced, but only intuitively. Most important is that students get a sense of electric and magnetic force fields (as well as of gravity) and of some simple relations between magnets and electric currents. Direction rules have little importance for general literacy. The priority should be on what conditions produce a magnetic field and what conditions induce an electric current. Diagrams of electric and magnetic fields promote some misconceptions about "lines of force," notably that the force exists only on those lines. Students should recognize that the lines are used only to show the direction of the field.

RESEARCH NOTES
page 340

ALSO SEE

Kindergarten through Grade 2

The focus should be on motion and on encouraging children to be observant about when and how things seem to move or not move. They should notice that things fall to the ground if not held up. They should observe motion everywhere, making lists of different kinds of motion and what things move that way. Even in the primary years, children should use magnets to get things to move without touching them, and thereby learn that forces can act at a distance with no perceivable substance in between.

By the end of the 2nd grade, students should know that

▶ Things near the earth fall to the ground unless something holds them up.

▶ Magnets can be used to make some things move without being touched. ■

Grades 3 through 5

The main notion to convey here is that forces can act at a distance. Students should carry out investigations to become familiar with the pushes and pulls of magnets and static electricity. The term *gravity* may interfere with students' understanding because it often is used as an empty label for the common (and ancient) notion of "natural motion" toward the earth. The important point is that the earth *pulls* on objects.

By the end of the 5th grade, students should know that

▶ The earth's gravity pulls any object toward it without touching it.

▶ Without touching them, a magnet pulls on all things made of iron and either pushes or pulls on other magnets.

▶ Without touching them, material that has been electrically charged pulls on all other materials and may either push or pull other charged materials. ■

Grades 6 through 8

The idea of gravity—up until now seen as something happening near the earth's surface—can be generalized to all matter everywhere in the universe. Some demonstration, in the laboratory or on film or videotape, of the gravitational force between objects may be essential to break through the intuitive notion that things just naturally fall. Students should make devices to observe the magnetic effects of current and the electric effects of moving magnets. At first, the devices can be simple electromagnets; later, more complex devices, such as motor kits, can be introduced.

By the end of the 8th grade, students should know that

▶ **Every object exerts gravitational force on every other object. The force depends on how much mass the objects have and on how far apart they are. The force is hard to detect unless at least one of the objects has a lot of mass.**

▶ **The sun's gravitational pull holds the earth and other planets in their orbits, just as the planets' gravitational pull keeps their moons in orbit around them.**

▶ **Electric currents and magnets can exert a force on each other. ■**

95

Grades 9 through 12

Students should now learn how well the principle of universal gravitation explains the architecture of the universe and much that happens on the earth. The principle will become familiar from many different examples (star formation, tides, comet orbits, etc.) and from the study of the history leading to this unification of earth and sky. The "inversely proportional to the square" aspect is not a high priority for literacy. Much more important is escaping the common adult misconceptions that the earth's gravity does not extend beyond its atmosphere or that it is caused by the atmosphere.

Study of the nature of electric and magnetic forces should be joined to the study of the atom. What is likely to surprise many students is how much more powerful electromagnetic forces are than the gravitational forces, which are negligible on an atomic scale. Some students may have trouble seeing mechanical forces, such as pushing on an object with a stick, as being produced by electric charges on the atomic scale. It may help for them to recognize that the electric forces they do observe commonly (such as "static cling") result from *extremely* slight imbalances of electric charges. As students come to believe in the action/reaction principle, they will expect forces to be mutual.

By the end of the 12th grade, students should know that

▶ **Gravitational force is an attraction between masses. The strength of the force is proportional to the masses and weakens rapidly with increasing distance between them.**

▶ **Electromagnetic forces acting within and between atoms are vastly stronger than the gravitational forces acting between the atoms. At the atomic level, electric forces between oppositely charged electrons and protons hold atoms and molecules together and thus are involved in all chemical reactions. On a larger scale, these forces hold solid and liquid materials together and act between objects when they are in contact—as in sticking or sliding friction.**

▶ **There are two kinds of charges—positive and negative. Like charges repel one another, opposite charges attract. In materials, there are almost exactly equal proportions of positive and negative charges, making the materials as a whole electrically neutral. Negative charges, being associated with electrons, are far more mobile in materials than positive charges are. A very small excess or deficit of negative charges in a material produces noticeable electric forces.**

▶ Different kinds of materials respond differently to electric forces. In conducting materials such as metals, electric charges flow easily, whereas in insulating materials such as glass, they can move hardly at all. At very low temperatures, some materials become superconductors and offer no resistance to the flow of current. In between these extremes, semiconducting materials differ greatly in how well they conduct, depending on their exact composition.

▶ Magnetic forces are very closely related to electric forces and can be thought of as different aspects of a single electromagnetic force. Moving electric charges produce magnetic forces and moving magnets produce electric forces. The interplay of electric and magnetic forces is the basis for electric motors, generators, and many other modern technologies, including the production of electromagnetic waves.

▶ The forces that hold the nucleus of an atom together are much stronger than the electromagnetic force. That is why such great amounts of energy are released from the nuclear reactions in the sun and other stars. ■

HENRI ROUSSEAU, *Tropical Forest With Monkeys,*

Chapter 5 THE LIVING ENVIRONMENT

People have long been curious about living things—how many different species there are, what they are like, where they live, how they relate to each other, and how they behave. Scientists seek to answer these questions and many more about the organisms that inhabit the earth. In particular, they try to develop the concepts, principles, and theories that enable people to understand the living environment better.

Living organisms are made of the same components as all other matter, involve the same kinds of transformations of energy, and move using the same basic kinds of forces. Thus, all of the physical principles discussed in Chapter 4: The Physical Setting apply to life as well as to stars, raindrops, and television sets. But living organisms also have characteristics that can be understood best through the application of other principles.

SCIENCE FOR ALL AMERICANS

What can be anywhere near as awe-inspiring as the vast array of living things that occupy every nook and cranny of the earth's surface, unless it is the array of extinct species that once occupied the planet? Biologists have already identified over a million living species, each with its own way of surviving, sometimes in the least likely places, each readily able to propagate itself in the next generation. Because only organisms with hard shells or skeletons are generally preserved, the fossil record does not preserve a good record of the even greater number of extinct species that have existed over the span of the earth's history.

continued

This sense of wonder at the rich diversity and complexity of life is easily fostered in children. They spontaneously respond to nature. However, attempts to give them explanations for that diversity before they are able to handle the abstractions, or before they see the need for explanations, can dampen their natural curiosity.

Nevertheless, the explanations must come, for scientists not only revel in nature but try to understand it. The challenge for educators is to capitalize on the interest that students have in living things while moving them gradually toward ideas that make sense out of nature. Familiarity with the phenomena should precede their explanation, and attention to the concrete object should precede abstract theory.

Perhaps this is another instance in which following the course of history pays off. Long before Darwin provided an entirely new framework for explaining evolution and before the microscope led scientists to cells and chemistry led them to protein and DNA, the earth was under close scrutiny.

Botanists, zoologists, geologists, surveyors, explorers, amateur collectors, and even fortune-hunters were busy finding out what was "out there." On every continent, indigenous people had intimate knowledge of the flora and fauna of their regions. Their very survival depended on acquiring this knowledge and passing it on from generation to generation. As information accumulated, interest in classification systems grew, and those systems became more complex, especially after the microscope revealed a whole new world to explore and catalogue. Eventually, scientists produced and tested the theories and models that are used to explain people's observations. They came to understand the living environment first through observations, then classifications, then theories. It's a useful model for students to follow in learning about the environment. Chapter 6: The Human Organism augments many of these ideas in the context of human beings. ■

5ᴀ Diversity of Life

General similarities and differences among organisms are easily observed. Most children enter kindergarten interested in living things and already able to distinguish among the common ones. Children know, for example, that fish resemble other fish, frogs resemble other frogs, and that fish and frogs are different. In the beginning, children can focus on any attribute—size, color, limbs, fins, or wings—but then should gradually be guided to realize that for purposes of understanding relatedness among organisms, some characteristics are more significant than others. The teacher's task is to move students toward a more sophisticated understanding of the features of organisms that connect or differentiate them: from external features and behavior patterns, to internal structures and processes, to cellular activity, to molecular structure.

Understanding and appreciating the diversity of life does not come from students' knowing bits of information or classification categories about many different species; rather it comes from their ability to see in organisms the patterns of similarity and difference that permeate the living world. Through these patterns, biologists connect the multitude of individual organisms to the theories of genetics, ecology, and evolution.

RESEARCH NOTES
page 340

◁ ALSO SEE

Kindergarten through Grade 2

All students, especially those who live in circumstances that limit their interaction with nature, must have the opportunity to observe a variety of plants and animals in the classroom, on the school grounds, in the neighborhood, at home, in parks and streams and gardens, and at the zoo. But observing is not enough. The students should have reasons for their observations—reasons that prompt them to do something with the information they collect. The reason can be to answer the students' own questions about how organisms live or care for their young. Some students may enjoy displaying, with drawings, photographs, or even real specimens, all the living things they can find where they live. The point is to encourage them to ask questions for which they can find answers by looking carefully (using hand lenses when needed) at plants and animals and then checking their observations and answers with one another.

The anthropomorphism embedded in most animal stories causes some worry. One suggestion is to ignore it. Stories sometimes give plants and animals attributes they do not have, but promoting student interest in reading is more important than giving students rigidly correct impressions in their reading. Students can be guided toward making distinctions between stories that portray animals the way they really are and those that do not. Differences among students over the correctness of the portrayal of animals or plants in books should lead the students to reference works, which are another source of information that students must start learning to use.

By the end of the 2nd grade, students should know that

▶ Some animals and plants are alike in the way they look and in the things they do, and others are very different from one another.

▶ Plants and animals have features that help them live in different environments.

▶ Stories sometimes give plants and animals attributes they really do not have. ■

Grades 3 through 5

Students should have the opportunity to learn about an increasing variety of living organisms, both the familiar and the exotic, and should become more precise in identifying similarities and differences among them. Although the emphasis can still be on external features, finer detail than before should be included. Hand lenses, introduced earlier, should now be routinely used by students. Microscopes should come into use, not to study cell structure but to begin exploring the world of organisms that cannot be seen by the unaided eye. Fortunately, a wealth of films exists to supplement direct observation.

As students become more familiar with the characteristics of more and more organisms, they should be asked to invent schemes for classifying them—but without using the Linnean classification system. Hopefully, their classification schemes will vary according to the uses made of them as well as according to gross anatomy, behavior patterns, habitats, and other features. The aim is to move students toward the realization that there are many ways to classify things but how good any classification is depends on its usefulness. A scheme is useful if it contributes either to making decisions on some matter or to a deeper understanding of the relatedness of organisms. Classification schemes will, of course, vary with purpose (pets/nonpets; edible/nonedible).

By the end of the 5th grade, students should know that

▶ A great variety of kinds of living things can be sorted into groups in many ways using various features to decide which things belong to which group.

▶ Features used for grouping depend on the purpose of the grouping. ■

Grades 6 through 8

Science in the middle grades should provide students with opportunities to enrich their growing knowledge of the diversity of life on the planet and to begin to connect that knowledge to what they are learning in geography. That is, whenever students study a particular region in the world, they should learn about the plants and animals found there and how they are like or unlike those found elsewhere. Tracing simple food webs in varied environments can contribute to a better understanding of the dependence of organisms (including humans) on their environment.

Students should begin to extend their attention from external anatomy to internal structures and functions. Patterns of development may be brought in to further illustrate similarities and differences among organisms. Also, they should move from their invented classification systems to those used in modern biology. That is not done to teach them the standard system but to show them what features biologists typically use in classifying organisms and why. Classification systems are not part of nature. Rather, they are frameworks created by biologists for describing the vast diversity of organisms, suggesting relationships among living things, and framing research questions. A provocative exercise is to have students try to differentiate between familiar organisms that are alike in many ways—for example, between cats and small dogs.

By the end of the 8th grade, students should know that

▶ One of the most general distinctions among organisms is between plants, which use sunlight to make their own food, and animals, which consume energy-rich foods. Some kinds of organisms, many of them microscopic, cannot be neatly classified as either plants or animals.

▶ Animals and plants have a great variety of body plans and internal structures that contribute to their being able to make or find food and reproduce.

▶ Similarities among organisms are found in internal anatomical features, which can be used to infer the degree of relatedness among organisms. In classifying organisms, biologists consider details of internal and external structures to be more important than behavior or general appearance.

▶ For sexually reproducing organisms, a species comprises all organisms that can mate with one another to produce fertile offspring.

▶ All organisms, including the human species, are part of and depend on two main interconnected global food webs. One includes microscopic ocean plants, the animals that feed on them, and finally the animals that feed on those animals. The other web includes land plants, the animals that feed on them, and so forth. The cycles continue indefinitely because organisms decompose after death to return food material to the environment. ■

Grades 9 through 12

Two aims dominate at this level. One is to advance student understanding of why diversity within and among species is important. The other is to take the study of diversity and similarity to the molecular level. Students can learn that it is possible to infer relatedness among organisms from DNA or protein sequences. An investigation of the DNA-fingerprinting controversy may provide an interesting way to approach the question of the nature and validity of molecular evidence.

By the end of the 12th grade, students should know that

▶ The variation of organisms within a species increases the likelihood that at least some members of the species will survive under changed environmental conditions, and a great diversity of species increases the chance that at least some living things will survive in the face of large changes in the environment.

▶ The degree of kinship between organisms or species can be estimated from the similarity of their DNA sequences, which often closely matches their classification based on anatomical similarities. ■

5B HEREDITY

Building an observational base for heredity ought to be the first undertaking. Explanations can come later. The organisms children recognize are themselves, their classmates, and their pets. And that is the place to start studying heredity. However, it is important to be cautious about having children compare their own physical appearance to that of their siblings, parents, and grandparents. At the very least, the matter has to be handled with great delicacy so no one is embarrassed. Direct observations of generational similarities and differences of at least some plants and animals are essential.

Learning the genetic explanation for how traits are passed on from one generation to the next can begin in the middle years and carry into high school. The part played by DNA in the story should wait until students understand molecules. The interaction between heredity and environment in determining plant and animal behavior will be of interest to students. Examining specific cases can help them grasp the complex interactions of genetics and environment.

RESEARCH NOTES
page 341

Kindergarten through Grade 2

Teachers should lead students to make observations about how the offspring of familiar animals compare to one another and to their parents. Children know that animals reproduce their own kind—rabbits have rabbits (but you can usually tell one baby rabbit from another), cats have kittens that have different markings (but cats never have puppies), and so forth. This idea should be strengthened by a large number of examples, both plant and animal, that the children can draw on.

By the end of the 2nd grade, students should know that

▶ There is variation among individuals of one kind within a population.

▶ Offspring are very much, but not exactly, like their parents and like one another. ∎

Grades 3 through 5

Students should move from describing individuals directly (she has blue eyes) to naming traits and classifying individuals with respect to those traits (eye color: blue). Students can be encouraged to keep lists of things that animals and plants get from their parents, things that they don't get, and things that the students are not sure about either way. This is also the time to start building the notion of a population whose members are alike in many ways but show some variation.

By the end of the 5th grade, students should know that

▶ Some likenesses between children and parents, such as eye color in human beings, or fruit or flower color in plants, are inherited. Other likenesses, such as people's table manners or carpentry skills, are learned.

▶ For offspring to resemble their parents, there must be a reliable way to transfer information from one generation to the next. ∎

Grades 6 through 8

Now is the time to begin the study of genetic traits—what offspring get from parents. This topic can be handled as a natural part of the study of human reproduction. Students should examine examples of lineages for which breeding has been used to emphasize or suppress certain features of organisms.

By the end of the 8th grade, students should know that

▶ In some kinds of organisms, all the genes come from a single parent, whereas in organisms that have sexes, typically half of the genes come from each parent.

▶ In sexual reproduction, a single specialized cell from a female merges with a specialized cell from a male. As the fertilized egg, carrying genetic information from each parent, multiplies to form the complete organism with about a trillion cells, the same genetic information is copied in each cell.

▶ New varieties of cultivated plants and domestic animals have resulted from selective breeding for particular traits. ■

Grades 9 through 12

DNA provides for both the continuity of traits from one generation to the next and the variation that in time can lead to differences within a species and to entirely new species. Understanding DNA makes possible an explanation of such phenomena as the similarities and differences between parents and offspring, hereditary diseases, and the evolution of new species. This understanding also makes it possible for scientists to manipulate genes and thereby create new combinations of traits and new varieties of organisms.

By the end of the 12th grade, students should know that

▶ Some new gene combinations make little difference, some can produce organisms with new and perhaps enhanced capabilities, and some can be deleterious.

▶ The sorting and recombination of genes in sexual reproduction results in a great variety of possible gene combinations from the offspring of any two parents.

▶ The information passed from parents to offspring is coded in DNA molecules.

▶ Genes are segments of DNA molecules. Inserting, deleting, or substituting DNA segments can alter genes. An altered gene may be passed on to every cell that develops from it. The resulting features may help, harm, or have little or no effect on the offspring's success in its environment.

▶ Gene mutations can be caused by such things as radiation and chemicals. When they occur in sex cells, the mutations can be passed on to offspring; if they occur in other cells, they can be passed on to descendant cells only. The experiences an organism has during its lifetime can affect its offspring only if the genes in its own sex cells are changed by the experience.

▶ The many body cells in an individual can be very different from one another, even though they are all descended from a single cell and thus have essentially identical genetic instructions. Different parts of the instructions are used in different types of cells, influenced by the cell's environment and past history. ■

5c CELLS

Students can get pretty far along in their study of organisms before they need to learn that all activities within those organisms are performed by cells and that organisms are mostly cells. The familiar description and depiction of cells in blood sometimes lead students to the notion that organisms *contain* cells rather than that organisms are mostly *made up of* cells. Imagining the large number of cells is also a problem for young students. Large organisms are composed of about a trillion cells, but this number means little to middle-school students. A million millions might have a better chance of making an impression.

Students may have even more difficulty with the idea that cells are the basic units in which life processes occur. Neither familiarity with functions of regular-sized organisms nor observation of single-celled organisms will reveal much about the chemical activity going on inside single cells. For most students, the story should be kept simple. The way to approach the idea of functioning microscopic units is to start with the needs of macroscopic organisms.

Information transfer and energy transformation are functions of nearly all cells. The molecular aspects of these processes should wait until students have observed the transformation of energy in a variety of physical systems and have examined more generally the requirements for the transfer of information. Information transfer may mean communication among cells within an organism or passing genetic codes from a cell to its descendants.

RESEARCH NOTES

page 342

Emphasis should be placed on examining a variety of familiar animals and plants and considering things and processes they all need to stay alive, such as food and getting rid of wastes. Students should use hand lenses to make things appear 3 to 10 times bigger and more detailed and should be encouraged to wonder what they might see with more powerful lenses.

By the end of the 2nd grade, students should know that

▶ **Magnifiers help people see things they could not see without them.**

▶ **Most living things need water, food, and air.** ∎

Students' experiences should expand to include the observation of microscopic organisms, so the scale of magnification should increase to 30- or 100-power (dissection scope or low power on microscopes). Watching microorganisms is always informative, but some events are so rare that prepared materials are a necessity. Students can observe films of living cells growing and dividing, taking in substances, and changing direction when they run into things. Some students may reason that because these tiny cells are alive, they probably have the same needs as other, larger organisms. That can stimulate discussions about how single-celled organisms satisfy their need for food, water, and air.

By the end of the 5th grade, students should know that

▶ **Some living things consist of a single cell. Like familiar organisms, they need food, water, and air; a way to dispose of waste; and an environment they can live in.**

▶ **Microscopes make it possible to see that living things are made mostly of cells. Some organisms are made of a collection of similar cells that benefit from cooperating. Some organisms' cells vary greatly in appearance and perform very different roles in the organism.** ∎

Grades 6 through 8

Once they have some "magnification sense," students can use photomicrographs to extend their observations of cells, gradually concentrating on cells that make up internal body structures. The main interest of youngsters at this level is the human body, so they can begin with as many different kinds of body cells as possible—nerve, bone, muscle, skin—and then move on to examining cells in other animals and plants. This activity can show students that cells are the fundamental building blocks of their own bodies and of other living things as well. Also, once students see that tissue in other animals looks pretty much the same as tissue in humans, two important claims of science will be reinforced: the ubiquity of cells and the unity of nature.

By the end of the 8th grade, students should know that

▶ All living things are composed of cells, from just one to many millions, whose details usually are visible only through a microscope. Different body tissues and organs are made up of different kinds of cells. The cells in similar tissues and organs in other animals are similar to those in human beings but differ somewhat from cells found in plants.

▶ Cells repeatedly divide to make more cells for growth and repair. Various organs and tissues function to serve the needs of cells for food, air, and waste removal.

▶ Within cells, many of the basic functions of organisms—such as extracting energy from food and getting rid of waste—are carried out. The way in which cells function is similar in all living organisms.

▶ About two thirds of the weight of cells is accounted for by water, which gives cells many of their properties. ■

The individual cell can be considered as a system itself and as part of larger systems, sometimes as part of a multicellular organism, always as part of an ecosystem. The cell membrane serves as a boundary between the cell and its environment, containing for its own use the proteins it makes, equipment to make them, and stockpiles of fuel. Students should be asked to consider the variety of functions cells serve in the organism and how needed materials and information get to and from the cells. It may help students to understand the interdependency of cells if they think of an organism as a community of cells, each of which has some common tasks and some special jobs.

The idea that protein molecules assembled by cells conduct the work that goes on inside and outside the cells in an organism can be learned without going into the biochemical details. It is sufficient for students to know that the molecules involved are different configurations of a relatively few kinds of amino acids, and that the different shapes of the molecules influence what they do.

Students should acquire a general picture of the functions of the cell and know that the cell has specialized parts that perform these functions. This can be accomplished without many technical terms. Emphasizing vocabulary can impede understanding and take the fun out of science. Discussion of what needs to be done in the cell is much more important than identifying or naming the parts that do it. For example, students should know that cells have certain parts that oxidize sugar to release energy and parts to stitch protein chains together according to instructions; but they don't need to remember that one type of part is a mitochondrion and the other a ribosome, or which is which.

By the end of the 12th grade, students should know that

▶ **Every cell is covered by a membrane that controls what can enter and leave the cell. In all but quite primitive cells, a complex network of proteins provides organization and shape and, for animal cells, movement.**

▶ **Within every cell are specialized parts for the transport of materials, energy transfer, protein building, waste disposal, information feedback, and even movement. In addition, most cells in multicellular organisms perform some special functions that others do not.**

continued

▶ The work of the cell is carried out by the many different types of molecules it assembles, mostly proteins. Protein molecules are long, usually folded chains made from 20 different kinds of amino-acid molecules. The function of each protein molecule depends on its specific sequence of amino acids and the shape the chain takes is a consequence of attractions between the chain's parts.

▶ The genetic information encoded in DNA molecules provides instructions for assembling protein molecules. The code used is virtually the same for all life forms. Before a cell divides, the instructions are duplicated so that each of the two new cells gets all the necessary information for carrying on.

▶ Complex interactions among the different kinds of molecules in the cell cause distinct cycles of activities, such as growth and division. Cell behavior can also be affected by molecules from other parts of the organism or even other organisms.

▶ Gene mutation in a cell can result in uncontrolled cell division, called cancer. Exposure of cells to certain chemicals or radiation increases mutations and thus increases the chance of cancer. of cancer.

▶ Most cells function best within a narrow range of temperature and acidity. At very low temperatures, reaction rates are too slow. High temperatures and/or extremes of acidity can irreversibly change the structure of most protein molecules. Even small changes in acidity can alter the molecules and how they interact. Both single cells and multicellular organisms have molecules that help to keep the cell's acidity within a narrow range.

▶ A living cell is composed of a small number of chemical elements mainly carbon, hydrogen, nitrogen, oxygen, phosphorous, and sulfur. Carbon atoms can easily bond to several other carbon atoms in chains and rings to form large and complex molecules. ◼

5_D   INTERDEPENDENCE OF LIFE

It is not difficult for students to grasp the general notion that species depend on one another and on the environment for survival. But their awareness must be supported by knowledge of the kinds of relationships that exist among organisms, the *kinds* of physical conditions that organisms must cope with, the *kinds* of environments created by the interaction of organisms with one another and their physical surroundings, and the complexity of such systems. Students should become acquainted with many different examples of ecosystems, starting with those near at hand.

RESEARCH NOTES
page 342

ALSO SEE

Chapter **1** C THE SCIENTIFIC ENTERPRISE   (ethics, effects on environment)
3 C ISSUES IN TECHNOLOGY   (impact of technology on ecosystem)
4 B THE EARTH   (habitat, needed resources)
C PROCESSES THAT SHAPE THE EARTH   (climate and habitat)
8 A AGRICULTURE   (ecosystems)
B MATERIALS AND MANUFACTURING   (loss and recovery of the earth's resources)
11 A SYSTEMS
B MODELS   (modeling ecosystems)
C CONSTANCY AND CHANGE   (ecosystems)

Kindergarten through Grade 2

Students should investigate the habitats of many different kinds of local plants and animals, including weeds, aquatic plants, insects, worms, and amphibians, and some of the ways in which animals depend on plants and on each other.

By the end of the 2nd grade, students should know that

▶ **Animals eat plants or other animals for food and may also use plants (or even other animals) for shelter and nesting.**

▶ **Living things are found almost everywhere in the world. There are somewhat different kinds in different places.** ■

Grades 3 through 5

Students should explore how various organisms satisfy their needs in the environments in which they are typically found. They can examine the survival needs of different organisms and consider how the conditions in particular habitats can limit what kinds of living things can survive. Their studies of interactions among organisms within an environment should start with relationships they can directly observe. By viewing nature films, students should see a great diversity of life in different habitats.

By the end of the 5th grade, students should know that

▶ **For any particular environment, some kinds of plants and animals survive well, some survive less well, and some cannot survive at all.**

▶ **Insects and various other organisms depend on dead plant and animal material for food.**

▶ **Organisms interact with one another in various ways besides providing food. Many plants depend on animals for carrying their pollen to other plants or for dispersing their seeds.**

▶ **Changes in an organism's habitat are sometimes beneficial to it and sometimes harmful.**

▶ **Most microorganisms do not cause disease, and many are beneficial.** ■

As students build up a collection of cases based on their own studies of organisms, readings, and film presentations, they should be guided from specific examples of the interdependency of organisms to a more systematic view of the kinds of interactions that take place among organisms. But a necessary part of understanding complex relationships is to know what a fair proportion of the possibilities are. The full-blown concept of ecosystem (and that term) can best be left until students have many of the pieces ready to put in place. Prior knowledge of the relationships between organisms and the environment should be integrated with students' growing knowledge of the earth sciences.

By the end of the 8th grade, students should know that

▶ In all environments—freshwater, marine, forest, desert, grassland, mountain, and others— organisms with similar needs may compete with one another for resources, including food, space, water, air, and shelter. In any particular environment, the growth and survival of organisms depend on the physical conditions.

▶ Two types of organisms may interact with one another in several ways: They may be in a producer/consumer, predator/prey, or parasite/host relationship. Or one organism may scavenge or decompose another. Relationships may be competitive or mutually beneficial. Some species have become so adapted to each other that neither could survive without the other.

The concept of an ecosystem should bring coherence to the complex array of relationships among organisms and environments that students have encountered. Students' growing understanding of systems in general can suggest and reinforce characteristics of ecosystems—interdependence of parts, feedback, oscillation, inputs, and outputs. Stability and change in ecosystems can be considered in terms of variables such as population size, number and kinds of species, and productivity.

By the end of the 12th grade, students should know that

▶ **Ecosystems can be reasonably stable over hundreds or thousands of years. As any population of organisms grows, it is held in check by one or more environmental factors: depletion of food or nesting sites, increased loss to increased numbers of predators, or parasites. If a disaster such as flood or fire occurs, the damaged ecosystem is likely to recover in stages that eventually result in a system similar to the original one.**

▶ **Like many complex systems, ecosystems tend to have cyclic fluctuations around a state of rough equilibrium. In the long run, however, ecosystems always change when climate changes or when one or more new species appear as a result of migration or local evolution.**

▶ **Human beings are part of the earth's ecosystems. Human activities can, deliberately or inadvertently, alter the equilibrium in ecosystems.** ■

5E FLOW OF MATTER AND ENERGY

Organisms are linked to one another and to their physical setting by the transfer and transformation of matter and energy. This fundamental concept brings together insights from the physical and biological sciences. But energy transfer in biological systems is less obvious than in physical systems. Tracing where energy comes from through its various forms is usually directly observable in physical systems. Fire heats water, falling water makes electricity. But energy stored in molecular configurations is difficult to show even with models.

The cycling of matter and flow of energy can be found at many levels of biological organization, from molecules to ecosystems. The study of food webs can start in the elementary grades with the transfer of matter, be added to in the middle grades with the flow of energy through organisms, and then be integrated in high school as students' understanding of energy storage in molecular configurations develops. The whole picture grows slowly over time for students. In their early years, the temptation to simplify matters by saying plants get food from the soil should be resisted.

RESEARCH NOTES

page 342

Kindergarten through Grade 2

Children should begin to be aware of the basic parts of the food chain: Plants need sunlight to grow, some animals eat plants, and other animals eat both plants and animals. The key step that *plants make their own food* is very difficult for elementary students and should be saved for middle school.

An awareness of recycling, both in nature and in human societies, may play a helpful role in the development of children's thinking. Familiarity with the recycling of materials fosters the notion that matter continues to exist even though it changes from one form to another.

By the end of the 2nd grade, students should know that

▶ **Plants and animals both need to take in water, and animals need to take in food. In addition, plants need light.**

▶ **Many materials can be recycled and used again, sometimes in different forms.** ■

Grades 3 through 5

Students should begin to notice that substances may change form and move from place to place, but they never appear out of nowhere and never just disappear. Questions should encourage students to consider where substances come from and where they go and to be puzzled when they cannot account for the origin or the fate of a substance.

It's all right to start students on chains of what eats what in various environments, but labeling the steps in the chain as *energy transfer* is not necessary. Transfers of energy at this level are better illustrated in physical systems; biological energy transfer is far too complicated.

By the end of the 5th grade, students should know that

▶ **Almost all kinds of animals' food can be traced back to plants.**

▶ **Some source of "energy" is needed for all organisms to stay alive and grow.**

▶ **Over the whole earth, organisms are growing, dying, and decaying, and new organisms are being produced by the old ones.** ■

Grades **6** through **8**

In the middle grades, the emphasis is on following matter through ecosystems. Students should trace food webs both on land and in the sea. The food webs that students investigate should first be local ones they can study directly. The use of films of food webs in other ecosystems can supplement their direct investigations but should not substitute for them. Most students see food webs and cycles as involving the creation and destruction of matter, rather than the breakdown and reassembly of invisible units. They see various organisms and materials as consisting of different types of matter that are not convertible into one another. Before they have an understanding of atoms, the notion of reusable building blocks common to plants and animals is quite mysterious. So following matter through ecosystems needs to be linked to their study of atoms.

Students' attention should be drawn to the transfer of energy that occurs as one organism eats another. It is important that students learn the differences between how plants and animals obtain food and from it the energy they need. The first stumbling block is *food*, which represents one of those instances in which differences between the common use of a term and the technical one cause persistent confusion. In popular language, *food* is whatever nutrients plants and animals must take in if they are to grow and survive (solutions of minerals that plants need traces of frequently bear the label "plant food"); in scientific usage, *food* refers only to those substances, such as carbohydrates, proteins, and fats, from which organisms derive the energy they need to grow and operate and the material of which they are made. It's important to emphasize that the sugars that plants make out of water and carbon dioxide are their *only* source of food. Water and minerals dissolved in it are not sources of energy for plants or for animals.

By the end of the 8th grade, students should know that

▶ **Food provides molecules that serves as fuel and building material for all organisms. Plants use the energy in light to make sugars out of carbon dioxide and water. This food can be used immediately for fuel or materials or it may be stored for later use. Organisms that eat plants break down the plant structures to produce the materials and energy they need to survive. Then they are consumed by other organisms.**

▶ **Over a long time, matter is transferred from one organism to another repeatedly and between organisms and their physical environment. As in all material systems, the total amount of matter remains constant, even though its form and location change.**

▶ **Energy can change from one form to another in living things. Animals get energy from oxidizing their food, releasing some of its energy as heat. Almost all food energy comes originally from sunlight. ■**

Grades 9 through 12

Now students have a sufficient grasp of atoms and molecules to link the conservation of matter with the flow of energy in living systems. Energy can be accounted for by thinking of it as being stored in molecular configurations constituted during photosynthesis and released during oxidation. Although there is no need to account for all the energy, students should observe heat generated by consumers and decomposers. Discussions of ecosystems can both contribute to and be reinforced by students' understanding of the systems concept in general. The difficulty of predicting the consequences of human tinkering with ecosystems can be illustrated with examples such as the ill-considered fire-prevention efforts in national forests.

This level is also a time to ask what this knowledge of the flow of matter and energy through living systems suggests for human beings. Issues such as the use of fossil fuels and the recycling of matter and energy are important enough to pay considerable attention to in high school.

By the end of the 12th grade, students should know that

▶ **At times, environmental conditions are such that plants and marine organisms grow faster than decomposers can recycle them back to the environment. Layers of energy-rich organic material have been gradually turned into great coal beds and oil pools by the pressure of the overlying earth. By burning these fossil fuels, people are passing most of the stored energy back into the environment as heat and releasing large amounts of carbon dioxide.**

▶ **The amount of life any environment can support is limited by the available energy, water, oxygen, and minerals, and by the ability of ecosystems to recycle the residue of dead organic materials. Human activities and technology can change the flow and reduce the fertility of the land.**

▶ **The chemical elements that make up the molecules of living things pass through food webs and are combined and recombined in different ways. At each link in a food web, some energy is stored in newly made structures but much is dissipated into the environment as heat. Continual input of energy from sunlight keeps the process going. ■**

121

5F EVOLUTION OF LIFE

In the twentieth century, no scientific theory has been more difficult for people to accept than biological evolution by natural selection. It goes against some people's strongly held beliefs about when and how the world and the living things in it were created. It hints that human beings had lesser creatures as ancestors, and it flies in the face of what people can plainly see— namely that generation after generation, life forms don't change; roses stay roses, worms stay worms. New traits arising by chance alone is a strange idea, unsatisfying to many and offensive to some. And its broad applicability is not appreciated by students, most of whom know little of the vast amount of biological knowledge that evolution by natural selection attempts to explain.

It is important to distinguish between *evolution*, the historical changes in life forms that are well substantiated and generally accepted as fact by scientists, and *natural selection*, the proposed mechanism for these changes. Students should first be familiar with the evidence of evolution so that they will have an informed basis for judging different explanations. This familiarity depends on knowledge from the life and physical sciences: knowledge of phenomena occurring at several different levels of biological organization and over very long time spans, and of how fossils form and how their ages are determined. Students may very well wonder why the fossil record has so many seeming holes in it. If so, the opportunity should be seized to show the value of mathematics. The probability of specimens of any species of organisms surviving is small—soft body parts are eaten or decomposed, and hard parts are crushed or dissolved. The probability of finding a specimen is small because most are buried or otherwise inexcavable. Mathematics holds that the probability of acquiring a specimen of an extinct species is *extremely* small—the product of the two probabilities.

Before natural selection is proposed as a mechanism for evolution, students must recognize the diversity and apparent relatedness of species. Students take years to acquire sufficient knowledge of living organisms and the fossil record. Natural selection should be offered as an explanation for familiar phenomena and then revisited as new phenomena are explored. To appreciate how natural selection can account for evolution, students have to understand the important distinction between the selection of an *individual* with a certain trait and the changing proportions of that trait in *populations*. Their being able to grasp this distinction requires some understanding of the mathematics of proportions and opportunities for them to reflect on the individual-versus-population distinction in other contexts.

Controversy is an important aspect of the scientific process. Students should realize that although virtually all scientists accept the general concept of evolution of species, scientists do have different opinions on how fast and by what mechanisms evolution proceeds. A separate issue altogether is how life itself began, a detailed mechanism for which has not yet emerged.

RESEARCH NOTES
page 343

Kindergarten through Grade 2

Students should begin to build a knowledge base about biological diversity. Student curiosity about fossils and dinosaurs can be harnessed to consider life forms that no longer exist. But the distinction between extinct creatures and those that still live elsewhere will not be clear for some time. "Long ago" has very limited meaning at this age level. Even as students make observations of organisms in their own environments, they can extend their experiences with other environments through film.

By the end of the 2nd grade, students should know that

▶ **Different plants and animals have external features that help them thrive in different kinds of places.**

▶ **Some kinds of organisms that once lived on earth have completely disappeared, although they were something like others that are alive today.** ■

Grades 3 through 5

Students can begin to look for ways in which organisms in one habitat differ from those in another and consider how some of those differences are helpful to survival. The focus should be on the consequences of different features of organisms for their survival and reproduction. The study of fossils that preserve plant and animal structures is one approach to looking at characteristics of organisms. Evidence for the similarity within diversity of existing organisms can draw upon students' expanding knowledge of anatomical similarities and differences.

By the end of the 5th grade, students should know that

▶ **Individuals of the same kind differ in their characteristics, and sometimes the differences give individuals an advantage in surviving and reproducing.**

▶ **Fossils can be compared to one another and to living organisms according to their similarities and differences. Some organisms that lived long ago are similar to existing organisms, but some are quite different.** ■

Grades 6 through 8

During middle school, several lines of evidence are further developed. The fossil evidence can be expanded beyond extinctions and survivals to the notion of evolutionary history. Sedimentation of rock can be brought in to show relative age. However, actual age, which requires an understanding of isotopic dating techniques, should wait until high school, when students learn about the structure of atoms. Breeding experiments can illustrate the heritability of traits and the effects of selection. It was familiarity with selective breeding that stimulated Darwin's thinking that differences between successive generations can naturally accumulate.

By the end of the 8th grade, students should know that

▶ **Small differences between parents and offspring can accumulate (through selective breeding) in successive generations so that descendants are very different from their ancestors.**

▶ **Individual organisms with certain traits are more likely than others to survive and have offspring. Changes in environmental conditions can affect the survival of individual organisms and entire species.**

▶ **Many thousands of layers of sedimentary rock provide evidence for the long history of the earth and for the long history of changing life forms whose remains are found in the rocks. More recently deposited rock layers are more likely to contain fossils resembling existing species. ■**

Grades 9 through 12

Knowing what evolutionary change is and how it played out over geological time, students can now turn to its mechanism. They need to shift from thinking in terms of selection of individuals with a trait to changing proportions of a trait in populations. Familiarity with artificial selection, coming from studies of pedigrees and their own experiments, can be applied to natural systems, in which selection occurs because of environmental conditions. Students' understanding of radioactivity makes it possible for them to comprehend isotopic dating techniques used to determine the actual age of fossils and hence to appreciate that sufficient time may have elapsed for successive changes to have accumulated. Knowledge of DNA contributes to the evidence for life having evolved from common ancestors and provides a plausible mechanism for the origin of new traits.

History should not be overlooked. Learning about Darwin and what led him to the concept of evolution illustrates the interacting roles of evidence and theory in scientific inquiry. Moreover, the concept of evolution provided a framework for organizing new as well as "old" biological knowledge into a coherent picture of life forms.

Finally there is the matter of public response. Opposition has come and continues to come from people whose interpretation of religious writings conflicts with the story of evolution. Schools need not avoid the issue altogether. Perhaps science courses can acknowledge the disagreement and concentrate on frankly presenting the scientific view. Even if students eventually choose not to believe the scientific story, they should be well informed about what the story is.

By the end of the 12th grade, students should know that

▶ The basic idea of biological evolution is that the earth's present-day species developed from earlier, distinctly different species.

▶ Molecular evidence substantiates the anatomical evidence for evolution and provides additional detail about the sequence in which various lines of descent branched off from one another.

▶ Natural selection provides the following mechanism for evolution: Some variation in heritable characteristics exists within every species, some of these characteristics give individuals an advantage over others in surviving and reproducing, and the advantaged offspring, in turn, are more likely than others to survive and reproduce. The proportion of individuals that have advantageous characteristics will increase.

▶ Heritable characteristics can be observed at molecular and whole-organism levels—in structure, chemistry, or behavior. These characteristics strongly influence what capabilities an organism will have and how it will react, and therefore influence how likely it is to survive and reproduce.

▶ New heritable characteristics can result from new combinations of existing genes or from mutations of genes in reproductive cells. Changes in other cells of an organism cannot be passed on to the next generation.

▶ Natural selection leads to organisms that are well suited for survival in particular environments. Chance alone can result in the persistence of some heritable characteristics having no survival or reproductive advantage or disadvantage for the organism. When an environment changes, the survival value of some inherited characteristics may change.

▶ The theory of natural selection provides a scientific explanation for the history of life on earth as depicted in the fossil record and in the similarities evident within the diversity of existing organisms.

▶ Life on earth is thought to have begun as simple, one-celled organisms about 4 billion years ago. During the first 2 billion years, only single-cell microorganisms existed, but once cells with nuclei developed about a billion years ago, increasingly complex multicellular organisms evolved.

▶ Evolution builds on what already exists, so the more variety there is, the more there can be in the future. But evolution does not necessitate long-term progress in some set direction. Evolutionary changes appear to be like the growth of a bush: Some branches survive from the beginning with little or no change, many die out altogether, and others branch repeatedly, sometimes giving rise to more complex organisms. ■

TAO-CHI, *Among Peaks and Pines (Mt. Huang) ca 1701.*

Chapter **6** THE HUMAN ORGANISM

As similar as human beings are in many ways to other species, we are unique among the earth's life forms in our ability to use language and thought. Having evolved a large and complex brain, our species has a facility to think, imagine, create, and learn from experience that far exceeds that of any other species. We have used this ability to create technologies and literary and artistic works on a vast scale and to develop a scientific understanding of ourselves and the world.

We are also unique in our profound curiosity about ourselves: How are we put together physically? How were we formed? How do we relate biologically to other life forms and to our ancestors? How are we as individuals like or unlike other humans? How can we stay healthy? Much of the scientific endeavor focuses on such questions.

SCIENCE FOR ALL AMERICANS

This chapter relates to many ideas in Chapter 5: The Living Environment. Many of the characteristics of the human organism covered in this chapter are common to all mammals, or all animals, or all life forms. They are presented in a human context because that is easiest to learn for most students. Still, some features of life may be less readily recognizable in human beings because they are covered by layers of socialization and language. People sometimes become aware of their own characteristics only when they see them in other animals.

6ᴀ   Hᴜᴍᴀɴ Iᴅᴇɴᴛɪᴛʏ

However much people may vary in appearance and behavior, the variations are minor when compared with the internal similarity of all human beings. Chemical similarities make it possible for people from anywhere in the world to donate blood and organs to others or to mate and produce offspring. Furthermore, as great as cultural differences among groups of people seem to be, it is their languages, technologies, and arts that distinguish human beings from other species. The theme of same/different is at the core of distinguishing what is human. Often individuals are very aware of *differences* between themselves and their family members, between family members and neighbors, between neighbors and foreigners, etc. Interest shifts to *similarities* among people mostly when human beings are contrasted with other species.

At this level, children should be finding out about themselves and other animals, developing ideas about how people and other animals live, grow, feed, move, and use their senses. They should concentrate mainly on external features. They may be able to identify some major internal organs and have simple views of their functions, but those should not be emphasized. Although children easily imagine animals acting like people, they may have difficulty seeing people as animals. They need not be coerced into this idea, but they should explore the similarities and differences. As they progressively find similarities and differences among human beings and between human beings and other animals, they see where the animal classification is usefully applied to people and where it is not.

By the end of the 2nd grade, students should know that

▶ **People have different external features, such as the size, shape, and color of hair, skin, and eyes, but they are more like one another than like other animals.**

▶ **People need water, food, air, waste removal, and a particular range of temperatures in their environment, just as other animals do.**

▶ **People tend to live in families and communities in which individuals have different roles.** ■

Grades 3 through 5

Children should study some aspects of feeding, sensing, defending, and reproducing in human beings and other animals. It's a good time to examine how people accomplish various tasks and how machines improve on what people can perceive and do. Some students at this level may be intrigued with sports, and others may delve into topics such as dinosaurs or fossils and learn everything they can about them. Such explorations should be encouraged and opportunities provided for students to share with classmates what they have learned.

Discussing sex differences and roles requires sensitivity. In most species, the roles are biologically determined, whereas in human beings, the roles can be socially decided. History provides good examples of how human roles changed from hunting and gathering to farming to living and working in villages, towns, and cities.

By the end of the 5th grade, students should know that

▶ Unlike in human beings, behavior in insects and many other species is determined almost entirely by biological inheritance.

▶ Human beings have made tools and machines to sense and do things that they could not otherwise sense or do at all, or as quickly, or as well.

▶ Artifacts and preserved remains provide some evidence of the physical characteristics and possible behavior of human beings who lived a very long time ago. ■

Grades 6 through 8

At this level, students are studying the details of animal digestion, respiration, and reproduction, and so, in learning how human beings carry out these same functions, they can understand some of the commonalities between human beings and other animals. Middle-school students are interested in machines that support or enhance life functions, so they should also look at ways in which human beings use various machines to improve speed, mobility, strength, hearing, seeing, etc. Whenever students learn something about the ways that technology helps human beings, they also learn something about human capabilities and limitations.

By the end of the 8th grade, students should know that

▶ **Like other animals, human beings have body systems for obtaining and providing energy, defense, reproduction, and the coordination of body functions.**

▶ **Human beings have many similarities and differences. The similarities make it possible for human beings to reproduce and to donate blood and organs to one another throughout the world. Their differences enable them to create diverse social and cultural arrangements and to solve problems in a variety of ways.**

▶ **Fossil evidence is consistent with the idea that human beings evolved from earlier species.**

continued

▶ Specialized roles of individuals within other species are genetically programmed, whereas human beings are able to invent and modify a wider range of social behavior.

▶ Human beings use technology to match or excel many of the abilities of other species. Technology has helped people with disabilities survive and live more conventional lives.

▶ Technologies having to do with food production, sanitation, and disease prevention have dramatically changed how people live and work and have resulted in rapid increases in the human population. ■

Similarities of complex molecules, as well as similarities in organ systems, can be used to illustrate human beings' kinships with one another and with their more distant relations. The relationships deduced from molecular evidence are highly consistent with those deduced from the fossil record.

Apparently all species have some capacity for communication, and it may interest students to speculate on some possibilities and then see if they can find anything in the literature to support their ideas. That can then be contrasted to the ability of human beings to communicate.

By the end of the 12th grade, students should know that

▶ The similarity of human DNA sequences and the resulting similarity in cell chemistry and anatomy identify human beings as a single species.

▶ Written records and photographic and electronic devices enable human beings to share, compile, use, and misuse great amounts of information and misinformation. No other species uses such technologies. ■

Human fertilization, followed by birth, growth and development, and finally aging and death continue in a cyclic fashion over generations. Birth and death are subjects that have awed and inspired people of all ages. Perhaps no other topic brings individuals closer to a sense of connectedness to people of all cultures and all times. Considerable amounts of society's resources go towards developing technology to control birth and death. The options opened up by technology raise ethical dilemmas for both individuals and society.

RESEARCH NOTES

page 344

Chapter 3 C ISSUES IN TECHNOLOGY (health technology)
5 A DIVERSITY OF LIFE (common patterns of development)
B HEREDITY (fertilization, cell differentiation)
7 A CULTURAL EFFECTS ON BEHAVIOR
8 F HEALTH TECHNOLOGY
11 A SYSTEMS (mother-embryo feedback)
C CONSTANCY AND CHANGE

ALSO SEE

Although not much may change in the lives of young children over a couple of years, they can certainly become aware of each human life's stages—infancy, childhood, adolescence, adulthood, and old age. They see evidence of a life cycle even though they may not think of it as such. Imagining their parents as children, or themselves as old, may be impossible, so short-cycle animal (even plant) examples may be the best first step in building this understanding.

By the end of the 2nd grade, students should know that

▶ **All animals have offspring, usually with two parents involved. People may prevent some animals from producing offspring.**

▶ **A human baby grows inside its mother until its birth. Even after birth, a human baby is unable to care for itself, and its survival depends on the care it receives from adults.** ■

Children are fascinated by films and stories about early stages of human development and they are particularly intrigued by comparisons of themselves now and earlier. It may be helpful at this level to inform students about changes that will take place in them during adolescence, since when they reach puberty, they may be too embarrassed to talk with adults about it. The importance for growth of adequate rest, proper food, regular checkups, and shots to prevent disease should be supported by some of the science behind the advice.

By the end of the 5th grade, students should know that

▶ **It takes about 9 months for a human embryo to develop. Embryos are nourished by the mother, so substances she takes in will affect how well or poorly the baby develops.**

▶ **Human beings live longer than most other animals, but all living things die.**

▶ **There is a usual sequence of stages in physical and mental development in human beings, although individuals differ in exactly when they reach each stage.**

▶ **People are usually able to have children before they are able to care for them properly.** ■

Conception, both natural and artificial, and the idea (introduced earlier without saying how) of birth control are important issues for some middle-school adolescents. Films showing how embryos develop are likely to interest them. The study of the life cycle ties in well with the social issues of parenthood. Most students are intrigued to learn about rites of passage in different cultures and compare them to their own.

By the end of the 8th grade, students should know that

▶ **Fertilization occurs when sperm cells from a male's testes are deposited near an egg cell from the female ovary, and one of the sperm cells enters the egg cell. Most of the time, by chance or design, a sperm never arrives or an egg isn't available.**

▶ **Contraception measures may incapacitate sperm, block their way to the egg, prevent the release of eggs, or prevent the fertilized egg from implanting successfully.**

▶ **Following fertilization, cell division produces a small cluster of cells that then differentiate by appearance and function to form the basic tissues of an embryo. During the first three months of pregnancy, organs begin to form. During the second three months, all organs and body features develop. During the last three months, the organs and features mature enough to function well after birth. Patterns of human development are similar to those of other vertebrates.**

▶ **The developing embryo—and later the newborn infant—encounters many risks from faults in its genes, its mother's inadequate diet, her cigarette smoking or use of alcohol or other drugs, or from infection. Inadequate child care may lead to lower physical and mental ability.**

▶ **Various body changes occur as adults age. Muscles and joints become less flexible, bones and muscles lose mass, energy levels diminish, and the senses become less acute. Women stop releasing eggs and hence can no longer reproduce. The length and quality of human life are influenced by many factors, including sanitation, diet, medical care, sex, genes, environmental conditions, and personal health behaviors.** ■

Grades 9 through 12

Students should know enough about atoms and molecules to make sense of the idea that DNA carries instructions for the assembly of proteins, determining their structure and the rates at which they are made. Students' growing notion of systems can help them understand how turning instructions on and off can sequence developments over a lifetime and that each cell's immediate environment can influence its development, even though nearly all cells carry the same DNA instructions. The use of health technologies raises many social issues—what certainty is necessary before a new drug is marketed, who benefits and who pays, and what constitutes a reasonable quality of life and who should decide. By now, students can take up such controversial issues and consider the trade-offs involved.

By the end of the 12th grade, students should know that

▶ As successive generations of an embryo's cells form by division, small differences in their immediate environments cause them to develop slightly differently, by activating or inactivating different parts of the DNA information.

▶ Using artificial means to prevent or facilitate pregnancy raises questions of social norms, ethics, religious beliefs, and even politics.

▶ The very long period of human development (compared to that of other species) is associated with the prominent role of the brain in human evolution. The ability to learn persists throughout life and may improve as people build a base of ideas and come to understand how to learn well. Human mental abilities that apparently evolved for survival are used for newly invented cultural purposes such as art, literature, ritual, and games.

▶ The development and use of technologies to maintain, prolong, sustain, or terminate life raise social, moral, ethical, and legal issues. ■

6c BASIC FUNCTIONS

Like other organisms, human beings are composed of specialized cells grouped in organs that have special functions. However, rather than focusing on distinct anatomical and physiological systems (circulatory, digestive, etc.), instruction should focus on the essential requirements for life—obtaining food and deriving energy from it, protecting against injury, providing internal coordination, and reproducing. These grand body systems and their subsystems illustrate important aspects of systems in general.

RESEARCH NOTES

page 344

◁ ALSO SEE

•

RELATIVITY

BIRTH
ENVIRON-
MENT

√PARENTS × SIBLING/RIVALRY + OBSESSION√EXPECTATIONS × JOB IN LAWS² SINGLES BARS × DESPERATION

SCHOOL = GRADES × DATING³ I.R.S. 1ST MARRIAGE CHILDREN ECONOMY ÷ INCOMES

ANGST ÷ JOY STEP CHILDREN

IF EINSTEIN HAD BEEN A SOCIAL SCIENTIST

©1993 WASHINGTON POST WRITERS GROUP

135

Kindergarten through Grade 2

Children at this level think each organ has its own independent function. The eyes are for seeing, the brain is for thinking, the stomach is for digesting food, and so forth. Only later will students be able to learn how organs work in coordinated ways to make systems. One can expose young children to some of the facts in response to their questions, but they cannot understand those facts until they are older.

By the end of the 2nd grade, students should know that

▶ **The human body has parts that help it seek, find, and take in food when it feels hunger—eyes and noses for detecting food, legs to get to it, arms to carry it away, and a mouth to eat it.**

▶ **Senses can warn individuals about danger; muscles help them to fight, hide, or get out of danger.**

▶ **The brain enables human beings to think and sends messages to other body parts to help them work properly.** ■

Grades 3 through 5

At this level, children can begin to view the body as a system, in which parts do things for other parts and for the organism as a whole. Models help children to see and touch the internal organs and to know where they are located in the body. Questions about familiar body systems can be useful in getting students to start thinking about systems generally. They can then begin to understand that each organ affects and is affected by others.

By the end of the 5th grade, students should know that

▶ **From food, people obtain energy and materials for body repair and growth. The undigestible parts of food are eliminated.**

▶ **By breathing, people take in the oxygen they need to live.**

▶ **Skin protects the body from harmful substances and other organisms and from drying out.**

▶ **The brain gets signals from all parts of the body telling what is going on there. The brain also sends signals to parts of the body to influence what they do.** ■

Students can now develop more sophisticated understandings of how organs and organ systems work together. The circulation of blood carries digested food to the cells and removes wastes from them. Nerves and hormones carry messages that contract muscles to help the organism respond to its environment. Asking "What if?" questions such as "What might happen if some other parts weren't there or weren't working?" can stimulate students to reflect on connections among organs.

Students can relate knowledge of organs and organ systems to their growing knowledge of cells. The specialization of cells serves the operation of the organs, and the organs serve the needs of cells.

By the end of the 8th grade, students should know that

▶ **Organs and organ systems are composed of cells and help to provide all cells with basic needs.**

▶ **For the body to use food for energy and building materials, the food must first be digested into molecules that are absorbed and transported to cells.**

▶ **To burn food for the release of energy stored in it, oxygen must be supplied to cells, and carbon dioxide removed. Lungs take in oxygen for the combustion of food and they eliminate the carbon dioxide produced. The urinary system disposes of dissolved waste molecules, the intestinal tract removes solid wastes, and the skin and lungs rid the body of heat energy. The circulatory system moves all these substances to or from cells where they are needed or produced, responding to changing demands.**

▶ **Specialized cells and the molecules they produce identify and destroy microbes that get inside the body.**

▶ **Hormones are chemicals from glands that affect other body parts. They are involved in helping the body respond to danger and in regulating human growth, development, and reproduction.**

▶ **Interactions among the senses, nerves, and brain make possible the learning that enables human beings to cope with changes in their environment.** ■

Grades 9 through 12

Students' understanding of the human organism can expand to encompass molecular energy release, protection by the immune and nervous systems, cognition, and some of the ways in which systems interact to maintain a fairly constant environment for cells. Although some concepts can be learned from print and video, students can have direct experiences examining the effects of exercise on biological rhythms, or of food on body measurements such as temperature, pulse, blood pressure, or oxygen consumption. These types of observations can be linked to mathematical description of changes, to physical and chemical measurements, to statistical summary, and to controlled experiments.

By the end of the 12th grade, students should know that

▶ **The immune system is designed to protect against microscopic organisms and foreign substances that enter from outside the body and against some cancer cells that arise within.**

▶ **The nervous system works by electrochemical signals in the nerves and from one nerve to the next. The hormonal system exerts its influences by chemicals that circulate in the blood. These two systems also affect each other in coordinating body systems.**

▶ **Communication between cells is required to coordinate their diverse activities. Some cells secrete substances that spread only to nearby cells. Others secrete hormones, molecules that are carried in the bloodstream to widely distributed cells that have special receptor sites to which they attach. Along nerve cells, electrical impulses carry information much more rapidly than is possible by diffusion or blood flow. Some drugs mimic or block the molecules involved in transmitting nerve or hormone signals and therefore disturb normal operations of the brain and body.**

▶ **Reproduction is necessary for the survival of any species. Sexual behavior depends strongly on cultural, personal, and biological factors.** ■

6D   LEARNING

Human behavior results from the interaction of inheritance and learning. Besides being a basic function of most animals, learning defines the most prominent way in which human beings are different from other species. The apparently unique human ability to transmit ideas and practices from one generation to the next and to invent new ones has resulted in the virtually unlimited variations in ideas and behavior that are associated with different cultures. There are multiple reasons for studying how people learn. By learning about how people learn, students may be able to learn more effectively themselves and to know what difficulties they may face. Finally, knowing about the limitations of human learning can help people to anticipate problems (their own and those of others) in learning how to teach children better.

RESEARCH NOTES

page 345

Chapter **1** B SCIENTIFIC INQUIRY   (beliefs depend on prior knowledge)
7 A CULTURAL EFFECTS ON BEHAVIOR
B GROUP BEHAVIOR
12 E CRITICAL-RESPONSE SKILLS   (reflecting on one's own thought)

ALSO SEE

This level is the time to be sure that all children learn that they can learn almost anything they want to. Children are most interested in learning about their surroundings and all the ways they can interact with these surroundings. They should be encouraged to notice how they learn by asking them how they learned something in the past or how they might learn to do something new or by having them teach a skill to someone else.

By the end of the 2nd grade, students should know that

▶ **People use their senses to find out about their surroundings and themselves. Different senses give different information. Sometimes a person can get different information about the same thing by moving closer to it or further away from it.**

▶ **Some of the things people do, like playing soccer, reading, and writing, must be deliberately learned. Practicing helps people to improve. How well one learns sometimes depends on *how* one does it and *how often* and *how hard* one tries to learn.**

▶ **People can learn from each other by telling and listening, showing and watching, and imitating what others do. ■**

At this level, children are more aware that they are separate from their surroundings. As their self-awareness increases, they want to know more about their personal capabilities, what they might be able to do and know. They should be given many opportunities to explore areas of personal interest and to develop new skills.

Students may select a topic they wish to delve into and then learn all about it, and they may also select a skill at which they wish to excel. Their concern with learning is how it can help them achieve in their areas of interest. Many also take pride in teaching things to younger children by reading them stories and asking them questions about what they've heard, showing them how to play new games, etc. Attention should turn to how learning can be improved, to factors that influence learning and help students to improve their own learning.

By the end of the 5th grade, students should know that

▶ **Human beings have different interests, motivations, skills, and talents.**

▶ **Human beings can use the memory of their past experiences to make judgments about new situations.**

▶ **Many skills can be practiced until they become automatic.**

Human beings tend to repeat behaviors that feel good or have pleasant consequences and avoid behaviors that feel bad or have unpleasant consequences.

Learning means using what one already knows to make sense out of new experiences or information, not just storing the new information in one's head. ■

Grades 6 through 8

Emphasis should now be on how to figure out what learning has taken place as a consequence of studying something. Students can design various tests and administer them to individuals and groups as practice for longer studies of learning. They can investigate different ways of learning different things and compare the results they get. Many students are interested in animal behavior experiments, which can help them understand learning and the nature of scientific inquiry.

By the end of the 8th grade, students should know that

Some animal species are limited to a repertoire of genetically determined behaviors; others have more complex brains and can learn a wide variety of behaviors. All behavior is affected by both inheritance and experience.

The level of skill a person can reach in any particular activity depends on innate abilities, the amount of practice, and the use of appropriate learning technologies.

Human beings can detect a tremendous range of visual and olfactory stimuli. The strongest stimulus they can tolerate may be more than a trillion times as intense as the weakest they can detect. Still, there are many kinds of signals in the world that people cannot detect directly.

Attending closely to any one input of information usually reduces the ability to attend to others at the same time.

continued

▶ Learning often results from two perceptions or actions occurring at about the same time. The more often the same combination occurs, the stronger the mental connection between them is likely to be. Occasionally a single vivid experience will connect two things permanently in people's minds.

▶ Language and tools enable human beings to learn complicated and varied things from others. ■

Students can reflect on and generalize from the particulars studied in previous grades. Now is the time to consider some explanations of how learning takes place. Claims of sophisticated learning by other animals, such as language in lower primates, can be considered in light of available evidence.

By the end of the12th grade, students should know that

▶ Differences in the behavior of individuals arise from the interaction of heredity and experience—the effect of each depends on what the other is. Even instinctive behavior may not develop well if the individual is exposed to abnormal conditions.

▶ The expectations, moods, and prior experiences of human beings can affect how they interpret new perceptions or ideas. People tend to ignore evidence that challenges their beliefs and to accept evidence that supports them. The context in which something is learned may limit the contexts in which the learning can be used.

▶ Human thinking involves the interaction of ideas, and ideas about ideas. People can produce many associations internally without receiving information from their senses. ■

Knowledge of health and knowledge about illness and disease are closely connected. Human beings' knowledge of diseases has helped them understand how the healthy body works, just as knowing about normal body functioning helps to define and detect diseases.

Knowledge of science can inform choices about nutrition and exercise, but that doesn't ensure healthy practices. Some adults have ideas about health that are contrary to scientific facts. Ideas about what constitutes good nutrition change somewhat as new information accumulates, but the basics are quite stable. Students should learn these basics.

The learning goals in this section are closely related to those of 8F: Health Technology. The connections are not always drawn explicitly here, but children should learn how to make and graph health-relevant measurements (body temperature, pulse), discuss trade-offs in using prescription drugs, and so on.

RESEARCH NOTES

page 345

◁ ALSO SEE

Chapter 5 B HEREDITY   (genetic disease)
          C CELLS   (cell roles)
       7 A CULTURAL EFFECTS ON BEHAVIOR   (health habits)
       8 F HEALTH TECHNOLOGY
      10 I DISCOVERING GERMS

Kindergarten through Grade 2

Children should learn how to keep healthy, although they may not understand why certain diets, exercise, and rest all help. They do know some of the ways one can be in poor health, and instruction should build on that.

By the end of the 2nd grade, students should know that

▶ Eating a variety of healthful foods and getting enough exercise and rest help people to stay healthy.

▶ Some things people take into their bodies from the environment can hurt them.

▶ Some diseases are caused by germs, some are not. Diseases caused by germs may be spread by people who have them. Washing one's hands with soap and water reduces the number of germs that can get into the body or that can be passed on to other people. ■

Grades 3 through 5

Children should explore ways in which good health can be promoted. Here, they can begin to understand some of the evidence, though not in great detail. They may get their first look at microorganisms through a microscope.

By the end of the 5th grade, students should know that

▶ Food provides energy and materials for growth and repair of body parts. Vitamins and minerals, present in small amounts in foods, are essential to keep everything working well. As people grow up, the amounts and kinds of food and exercise needed by the body may change.

▶ Tobacco, alcohol, other drugs, and certain poisons in the environment (pesticides, lead) can harm human beings and other living things.

▶ If germs are able to get inside one's body, they may keep it from working properly. For defense against germs, the human body has tears, saliva, skin, some blood cells, and stomach secretions. A healthy body can fight most germs that do get inside. However, there are some germs that interfere with the body's defenses.

▶ There are some diseases that human beings can catch only once. After they've recovered they don't get sick from them again. There are many diseases that can be prevented by vaccination, so that people don't catch them even once. ■

Grades 6 through 8

Students should extend their study of the healthy functioning of the human body and ways it may be promoted or disrupted by diet, lifestyle, bacteria, and viruses. Students should consider the effects of tobacco, alcohol, and other drugs on the way the body functions. They should start reading the labels on food products and considering what healthful diets could be like.

By the end of the 8th grade, students should know that

▶ The amount of food energy (calories) a person requires varies with body weight, age, sex, activity level, and natural body efficiency. Regular exercise is important to maintain a healthy heart/lung system, good muscle tone, and bone strength.

▶ Toxic substances, some dietary habits, and some personal behavior may be bad for one's health. Some effects show up right away, others may not show up for many years. Avoiding toxic substances, such as tobacco, and changing dietary habits to reduce the intake of such things as animal fat increases the chances of living longer.

▶ Viruses, bacteria, fungi, and parasites may infect the human body and interfere with normal body functions. A person can catch a cold many times because there are many varieties of cold viruses that cause similar symptoms.

▶ White blood cells engulf invaders or produce antibodies that attack them or mark them for killing by other white cells. The antibodies produced will remain and can fight off subsequent invaders of the same kind.

▶ The environment may contain dangerous levels of substances that are harmful to human beings. Therefore, the good health of individuals requires monitoring the soil, air, and water and taking steps to keep them safe. ∎

Grades 9 through 12

Students should relate their knowledge of normal body functioning to situations, both hereditary and environmental, in which functioning is impaired. As they come across medical news in the media, students can identify new ways of detection, diagnosis, treatment, prevention, or monitoring. They should routinely try to find explanations for various disease conditions in physiological, molecular, or systems terms.

By the end of the 12th grade, students should know that

▶ Some allergic reactions are caused by the body's immune responses to usually harmless environmental substances. Sometimes the immune system may attack some of the body's own cells.

▶ Faulty genes can cause body parts or systems to work poorly. Some genetic diseases appear only when an individual has inherited a certain faulty gene from both parents.

▶ New medical techniques, efficient health care delivery systems, improved sanitation, and a fuller understanding of the nature of disease give today's human beings a better chance of staying healthy than their forebears had. Conditions now are very different from the conditions in which the species evolved. But some of the differences may not be good for human health.

▶ Some viral diseases, such as AIDS, destroy critical cells of the immune system, leaving the body unable to deal with multiple infection agents and cancerous cells. ■

6F MENTAL HEALTH

Sound mental health involves the interaction of psychological, physiological, and cultural systems. It is generally regarded as the ability to cope with the circumstances people encounter in their personal, professional, and social lives. Ideas about what constitutes good mental health vary from one culture to another and from one time period to another. This fact is probably the most important insight students can gain about mental health.

Chapter 7 A CULTURAL EFFECTS ON BEHAVIOR
        8 F HEALTH TECHNOLOGY

◁ ALSO SEE

Kindergarten through Grade 2

Children should be helped to identify internal feelings and distinguish them from external sensations. Through discussion, they can appreciate that everybody has both pleasant and unpleasant feelings.

By the end of the 2nd grade, students should know that

▶ **People have many different feelings—sadness, joy, anger, fear, etc.—about events, themselves, and other people.**

▶ **People react to personal problems in different ways. Some ways are more likely to be helpful than others.**

▶ **Talking to someone (a friend, relative, teacher, or counselor) may help people understand their feelings and problems and what to do about them.** ■

Grades 3 through 5

Children at this level are less concerned about identifying emotions than about knowing what to do with them. They know that everyone has emotions, and they may even realize that people respond differently to their own emotions. But it might seem strange that other people don't always feel the same way they do in similar situations. Students should be encouraged to wonder why they and others have certain emotions. They should learn that different ways of dealing with emotions have different consequences, and that it is normal for feelings to fluctuate. Role playing offers opportunities for students to explore ways of dealing with emotional situations. They can begin to make a connection between their physical and emotional well-being and see how one can affect the other.

By the end of the 5th grade, students should know that

▶ **Different individuals handle their feelings differently, and sometimes they have different feelings in the same situation.**

▶ **Often human beings don't understand why others act the way they do, and sometimes they don't understand their own behavior and feelings.**

▶ **Physical health can affect people's emotional well-being and vice versa.**

▶ **One way to respond to a strong feeling, either pleasant or unpleasant, is to think about what caused it and then consider whether to seek out or avoid similar situations.** ■

Grades 6 through 8

Students at this level are often drawn to situations involving intense emotions, such as those they encounter at sports events, concerts, playground fights, and in horror stories and movies. They may connect these extremes of emotion to their own thoughts and feelings. The consequences of unbridled emotion, such as violence, death, drug use, etc., are now so serious that it is important for students to understand the connection between emotion and risky behavior.

By the end of the 8th grade, students should know that

▶ **Individuals differ greatly in their ability to cope with stressful situations. Both external and internal conditions (chemistry, personal history, values) influence how people behave.**

▶ **Often people react to mental distress by denying that they have any problem. Sometimes they don't know why they feel the way they do, but with help they can sometimes uncover the reasons.** ■

Grades 9 through 12

Students at this level begin to search for their place in a complex world. Some researchers report that adolescents want to understand why people behave as they do and will seek general truths about social and psychological processes. Students are not so interested in the particulars as in the meaning of things in general.

By the end of the 12th grade, students should know that

▶ **Stresses are especially difficult for children to deal with and may have long-lasting effects.**

▶ **Biological abnormalities, such as brain injuries or chemical imbalances, can cause or increase susceptibility to psychological disturbances.**

▶ **Reactions of other people to an individual's emotional disturbance may increase its effects.**

▶ **Human beings differ greatly in how they cope with emotions and may therefore puzzle one another.**

▶ **Ideas about what constitutes good mental health and proper treatment for abnormal mental states vary from one culture to another and from one time period to another.** ■

GEORGE CATLIN, *A Little Sioux Village, 1857/1869.*

Chapter 7 HUMAN SOCIETY

As a species, we are social beings who live out our lives in the company of other humans. We organize ourselves into various kinds of social groupings, such as nomadic bands, villages, cities, and countries, in which we work, trade, play, reproduce, and interact in many other ways. Unlike other species, we combine socialization with deliberate changes in social behavior and organization over time. Consequently, the patterns of human society differ from place to place and era to era and across cultures, making the social world a very complex and dynamic environment.

Insight into human behavior comes from many sources. The views presented here are based principally on scientific investigation, but it should also be recognized that literature, drama, history, philosophy, and other nonscientific disciplines contribute significantly to our understanding of ourselves. Social scientists study human behavior from a variety of cultural, political, economic, and psychological perspectives, using both qualitative and quantitative approaches. They look for consistent patterns of individual and social behavior and for scientific explanations of those patterns. In some cases, such patterns may seem obvious once they are pointed out, although they may not have been part of how most people consciously thought about the world. In other cases, the patterns—as revealed by scientific investigation—may show people that their long-held beliefs about certain aspects of human behavior are incorrect.

Science for All Americans

Social science is a collection of disciplines, each of which examines human behavior from a different perspective and has its own particular techniques, modes of expression, and history. The social-science disciplines employ the evidence-based, hypothesis-testing, model-building approach of science in general to investigate social phenomena, using both qualitative and quantitative approaches, and they share the values characteristic of all science. Being literate in science does not require expertise in anthropology, economics, psychology, sociology, and the other social sciences individually, any more than it requires expertise in astronomy, zoology, ecology, and the other natural sciences individually. Scientific literacy does require, however, a broad-based knowledge across all of these disciplines. Therefore, instead of focusing on the special features of the individual social-science disciplines, the benchmarks in this chapter delineate what students at various levels of development ought to know about how social science illuminates human behavior.

The benchmarks that follow are aimed at getting students to understand human social behavior, not at fostering social action or socializing future citizens. In everyday life, scientific knowledge is seldom essential or sufficient for guiding social action. Nevertheless, if scientific knowledge of how human society works is widely shared, it can contribute to better personal and public decision making.

Exploring the findings of social science requires teachers and students to be both the observers and the observed. Students are being asked to observe the human drama critically and dispassionately, but at the same time they—and their family members, friends, neighbors, schoolmates, teachers, shopkeepers, and local authority figures—are also subject matter. It takes time for students to learn how to shift roles. And there is a natural tendency on the part of students of almost any age to reject or ridicule ideas that seem to violate the mores of their major peer group, and that tendency needs to be taken into account.

For their part, teachers should try to lead students to the kind of understanding of human behavior that derives from science. The role of the teacher is to provide a context for multiple perspectives in a democratic society. One useful approach, especially in the upper grades, is for teachers of different subjects to participate in seminars in which students and faculty explore social behavior from a variety of perspectives. In the lower grades, organizing learning around broad questions rather than subject-matter domains can make it easier to meet this objective. Reasoning with statistics is especially important for making sense out of social phenomena. Therefore it is useful to get mathematics and social studies teachers to work together to help all students understand social applications of probability and statistics. Students should also come to realize that insight into human behavior derives from many sources, including the biological, behavioral, and social sciences and also literature, history, art, philosophy, and religion. ■

7A Cultural Effects on Behavior

One of the central questions of human existence concerns how genetic inheritance, social inheritance, and life experience interact in making people what they are. How much control do individuals have over what they become? The conclusion reached in science is that genetics and culture interact in complex ways to influence what individuals can become but that the range of possibilities is so great that each person can shape his or her own life to a significant degree.

In considering the impact of culture on human behavior, one of the most sensitive issues that arises is social class. Historically, the class into which people are born has determined what their lives can be, although in today's world, skill, wealth, profession, and other factors may be more important than lineage in determining social status. Even where there is greater class mobility, the influence of birth status continues to be a significant factor. Analysts often divide citizens into lower, middle, upper middle, and upper classes (and sometimes use even finer distinctions) based on parent income, occupation, and education. Students in the upper grades should actually examine how those and other socioeconomic indices are determined and become familiar with their uses and limitations. School can help students understand that one can be proud of one's own cultural origins without having to denigrate other cultures. Religious, racial, language, and national prejudices are deep, generational, and not easily eliminated, but at least students can realize that those attitudes are part of everyone's cultural inheritance, and they can become familiar with the effects such attitudes can have on human behavior.

The benchmarks treat standards for defining crime and assigning punishment as cultural variables. This approach is controversial, for many people believe that there are absolute standards for acceptable and unacceptable human behavior. And, of course, students themselves are learning what is and is not acceptable behavior in different settings and are having to face the consequences of their own, sometimes unacceptable, behavior. This fact may be used to enhance the relevance of studying behavior in the context of culture, as long as the scientific study of behavior is not turned into an occasion for moralizing.

RESEARCH NOTES
page 346

◁ ALSO SEE

Chapter 3 C Issues in Technology (technology and safety)
5 B Heredity
6 The Human Organism
8 D Communication (influence of media, transportation)
12 A Values and Attitudes

Kindergarten through Grade 2

The emphasis in the first years of schooling should be on helping children to become aware of the range of society's implicit rules. Students can begin by finding out what the rules are in different classrooms and families, observing how children respond to the rules and recording their findings in drawings and notes. Discussions can focus on how the rules and behaviors resemble or differ from those in their own classroom or family. Such observations should introduce students to the idea of cultural diversity (though of course no such term need be used at this stage), and this impression should be strongly reinforced by the stories they read.

By the end of the 2nd grade, students should know that

▶ People are alike in many ways and different in many ways.

▶ Different families or classrooms have different rules and patterns of behavior. Some behaviors are not accepted in most families or schools.

▶ People often choose to dress, talk, and act like their friends, do the same things they do, and have the same kinds of things they have. They also often choose to do certain things their own way. ■

Grades 3 through 5

The curriculum should broaden the perception that students have of cultural diversity and shift their attention from observing and describing rules and behavior to considering, from a scientific viewpoint, some reasons for such rules and behavior to exist. Contrasting the common national culture of the United States with other national cultures, or with American subcultures, or with the American culture in former times can be particularly helpful. Research suggests that students are often surprised and puzzled by the actions of people in the past and so become interested in why people behave as they do in the present.

By the end of the 5th grade, students should know that

▶ People can learn about others from direct experience, from the mass communications media, and from listening to other people talk about their work and their lives. People also sometimes imitate people—or characters— in the media.

▶ People tend to feel uncomfortable with other people who dress, talk, or act very differently from themselves. What is considered to be acceptable human behavior varies from culture to culture and from one time period to another, but there are some behaviors that are unacceptable in almost all cultures, past and present. ■

Grades 6 through 8

Students should begin to understand more fully why people in different situations and other cultures, past and present, might behave or have behaved differently. Data on economics, education, employment and other demographic variables should now be studied and interrelated to deepen student understanding of the effects of social forces on behavior. Students should use graphs, probability theory, and mathematical modeling to make projections about graduation, jobs, family size, etc. Student projections can be compared to official projections as a basis for discussion of both methods and results.

Data for student analysis can come from census and other statistical databases and from their own collection efforts. Students can conduct careful surveys and, what may be especially interesting to them, they can interview older members of their families to collect their notions of what social, personal, or technological factors influenced their lives the most. Based on the testimony of grandparents, parents, and children their own age, students may get a sense of how much change can occur in a lifetime; they may then attempt to predict 20 years or so into the future and then compare their predictions with one another.

By the end of the 8th grade, students should know that

▶ **Each culture has distinctive patterns of behavior, usually practiced by most of the people who grow up in it.**

▶ **Within a large society, there may be many groups, with distinctly different subcultures associated with region, ethnic origin, or social class.**

▶ **Although within any society there is usually broad general agreement on what behavior is unacceptable, the standards used to judge behavior vary for different settings and different subgroups, and they may change with time and different political and economic conditions. Moreover, the punishments vary widely among, and even within, different societies.**

▶ **Technology, especially in transportation and communication, is increasingly important in spreading ideas, values, and behavior patterns within a society and among different societies. New technology can change cultural values and social behavior.** ■

155

Grades 9 through 12

Descriptive and statistical information about different cultures can be used to stimulate discussion about how circumstances, beliefs, and patterns of behavior are linked. The challenge is to help students make sense of behavior patterns that may seem puzzling out of the context of cultural diversity. Although students might be able to describe cultural influences on other people's thinking, the tougher goal is for them to see what influences have an effect on their own ideas and behavior.

By the end of the 12th grade, students should know that

▶ Cultural beliefs strongly influence the values and behavior of the people who grow up in the culture, often without their being fully aware of it. Response to these influences varies among individuals.

▶ The ways that unacceptable social behavior is punished depend partly on beliefs about the purposes of punishment and about its effectiveness. Effectiveness is difficult to test scientifically because circumstances vary greatly and because legal and ethical barriers interfere.

▶ Social distinctions are a part of every culture, but take many different forms, ranging from rigid classes based solely on parentage to gradations based on the acquisition of skill, wealth, or education. Differences in speech, dress, behavior, or physical features are often taken by people to be signs of social class. The difficulty of moving from one social class to another varies greatly with time, place, and economic circumstances.

▶ Heredity, culture, and personal experience interact in shaping human behavior. Their relative importance in most circumstances is not clear. ■

7B   GROUP BEHAVIOR

Students may have trouble accommodating the many meanings of social group. The term includes any set of people who regularly spend time together for any reason, invented classifications of people who have mutual interests (such as blue-collar workers or Southerners), and groups that people deliberately and formally join or are assigned to (such as college sororities, military units, scouts, street gangs, or the Shriners). Clearly these are groups in very different senses. Nonetheless, for the purpose of understanding the general consequences of group affiliation, these distinctions make little difference. Indeed, the main point may be that no matter how groups are defined, there are certain common patterns in the relationships between those who belong and those who do not.

Group membership implies some sense of commonality for members and thereby some sense of difference from nonmembers. Often members of a group tend to stereotype outsiders, and nonmembers tend to stereotype the group's members. Group membership does reveal something useful about individuals, but it is a mistake to attribute all of a group's real and imagined properties to every individual who belongs to it. The task for science education is to alert students to the prevalence and error of stereotyping, without disparaging the value of group membership.

RESEARCH NOTES
page 347

◁ ALSO SEE

Kindergarten through Grade 2

Instruction should use children's experiences to help them make explicit their intuitive notions about behavior in groups. Students can identify their own groups (family, classroom, scouts, or sports team) and indicate how one becomes a member of each. They should also know some of the things that members have in common with each other. Issues of adoption in families, exclusion from groups based on race or sex, and other potentially difficult issues might arise in these discussions, and they should be handled deftly. The school should be a model of inclusiveness, reinforcing the openness that children this age generally come to school with.

By the end of the 2nd grade, students should know that

▶ **People belong to some groups by birth and belong to some groups because they join them.**

▶ **The way people act is often influenced by the groups to which they belong.** ∎

Grades 3 through 5

Students can now identify the value that various kinds of social groupings have for their members, including the shared understandings of what behavior is and is not acceptable. Although the emphasis of their study of group behavior should be generally positive, students at this level are beginning to form cliques and should be aware of what it is like to be excluded as well as to be included. Students should be introduced to the idea of crowd behavior and asked to identify examples from their own experience of individuals in groups saying and doing things they never would say or do alone.

By the end of the 5th grade, students should know that

▶ **People often like or dislike other people because of membership in or exclusion from a particular social group. Individuals tend to support members of their own group and perceive them as being like themselves.**

▶ **Different groups have different expectations for how their members should act. Sometimes the rules are written down and strictly enforced, sometimes they are just understood from example.**

▶ **When acting together, members of a group and even people in a crowd sometimes do and say things, good or bad, that they would not do or say on their own.** ∎

Grades 6 through 8

Case studies and simulations can be used to examine how groups can influence attitudes and cause change. Literature also can contribute to students' understanding of group behavior. Appropriate novels, short stories, and plays can be read and discussed, and of course students can themselves write and perform short stories or plays that illustrate some of the ideas being studied. Source material is available on videotape as well.

By the end of the 8th grade, students should know that

▶ **Affiliation with a group can increase the power of members through pooled resources and concerted action. Joining a group often has personal advantages, such as companionship, a sense of identity, and recognition by others inside and outside the group. Group identity may create a feeling of superiority, which increases group cohesion but may also entail hostility toward other groups.**

▶ **People sometimes react to all members of a group as though they were the same and perceive in their behavior only those qualities that fit preconceptions of the group. Such stereotyping leads to uncritical judgments, such as showing blind respect for members of some groups and equally blind disrespect for members of other groups. ■**

Grades 9 through 12

Students should have the skills to survey community groups to identify types of organizations and their compositions and to interview members to find out what they believe are the benefits of belonging to the groups they do. Historical and contemporary cases can be explored for extraordinary results of group affiliation—say, the Bolshevik Revolution, the Montgomery boycotts, or the Vietnam War. At some point it is important that students examine science itself as a group enterprise—including pooling of resources, use of mutual critique, and shared biases— and all that implies for the nature of science.

By the end of the 12th grade, students should know that

▶ **The behavior of a group may not be predictable from an understanding of each of its members.**

▶ **Social organizations may serve business, political, or social purposes beyond those for which they officially exist, including unstated ones such as excluding certain categories of people from activities.** ■

7c   SOCIAL CHANGE

Social change happens sometimes in a flash, but more often slowly. The point to raise with students is not whether change is good or bad (it is usually some of each, and in any case different people are apt to judge it differently) or whether it is needed or not (societies need to change over time but they also need stability). What is sought here is an understanding of what kinds of internal and external factors foster social change or influence its character. Another major aim should be to help students to recognize that unless it is imposed by force, social change involves negotiation among different interests—on every level from deciding who does the dishes to organizing a neighborhood activity to working out international treaties. Developing such an understanding takes time—time for students to encounter and examine social change in a variety of present and historical contexts.

RESEARCH NOTES
page 347

ALSO SEE

Kindergarten through Grade 2

The way children adapt to change in their own lives can influence how they understand and relate to social change in later life. They can be helped to examine changes that affect their lives, including those they expect (moving to the next grade) and those they don't (moving away).

By the end of the 2nd grade, students should know that

▶ **Changes happen in everyone's life, sometimes suddenly, more often slowly. People cannot control some changes, but they can usually learn to cope with them.** ■

Grades 3 through 5

To a large extent, social structure can be characterized by its rules, formal and informal. Social change typically entails negotiation of some rules. Children know that breaking a rule carries a penalty, even though they may also think that some rules are unfair. They often regard all rules as unchangeable and don't know that rules can be changed through negotiation. Negotiation should be emphasized in classroom management and after-school activities. Students, to some degree, can take part in making school and classroom rules that relate to procedures, compliance, and rewards or penalties.

By the end of the 5th grade, students should know that

▶ **Although rules at home, school, church, and in the community stay mostly the same, sometimes they change. Changes in social arrangements happen because some rules do not work or new people are involved or outside circumstances change.**

▶ **Rules and laws can sometimes be changed by getting most of the people they affect to agree to change them.** ■

Grades 6 through 8

Middle-school students can imagine themselves in situations different from their own. Interviews with senior citizens, literary and media accounts of life in times past, simulations, and role playing all provide raw material for discussions about social change. Students can be helped to see that cultural patterns change because of technological innovations, scientific discoveries, and population changes. They can identify social changes that happen gradually as well as those that happen quickly because of natural disasters and wars. Students should also begin to identify aspects of family and community life that have remained relatively constant over generations.

By the end of the 8th grade, students should know that

▶ Some aspects of family and community life are the same now as they were a generation ago, but some aspects are very different. What is taught in school and school policies toward student behavior have changed over the years in response to family and community pressures.

▶ By the way they depict the ideas and customs of one culture, communications media may stimulate changes in others.

▶ Migration, conquest, and natural disasters have been major factors in causing social and cultural change. ■

Grades 9 through 12

Students at this level can consider how technology has affected mobility and the capacity for crowding in cities. They can use case studies to examine the causes and results of social changes occurring under conditions of close and prolonged contact between two cultures. Students become aware of the complexity of explaining human population patterns such as urban crowding or mobility and can search databases and identify and display trends and relationships.

By the end of the 12th grade, students should know that

▶ The size and rate of growth of the human population in any location is affected by economic, political, religious, technological, and environmental factors. Some of these factors, in turn, are influenced by the size and rate of growth of the population.

▶ The decisions of one generation both provide and limit the range of possibilities open to the next generation.

▶ Mass media, migrations, and conquest affect social change by exposing one culture to another. Extensive borrowing among cultures has led to the virtual disappearance of some cultures but only modest change in others.

▶ To various degrees, governments try to bring about social change or to impede it through policies, laws, incentives, or direct coercion. Sometimes such efforts achieve their intended results and sometimes they do not. ■

7D   SOCIAL TRADE-OFFS

Gaining an understanding of the concept of social trade-offs may be one of the most important components of a comprehensive education. Because social problems typically involve a variety of factors and interests, it is rare that one solution to a problem will carry all the benefits and avoid all disadvantages. Because increasing some advantages is likely to decrease others, most realistic solutions involve compromise among advantages and disadvantages. If the habit of considering alternatives and their consequences is to be functional for students, they should exercise it in a rich variety of contexts.

Thus the concept of trade-offs should show up in every part of the curriculum, including social studies, literature, physical education, technology, and science. Student decision making in the classroom, student government, clubs, newspaper, yearbook, community service, etc. should help them learn the inevitability of trade-offs and the need to take benefits and costs into account in any proposed action. When people with different interests are involved in solving social problems, compromise is also needed to accommodate their different perceptions of advantages and disadvantages. As students mature, they can consider social trade-offs in broader and more sophisticated situations. The principles continue to apply, but there is an increase in the range and complexity of contexts and the difficulty of making decisions.

RESEARCH NOTES

page 347

ALSO SEE ☞

Kindergarten through Grade 2

Life is full of choices, even for very young children. Many choices are made for them, of course, by parents and teachers. Telling children some of the reasons that particular choices were made can show them what counts as explanation and get them into the habit of looking for reasoned decisions. But children do make many decisions for themselves—and should be encouraged to anticipate the possible consequences of their choices.

By the end of the 2nd grade, students should know that

▶ **Getting something one wants may mean giving up something in return.**

▶ **Different people may make different choices for different reasons.**

▶ **Choices have consequences, some of which are more serious than others.** ∎

Grades 3 through 5

Children can improve their decision-making skills and apply them in new and varied situations. The decisions they make and the decisions others make in their behalf can serve as the subject for discussions about trade-offs. Discussions should include examining possible options, considering how various options will affect others, identifying possible risks, and deciding which risks, if any, are worth taking.

By the end of the 5th grade, students should know that

▶ **In making decisions, it helps to take time to consider the benefits and drawbacks of alternatives.**

▶ **In making decisions, benefits and drawbacks of alternatives can be taken into account more effectively if the people who will be affected are involved.**

▶ **Sometimes social decisions have unexpected consequences, no matter how carefully the decisions are made.** ∎

Grades 6 through 8

Students at this level are very preoccupied by personal and social relationships. Their greatest concerns are usually peer approval and popularity, sexual development and feelings, personal appearance, and the struggle to separate from family and become an individual. They can consider personal and social consequences of individual choices in health (sexual activity, immunization), education (how different course choices limit future options), and popularity (how affiliation with one group might exclude a person from others). Students should assess trade-offs that occur in the lives of their friends (or their own) and that offer only unwanted choices (such as sexual abstinence and venereal disease).

By the end of the 8th grade, students should know that

▶ There are trade-offs that each person must consider in making choices—about personal popularity, health, family relations, and education, for example—that often have life-long consequences.

▶ One common aspect of all social trade-offs pits personal benefit and the rights of the individual, on one side, against the social good and the rights of society, on the other.

▶ Trade-offs are not always between desirable possibilities. Sometimes social and personal trade-offs require accepting an unwanted outcome to avoid some other unwanted one. ■

Grades 9 through 12

Students at this age can examine the complexities of decision making and take into account different types of costs (direct, indirect, economic, social, emotional, etc.). They can examine trade-offs across generations and over great distances (actions taken in one place or time can have costly outcomes elsewhere). They can realize that the people who receive benefits and the people who bear the costs of those benefits are often not the same. Social, environmental, political, technological, and scientific case studies offer a rich foundation for developing decision-making skills.

By the end of the 12th grade, students should know that

▶ Benefits and costs of proposed choices include consequences that are long-term as well as short-term, and indirect as well as direct. The more remote the consequences of a personal or social decision, the harder it usually is to take them into account in considering alternatives. But benefits and costs may be difficult to estimate.

▶ In deciding among alternatives, a major question is who will receive the benefits and who (not necessarily the same people) will bear the costs.

▶ Social trade-offs are often generational. The cost of benefits received by one generation may fall on subsequent generations. Also, the cost of a social trade-off is sometimes borne by one generation although the benefits are enjoyed by their descendants. ■

7ᴇ POLITICAL AND ECONOMIC SYSTEMS

Political and economic systems evolve to become quite complex, so it's easy to convince ourselves that children cannot possibly understand them. Yet these systems are based upon very simple and basic premises that children can grasp, first mostly in personal terms and later in societal terms. It is important that they do so if they are to be expected to grasp the more complex refinements of these evolved systems as they become older.

When reduced to their simplest terms, political and economic models describe how people organize themselves to satisfy their needs: People tend to live in groups, and so must devise means for living and working together; people have needs, and so must devise means for allowing everyone a fair chance at meeting these needs; people can choose how they will live and work together and how they will go about meeting their basic needs.

Students can participate in student government and take some part in making, interpreting, and even enforcing rules of fair play. Students can visit local, state, and federal government institutions—city council, state legislature, Congress—and compare what these bodies do with what their student council does.

These benchmarks focus on two theoretical economic/political models, here called free-market and central-planning (in lieu of the more usual "market" and "command") to emphasize their theoretical nature. Neither exists in the pure form, though people often speak as though they did. The models are useful for promoting thoughtful analysis of government, in terms of principles and in terms of what works in practice. Discussions of what should be planned and what should be left to market forces can focus attention on issues of efficiency and fairness. As students study local, state, and federal governments and U.S. and world history, they can see that societies have very different ways of deciding what their needs are and how to meet them.

RESEARCH NOTES
page 347

◁ ALSO SEE

Kindergarten through Grade 2

Young children can engage in discussions about rules and fair play. They have a sense of what is and is not fair and are interested in discussion about it. They know about negotiation and compromise when expressed in terms of sharing in decisions. Although still focused primarily on self, they can learn to consider the interest of all in certain circumstances.

At this age, children encounter government systems mostly through the services they receive, such as mail service and police and fire protection. But at this age they are only dimly aware of what a government is. (They might think, for example, that the postman does his job because he likes people to get mail.) Children know that people have jobs, make money to buy things, and that jobs involve making or selling things or performing services. They can make connections between money and goods on the one hand and money and work on the other. They are not ready, however, to conceive of the whole system that relates production to consumption. Children should experience producing, buying, and selling, which can provide the foundation for discussions of more abstract ideas later on. They can learn about the range of jobs that parents and neighbors have through the communications media, invited speakers, and field trips.

By the end of the 2nd grade, students should know that

▶ **Money can buy things that people need or want. People earn money by working at a job making or growing things, selling things, or doing things to help other people.**

▶ **Everyone wants to be treated fairly, and some rules can help to do that.** ∎

Grades 3 through 5

Children can develop an understanding of the relationship between pay received and the quantity and quality of work done in any one job. But they have trouble making the connection between the pay received for different kinds of jobs and the amount of goods and services produced. Experience with producing goods from raw materials may be helpful for eventually understanding the larger system. Simulations of economic systems can help students explore money systems and trade. Students should also learn about systems of trade used in different places and by different groups, both past and present.

The free-market and central-planning models provide alternative ways of allocating resources. Although it's too early to ask students to compare the systems, they can have concrete experiences with the problems each model tries to solve—what is to be produced, for whom, by whom, and how? Young children have a strong sense of fairness, which should be taken seriously and used in discussions about rules, jobs, and money.

By the end of the 5th grade, students should know that

▶ **People tend to live together in groups and therefore have to have ways of deciding who will do what.**

▶ **Services that everyone gets, such as schools, libraries, parks, mail service, and police and fire protection, are usually provided by government.**

▶ There are not enough resources to satisfy all of the desires of all people, and so there has to be some way of deciding who gets what.

▶ Some jobs require more (or more expensive) training than others, some involve more risk, and some pay better. ■

Focusing on particular issues, such as health care or crime control, students can investigate how various countries answer the basic economic questions: Who decides the answers to basic economic questions, and who decides who gets to decide?

By the end of the 8th grade, students should know that

▶ Government provides some goods and services through its own agencies and some through contracts with private individuals or businesses. To pay for the goods and services, government must obtain money by taxing people or by borrowing the money.

▶ Government leaders come into power by election, appointment, or force.

▶ However they are formed, governments usually have most of the power to make, interpret, and enforce the rules and decisions that determine how a community, state, or nation will be run. Many of the rules established by governments are designed to reduce social conflict. The rules affect a wide range of human affairs, from marriage and education to scientific research and commerce.

▶ In a central-planning model, a single authority, usually a national government, decides what to produce, how to produce it, and for whom. In a free-market model, consumers and producers (individually or in organizations) make these decisions based on what they believe will benefit themselves. No real-world economy is a pure example of either model; all economies have some features of each kind. ■

169

Grades 9 through 12

As students examine more complicated cases, they should be challenged to imagine what the pure economic system would be like and to account for why all real-world systems are a mix. They can collect and analyze data from different economic systems. Games and simulations can show them how changing various conditions or making different assumptions would play out over time.

By the end of the 12th grade, students should know that

▶ **In the free-market model, the control of production and consumption is mainly in private hands. The best allocation of resources is believed to be achieved by individuals and organizations competing in the marketplace. Individual initiative, talent, and hard work are expected to be rewarded with success and wealth. Government's role is primarily to protect political and economic freedoms for society as a whole— even at the cost of some individual or group material benefits.**

▶ **In the central-planning model, production and consumption are controlled by the government. The best allocation of resources is thought to be achieved through government planning by experts. Dedication to the good of the society as a whole is expected to motivate initiative, talent, and hard work. The main purpose of government is to promote comparable welfare for all individuals and groups—even at the cost of some individual and group freedoms.**

▶ **In practice, countries make compromises with regard to economic models. Central planning has to allow for some individual initiative, and markets have to provide some protection for unsuccessful competitors. The countries of the world use elements of both systems and are neither purely free-market nor entirely centrally controlled. Countries change, some adopting more free-market policies and practices, others more central-planning ones, and still others doing some of each.** ■

7F SOCIAL CONFLICT

Human beings generally live in groups, voluntarily cooperating and competing with one another. Within the general context of interdependence, however, some social conflict is inevitable—from infants squabbling over toys to adolescent rivalries for friendship to adult constitutional debates and war. Human societies spend much time in negotiating differences— peacefully, heatedly, or violently.

Children are not strangers to conflict. One purpose of education is to help children learn how to deal thoughtfully and constructively with conflict in their own lives. Understanding the nature of social conflict and ways of moderating it can contribute to that end. Such understanding is also important for confronting community, religious, international, interracial, and intercultural conflict. School provides opportunities for students to obtain experience in resolving conflict in positive ways. Social conflict that occurs in school can be examined in ways that will lead to understanding and help students resolve conflict in constructive ways.

RESEARCH NOTES
page 348

Chapter **3**	B	DESIGN AND SYSTEMS (unequal impacts of different groups)
	C	ISSUES IN TECHNOLOGY

ALSO SEE

Young children experience many conflicts as they go about distinguishing their rights from those of others around them. The simple recognition that others have rights is a significant part of the socialization process. Children need considerable guidance to help them identify when their actions are likely to result in conflict with others and to learn strategies for avoiding it. Providing opportunities for students to explain their point of view can lead to new insights for children and teachers alike. Role playing and discussion of various situations involving conflict may also help.

By the end of the 2nd grade, students should know that

▶ **Disagreements are common, even between family members or friends. Some ways of dealing with them work better than others. People who are not involved in an argument may be helpful in solving it.**

▶ **Rules at home, at school, and in the community let individuals know what to expect and so can reduce the number of disputes.** ■

Children's own conflicts (and stories about those of others) continue to provide a rich source of material for discussion, which should include possible consequences of various ways of resolving conflict. Students can understand the fairness of majority votes, although they may not entirely accept that others' interests are as important as their own. Even when they are on the losing side, students can accept the results of an either/or choice if they believe it was fair, particularly in a class or school election. They want to take part in decision making and are able to negotiate compromises. They also can learn to negotiate from a minority position and try to lobby peers and even adults into supporting their position.

By the end of the 5th grade, students should know that

▶ **Communicating the different points of view in a dispute can often help people to find a satisfactory compromise.**

▶ **Resolving a conflict by force rather than compromise can lead to more problems.**

▶ **One person's exercise of freedom may conflict with the freedom of others. Rules can help to resolve conflicting freedoms.**

▶ **If a conflict cannot be settled by compromise, it may be decided by a vote—if everyone agrees to accept the results.** ■

Conflict has more serious consequences now than before and has immediate, as well as long-term, significance. Emphasizing conflict resolution is more important than focusing on the dangers of conflict.

By the end of the 8th grade, students should know that

▶ Being a member of a group can increase an individual's social power or hostile actions against other groups or individuals. It may also subject that person to the hostility of people who are outside the group.

▶ Most groups have formal or informal procedures for arbitrating disputes among their members. ∎

History provides endless examples of conflict, its causes, and its consequences. Adding social and economic analysis to historical narrative helps students better understand both. There is value in studying major episodes of conflict in history, but they should be made more relevant by relating them to more current episodes that have some personal meaning for students.

By the end of the 12th grade, students should know that

▶ Conflict between people or groups arises from competition over ideas, resources, power, and status. Social change, or the prospect of it, promotes conflict because social, economic, and political changes usually benefit some groups more than others. That, of course, is also true of the status quo.

▶ Conflicts are especially difficult to resolve in situations in which there are few choices and little room for compromise. Some informal ways of responding to conflict—use of pamphlets, demonstrations, cartoons, etc.—may sometimes reduce tensions and lead to compromise but at other times they may be inflammatory and make agreement more difficult to reach.

continued

▶ Conflict within a group may be reduced by conflict between it and other groups.

▶ Intergroup conflict does not necessarily end when one segment of society gets a decision in its favor, for the "losers" may then work all the harder to reverse, modify, or circumvent the change. Even when the majority of the people in a society agree on a social decision, the minority who disagree must be protected from oppression, just as the majority may need protection against unfair retaliation from the minority. ■

7ɢ Global Interdependence

The purpose here is not to promote any particular view of how nations should work together or to suggest what the balance between national interests and global ones ought to be for the United States or any other country. Rather, students need to become aware of the growing number of ways in which each nation is part of larger political, economic, military, environmental, biological, and technological systems.

In history, geography, and social studies, students ought to discuss the consequences of citizens' ties to religious and ethnic groups that transcend national borders and go farther back in history than the country itself. Whether in the United States or in other nations, nonnative groups are increasingly important elements of society.

RESEARCH NOTES
page 348

◁ **ALSO SEE**

Kindergarten through Grade 2

Children should have many opportunities to identify meaningful roles for themselves through a variety of personal interactions—in school, at home, and in the community. They can be "helpers" in all of these places—being responsible for classroom duties, picking up after themselves at home, recycling or picking up litter in their environment (and participating in other community-service activities). Children need to see themselves and what they do as important to others— enabling them to notice and appreciate how what others do affects them. They should have experiences in which "everyone must do their part" in order to achieve success—bringing an item for a class project or party, having a role in a class play or other performance, etc. Through interviews and other encounters with community, business, and government workers, children can develop a wider view of the many ways in which people can affect one another.

Children should be encouraged to ask where various products they use come from. Role playing or simulations involving trading goods and services can lead to discussions about problems that arise when bartering is the only way to get what one wants.

By the end of the 2nd grade, students should know that

▶ **For many things they need, people rely on others who are not part of the family and maybe not even part of their local community.** ■

Grades 3 through 5

Students should continue to have experiences that show their impact on the world around them, such as community-service projects. Students may not know exactly where or how far away various countries are, but they certainly know, through television or even the labels on their clothing, that many different countries exist. Books, films, other media, and direct personal experiences can expand the students' world beyond the borders of their community, state, and country. A variety of activities can familiarize (and fascinate) students with products grown or manufactured elsewhere in the world—many of which they see and use in their everyday lives. And television can show how one country helps another deal with the consequences of a natural (or other) disaster. Schools can provide opportunities to reflect on this information.

By the end of the 5th grade, students should know that

▶ **Many of the things people eat and wear come from other countries, and people in those countries use things from this country. Trade occurs between nations, between different people, and between regions in the same nation. Decisions made in one country about what is produced there may have an effect on other countries.** ■

Grades 6 through 8

The study of geographical differences in climate and natural resources will make the advantages of trade evident. Simulations can involve students in planning the use of available resources for the greatest benefit and making choices about what, how much, and how to produce things to meet wants and needs. Students can trace how policies of market participants or government agencies affect the production and distribution of resources.

By the end of the 8th grade, students should know that

▶ Trade between nations occurs when natural resources are unevenly distributed and the costs of production are very different in different countries. A nation has a trade opportunity whenever it can create more of a product or service at lower cost than another.

▶ The major ways to promote economic health are to encourage technological development, to increase the quantity or quality of a nation's productive resources—more or better-trained workers, better equipment and methods—and to engage in trade with other nations.

▶ The purpose of treaties being negotiated directly between individual countries or by international organizations is to bring about cooperation among countries.

▶ Scientists are linked to other scientists worldwide both personally and through international scientific organizations.

▶ The global environment is affected by national policies and practices relating to energy use, waste disposal, ecological management, manufacturing, and population. ■

177

Grades 9 through 12

The daily newspaper and news weeklies provide good raw material for stimulating fruitful discussions on economic and political models. Trade negotiations, worker migration, balance of payments, productivity, and the like are the focus of much international tension. In analyzing these matters, students should try to understand what is going on rather than judge what is desirable.

By the end of the 12th grade, students should know that

▶ The wealth of a country depends partly on the effort and skills of its workers, its natural resources, and the capital and technology available to it. It also depends on the balance between how much its products are sought by other nations and how much of other nations' products it seeks. Even if a country could produce everything it needs for itself, it would still benefit from trade with other countries.

▶ Because of increasing international trade, the domestic products of any country may be made up in part by parts made in other countries. The international trade picture is often complicated by political motivations taking priority over economic ones.

▶ Migration across borders, temporary and permanent, legal and illegal, plays a major role in the availability and distribution of labor in many nations. It can bring both economic benefits and political problems.

▶ The growing interdependence of world social, economic, and ecological systems does not always bring greater worldwide stability and often increases the costs of conflict. ■

GRANT WOOD, *Fall Plowing, 1931.*

Chapter 8 THE DESIGNED WORLD

The world we live in has been shaped in many important ways by human action. We have created technological options to prevent, eliminate, or lessen threats to life and the environment and to fulfill social needs. We have dammed rivers and cleared forests, made new materials and machines, covered vast areas with cities and highways, and decided—sometimes willy-nilly—the fate of many other living things.

In a sense, then, many parts of our world are designed—shaped and controlled, largely through the use of technology—in light of what we take our interests to be. We have brought the earth to a point where our future well-being will depend heavily on how we develop and use and restrict technology. In turn, that will depend heavily on how well we understand the workings of technology and the social, cultural, economic, and ecological systems within which we live.

SCIENCE FOR ALL AMERICANS

Here the focus is on particular technological systems, such as agriculture and manufacturing, and the benchmarks indicate what particular engineering, scientific, social, and historical understandings students should gain. In the companion Chapter 3: The Nature of Technology, the benchmarks deal with general principles of technology and engineering, with the relationships between technology and science, and with the effects of technology on society.

The sections in this chapter are not intended to cover all major areas of technology. Other areas—such as the technology of warfare, transportation, or architecture— might also have been included. The areas covered here should supply an ample sampling of major ideas to serve as a basis for understanding various key technologies of today—and those that will come tomorrow. For many of the ideas in this chapter, students will need a background understanding of the physical setting and the living environment, for which benchmarks are given in Chapter 4: The Physical Setting and Chapter 5: The Living Environment.

RESEARCH NOTES

page 349

The content should not be taught solely in the technology curriculum. The responsibility needs to be shared by science, mathematics, social studies, and history. Some of the instruction can be didactic but much of it should be done through student projects. Technology projects should be part of the curriculum from the earliest grades, gradually becoming longer and more complex. Most projects should be done by small student groups with teachers acting as advisers. Classroom visits by people involved in technology-related fields—such as architecture, transportation, and textiles—may help to acquaint students with occupational opportunities in technology. ■

8A AGRICULTURE

A majority of people never see food or fiber before those products get to retail stores, and primary-school children may have only vague ideas about where their foods and fabrics come from. So the first steps in teaching children about agriculture are to acquaint them with basics: what grows where, what is required to grow and harvest it, how it gets to the stores, and how modern-day U.S. agriculture compares with agriculture in other places and other times. Such comparisons prepare students to consider how agriculture can be improved, what resources are needed, and the consequences for society and the environment.

For most students, media resources about agricultural production in the United States and elsewhere may have to supplement firsthand experiences. Projects to trace locally available food and fiber back to their origins are helpful in providing at least some personal experience. As students become better able to handle complexity, they can undertake projects that require planting, fertilizing, selecting desirable features, and adjusting the amount of light, water, and warmth.

Projects for older students can involve the preservation of food and fiber, requirements for good nutrition, comparing energy efficiency of different products, and long-term changes in water, soil, and forest resources. They should expand their sense of what agriculture is to include the planting and harvesting of materials for use as fibers and fuel and for building shelters. When students are able to grasp the interdependent elements of the agricultural system, including fuel, roads, communications, weather, and prices, they may assess what disasters do to an agricultural system and possible ways of recovering or even reducing their likelihood.

ALSO SEE

Kindergarten through Grade 2

The basic experiences for primary-school children include seeing plants grow from seeds they have planted, eating the edible portions of the mature plants, and noticing what plants and other things animals eat. Comparisons can be made to see what happens if some plants don't get water or light, but carefully controlled experiments should be delayed until later, when students will know better how to conduct scientific investigations. Some of the earliest stories to be read to and by small children can tell about life on the farm and what happens to food between the farm and the store.

By the end of the 2nd grade, students should know that

▶ **Most food comes from farms either directly as crops or as the animals that eat the crops. To grow well, plants need enough warmth, light, and water. Crops also must be protected from weeds and pests that can harm them.**

▶ **Part of a crop may be lost to pests or spoilage.**

▶ **A crop that is fine when harvested may spoil before it gets to consumers.**

▶ **Machines improve what people get from crops by helping in planting and harvesting, in keeping food fresh by packaging and cooling, and in moving it long distances from where it is grown to where people live.** ■

Grades 3 through 5

Students should enhance their earlier experiences by following plants through the production of new seeds and offspring. They can design experiments to see the effects of water, light, and fertilizer, although their experiments should involve only one variable at a time.

They should study what crops are found in different environments, including oceans, and trace the paths that various foods and fibers take as they move from growers to consumers. Storage, transportation, preservation, processing, and packaging should be considered. Where possible, students should visit markets, farms, grain elevators, and processing plants and examine trucks, trains, cargo planes, and as many other parts of the "technological food chain" as possible.

To appreciate the rigors of agriculture, students should learn about life in earlier times and the great effort that went into planting, nurturing, harvesting, and using crops. It is important that they know some of the hazards that food encounters from the time it is a seed until it reaches the kitchen. Food preservation and sanitation can be explored in early grades, but explanation of spoilage as the result of microorganisms should wait until 6th through 8th grades.

By the end of the 5th grade, students should know that

▶ **Some plant varieties and animal breeds have more desirable characteristics than others, but some may be more difficult or costly to grow. The kinds of crops that can grow in an area depend on the climate and soil. Irrigation and fertilizers can help crops grow in places where there is too little water or the soil is poor.**

▶ The damage to crops caused by rodents, weeds, and insects can be reduced by using poisons, but their use may harm other plants or animals as well, and pests tend to develop resistance to poisons.

▶ Heating, salting, smoking, drying, cooling, and airtight packaging are ways to slow down the spoiling of food by microscopic organisms. These methods make it possible for food to be stored for long intervals before being used.

▶ Modern technology has increased the efficiency of agriculture so that fewer people are needed to work on farms than ever before.

▶ Places too cold or dry to grow certain crops can obtain food from places with more suitable climates. Much of the food eaten by Americans comes from other parts of the country and other places in the world. ■

In middle school, students can examine how changes in climate, fashion, or ecosystems affect agriculture. The news media, even in the cities, often report how well particular crops are doing in response to weather, pestilence, market demand, federal policies, and the like. Students' discussions of such current events can lead them to raise technological, scientific, economic, and political questions for further study.

Students should continue to be engaged in gardening and experimentation. As an addition to traditional seeds-in-soil activities, hydroponics is an inexpensive and relatively rapid way to help students understand modern agriculture because it allows them to monitor and control many of the variables that contribute to plant growth and development. Students at this level also study geography and the early history of the human species, including the transformation from hunting and gathering to farming. This agricultural revolution provides a dramatic instance of social change made possible by technological advances and, conversely, of technological advance promoted by social change.

By the end of the 8th grade, students should know that

▶ Early in human history, there was an agricultural revolution in which people changed from hunting and gathering to farming. This allowed changes in the division of labor between men and women and between children and adults, and the development of new patterns of government.
continued

185

▶ **People control the characteristics of plants and animals they raise by selective breeding and by preserving varieties of seeds (old and new) to use if growing conditions change.**

▶ **In agriculture, as in all technologies, there are always trade-offs to be made. Getting food from many different places makes people less dependent on weather in any one place, yet more dependent on transportation and communication among far-flung markets. Specializing in one crop may risk disaster if changes in weather or increases in pest populations wipe out that crop. Also, the soil may be exhausted of some nutrients, which can be replenished by rotating the right crops.**

▶ **Many people work to bring food, fiber, and fuel to U.S. markets. With improved technology, only a small fraction of workers in the United States actually plant and harvest the products that people use. Most workers are engaged in processing, packaging, transporting, and selling what is produced.** ■

Grades 9 through 12

Students' understanding of agricultural technology can increasingly draw upon their understanding of underlying science concerning the interaction of living things with their environments in ecosystems, the inheritance of traits, mutations, and natural selection. Their growing familiarity with systems concepts should be exploited in agricultural contexts to study the interactions among production, preservation, transportation, communications, government regulations, subsidies, and world markets. Social side-effects and trade-offs of agricultural strategies should be discussed in both local and world contexts.

By the end of the 12th grade, students should know that

▶ **New varieties of farm plants and animals have been engineered by manipulating their genetic instructions to produce new characteristics.**

▶ **Government sometimes intervenes in matching agricultural supply to demand in an attempt to ensure a stable, high-quality, and inexpensive food supply. Regulations are often also designed to protect farmers from abrupt changes in farming conditions and from competition by farmers in other countries.**

▶ **Agricultural technology requires trade-offs between increased production and environmental harm and between efficient production and social values. In the past century, agricultural technology led to a huge shift of population from farms to cities and a great change in how people live and work.** ■

8B MATERIALS AND MANUFACTURING

Most children like to make things. Over the school years, students should study and manipulate (shape, cut, drill, pound, bake, soak, radiate, join, grind, etc.) many different kinds of materials, from mud, clay, and paper to chemical reagents, alloys, and plastics. In doing so, they learn about the physical and chemical properties of materials as well as about manufacturing. In their building activities, students should progress from using simple tools (scissors, paste, string, rulers) to standard hand tools and cooking utensils to sensitive measuring instruments and power tools.

Students should also move from designing and making simple objects to designing, assembling, and operating a manufacturing system. The importance of planning, coordination, and control should become as evident as the importance of selecting the most appropriate materials and processes. Also evident will be the need for financing, sales, and follow-up (including maintenance, repair, and handling complaints).

Historical, social, cultural, and scientific perspectives, involving readings and films to focus class discussion and student papers, can help to fill in the picture of materials and manufacturing as essential components of human society.

ALSO SEE

Chapter 3		THE NATURE OF TECHNOLOGY
4	B	THE EARTH (effects on resources)
	D	STRUCTURE OF MATTER (synthesis of new material)
	G	FORCES OF NATURE (properties of materials)
5	D	INTERDEPENDENCE OF LIFE (loss and recovery of resources)
6	A	HUMAN IDENTITY
7	G	GLOBAL INTERDEPENDENCE (trade)
10	J	HARNESSING POWER
11	A	SYSTEMS

| Kindergarten through Grade 2 | Grades 3 through 5 |

Young children should have many experiences in working with different kinds of materials, identifying and composing their properties and figuring out their suitability for different purposes. *(The Three Little Pigs* is a familiar introduction to the world of materials for very young children.). It is not too early for children to begin to wonder what happens to something after it has been thrown away. They can monitor the amount of waste that people produce or take part in community recycling projects.

By the end of the 2nd grade, students should know that

▶ **Some kinds of materials are better than others for making any particular thing. Materials that are better in some ways (such as stronger or cheaper) may be worse in other ways (heavier or harder to cut).**

▶ **Several steps are usually involved in making things.**

▶ **Tools are used to help make things, and some things cannot be made at all without tools. Each kind of tool has a special purpose.**

▶ **Some materials can be used over again.** ■

Many interesting activities enable children to experience how people process materials. Cooking can help young people develop concepts about the effects of combining various ingredients and treating mixtures to change their properties. Weaving cloth and straw, shaping metal and plastic, cutting wood, and stamping leather can help students discover the properties of various materials and experience how people transform materials into useful objects.

Teachers can channel students' inclination to make things into assembly activities that benefit from teamwork and go beyond producing a single product. Students can develop and use a series of simple workstations to make sandwiches or fold paper into objects. Students should consider how to improve product uniformity, quantity, and quality and reduce the costs of manufacturing products.

By the end of the 5th grade, students should know that

▶ **Naturally occurring materials such as wood, clay, cotton, and animal skins may be processed or combined with other materials to change their properties.**

▶ **Through science and technology, a wide variety of materials that do not appear in nature at all have become available, ranging from steel to nylon to liquid crystals.**

Grades 6 through 8

▶ Discarded products contribute to the problem of waste disposal. Sometimes it is possible to use the materials in them to make new products, but materials differ widely in the ease with which they can be recycled.

▶ Through mass production, the time required to make a product and its cost can be greatly reduced. Although many things are still made by hand in some parts of the world, almost everything in the most technologically developed countries is now produced using automatic machines. Even automatic machines require human supervision. ■

Recycling activities take on added value when students learn about a material's origins and history. Students at this level can trace the production cycle of common materials such as paper, lumber, rubber, steel, aluminum, glass, petroleum, and plastics. Their investigation should begin with the natural formation of raw materials and examine the techniques employed to gather these raw materials, process them into workable materials, transform them into industrial and consumer products, and dispose of the products when they are no longer useful. Students should identify points in the production and disposal cycle where used materials can be collected, sorted, and reprocessed into usable materials. Once students have a sense of the whole cycle, they can understand how recycling can conserve energy and natural resources. Students can reflect on the influences that their own consumption choices can have on what products are made and how they are packaged. (Later, they can find out that sometimes recycling may use more energy and other resources than it saves.)

It is appropriate in the middle grades for students to undertake one or more manufacturing initiatives of some magnitude and complexity. At this level, students should address the challenges of conducting efficiency studies, designing production tooling, engineering a production facility, maintaining quality-control standards, and marketing their final product. The emphasis at this level should be on efficiency by maximizing production while minimizing losses (for example, of time, material, energy, and effort).

189

By the end of the 8th grade, students should know that

▶ **The choice of materials for a job depends on their properties and on how they interact with other materials. Similarly, the usefulness of some manufactured parts of an object depends on how well they fit together with the other parts.**

▶ **Manufacturing usually involves a series of steps, such as designing a product, obtaining and preparing raw materials, processing the materials mechanically or chemically, and assembling, testing, inspecting, and packaging. The sequence of these steps is also often important.**

▶ **Modern technology reduces manufacturing costs, produces more uniform products, and creates new synthetic materials that can help reduce the depletion of some natural resources.**

▶ **Automation, including the use of robots, has changed the nature of work in most fields, including manufacturing. As a result, high-skill, high-knowledge jobs in engineering, computer programming, quality control, supervision, and maintenance are replacing many routine, manual-labor jobs. Workers therefore need better learning skills and flexibility to take on new and rapidly changing jobs.** ■

The study and design of materials involves several disciplines and issues. An effort should be made to explore how scientific knowledge fuels technological advances and how technology creates new scientific knowledge. Chemistry, physics, biology, and geology provide many clear examples of this interactive relationship between science and technology. As students understand better how atoms are configured in molecules and crystals (and less-well-defined arrangements), they can begin to see the connections to large-scale properties of materials. This understanding leads naturally to laboratory tests that measure a material's physical properties (such as tensile strength, hardness, and absorbency). Such tests can be included in problems that require students to select and process materials to give the optimum compromise between properties available and properties needed. Students should see some automated production process firsthand, if possible, or at least they should see some media presentations of several automated processes.

To develop an understanding of how modern manufacturing works, students need to study and experience the role of automation in freeing people from tasks that are typically "dull, dirty, or dangerous." Students should have opportunities to manipulate and program automated devices such as tabletop robots. Students generally have a lot of misconceptions and negative attitudes about industrial robots, often based on television and movie depictions of robots. Without concrete experience, they tend to think robots are

continued

intelligent and evil machines that take jobs away from people. After a little experience playing with an industrial robot, they often report that robots are very stupid machines that are dependent on people for all the brain work and can perform only the very simplest tasks.

By the end of the 12th grade, students should know that

▶ **Manufacturing processes have been changed by improved tools and techniques based on more thorough scientific understanding, increases in the forces that can be applied and the temperatures that can be reached, and the availability of electronic controls that make operations occur more rapidly and consistently.**

▶ **Waste management includes considerations of quantity, safety, degradability, and cost. It requires social and technological innovations, because waste-disposal problems are political and economic as well as technical.**

▶ **Scientific research identifies new materials and new uses of known materials.**

▶ **Increased knowledge of the molecular structure of materials helps in the design and synthesis of new materials for special purposes.** ■

8c ENERGY SOURCES AND USE

Here the focus is on what practical knowledge students should have about energy, for which benchmarks are presented in Chapter 4: The Physical Setting and Chapter 5: The Living Environment. Students will use the term energy long before they have much of an idea of what energy is. In the elementary grades, students can simply associate energy with getting things done and with heat. Students should have experience in using a variety of energy-transforming devices and considering what their inputs and outputs are. Understanding of the science and technology of energy can grow together and lead to a better grasp of this elusive term. It also can lead to understandings needed to inform decisions about energy use.

Kindergarten through Grade 2

Young children tend to associate the term *energy* with moving around a lot. They are likely to know sources of energy by what they are used for—electricity gives people lights or cooks their food, the sun melts snow or makes some calculators work, and moving air makes a pinwheel turn and helps some boats move. But young children probably don't see heat and light as forms of energy and need not be asked to. The emphasis should be on familiarizing them with a wide variety of phenomena that result from moving water, wind, burning fuel, or connecting to batteries and wall sockets.

By the end of the 2nd grade, students should know that

▶ **People can save money by turning off machines when they are not using them.**

▶ **People burn fuels such as wood, oil, coal, or natural gas, or use electricity to cook their food and warm their houses.** ■

Grades 3 through 5

The emphasis here is on energy sources. Students should have many opportunities to observe and talk about what the sun's energy is used for. They can see moving water as an energy source for "running" mills but its conversion to electricity should probably wait until they have some familiarity with the relation between electricity and magnetism.

Students may be intrigued with the story of fossil fuels, particularly if it is linked to the era of the dinosaurs. Some students may wonder why the plants that died so long ago didn't just turn into soil the way the plants in their garden do; wondering like this should be encouraged. Just realizing that fossil fuels formed under very special conditions can help students to appreciate that these fuels are not easily replaced.

For the more easily observed sources of energy, students can start to consider inputs and outputs; what it takes for something to work and what all the effects are.

By the end of the 5th grade, students should know that

▶ **Moving air and water can be used to run machines.**

▶ **The sun is the main source of energy for people and they use it in various ways. The energy in fossil fuels such as oil and coal comes from the sun indirectly, because the fuels come from plants that grew long ago.**

▶ **Some energy sources cost less than others and some cause less pollution than others.**

▶ **People try to conserve energy in order to slow down the depletion of energy resources and/or to save money.** ■

Grades 6 through 8

The emphasis here is on energy transformation. Students at this level usually respond enthusiastically to design challenges in which teams of students are called upon to create energy-conversion systems using readily available mechanical, electrical, and electronic devices. Ingenuity, simplicity, and complexity can all be rewarded but only for those teams that also can describe correctly the science of what is happening as energy goes through its transformation(s) in their machines.

At this level, students enjoy making and testing simple energy-conversion devices such as tabletop wind generators and model solar collectors. During the testing process, students can monitor the energy-conversion process by making input versus output comparisons. The data they gather can inspire hypotheses that subsequently inspire modifications. These modifications might include altering the pitch of a wind turbine's blades to increase their speed or adding reflector panels to a solar collector to increase the amount of radiant energy entering the device. Such modifications can result in a higher output voltage in the case of the wind generator or a greater temperature gain in the case of the solar collector.

Such tinkering experiences typically create a genuine desire and readiness on the part of students to understand the laws of nature that can help them explain why their devices behave the way they do. Alternative and appropriate energy-utilization systems are typically easy to understand because they are relatively simple. Because of the simplicity of such systems, almost all students can experience some degree of success in designing, building, and testing a model alternative-energy device.

By the end of the 8th grade, students should know that

▶ Energy can change from one form to another, although in the process some energy is always converted to heat. Some systems transform energy with less loss of heat than others.

▶ Different ways of obtaining, transforming, and distributing energy have different environmental consequences.

▶ In many instances, manufacturing and other technological activities are performed at a site close to an energy source. Some forms of energy are transported easily, others are not.

▶ Electrical energy can be produced from a variety of energy sources and can be transformed into almost any other form of energy. Moreover, electricity is used to distribute energy quickly and conveniently to distant locations.

▶ Energy from the sun (and the wind and water energy derived from it) is available indefinitely. Because the flow of energy is weak and variable, very large collection systems are needed. Other sources don't renew or renew only slowly.

▶ Different parts of the world have different amounts and kinds of energy resources to use and use them for different purposes. ■

Grades 9 through 12

Students can compare industrial and nonindustrial societies by their standards of living and energy consumption. They can examine the consequences of the world's dependence on fossil fuels, explore a wide range of alternative energy resources and technologies, and consider trade-offs in each. They might evaluate such matters as the use of high-quality energy resources such as natural gas for such applications as heating homes. They can even propose policies for conserving and managing energy resources.

By the end of the 12th grade, students should know that

▶ **A central factor in technological change has been how hot a fire could be made. The discovery of new fuels, the design of better ovens and furnaces, and the forced delivery of air or pure oxygen have progressively increased the available temperature. Lasers are a new tool for focusing radiation energy with great intensity and control.**

▶ **At present, all fuels have advantages and disadvantages so that society must consider the trade-offs among them.**

▶ **Nuclear reactions release energy without the combustion products of burning fuels, but the radioactivity of fuels and by-products poses other risks, which may last for thousands of years.**

▶ **Industrialization brings an increased demand for and use of energy. Such usage contributes to the high standard of living in the industrially developing nations but also leads to more rapid depletion of the earth's energy resources and to environmental risks associated with the use of fossil and nuclear fuels.**

▶ **Decisions to slow the depletion of energy sources through efficient technology can be made at many levels, from personal to national, and they always involve trade-offs of economic costs and social values. ■**

8D COMMUNICATION

Communication is the transfer of information and some means of ensuring that what is sent is also received. Technology increases the ways in which information can be communicated, the speed of transmission, and the total volume that can be handled at any one time. The spread of communication technologies brings social change, affects people's attitudes toward others, and influences behavior.

Nearly everyone is interested in audio and television systems, radar, and communications satellites, yet they need also to realize that earlier communication technologies, such as writing and moveable type, revolutionized civilization. And before that, the development of spoken language, coupled with mobility, was an important step forward in communication technology.

People are a part of every communications system, in both its design and operation. Many students see the communications industry as important for entertainment and job prospects. Students can move from being users of various communication devices to understanding general communications principles and appreciating opportunities and problems that come with these technologies.

Kindergarten through Grade 2

Even before children master the alphabet, they know that various shapes, symbols, and colors have special meanings in society (for example, red means danger, a red octagon means stop, green means go, arrows show direction, a circle with a slash means no). Young children are fascinated by various forms of giving messages, including sign language, road signs, recycling symbols, and company logos, and they should have opportunities to invent forms of their own. Their symbols can be used in classroom routines, illustrating the need to have common meanings for signs, symbols, and gestures. They should learn that writing things down and drawing pictures can help them tell their ideas to others accurately. (Second-graders need not be burdened yet with "communicating information"— they can tell and hear and send and get messages.). Students can discuss what the best ways are to convey different kinds of messages—not to decide on right answers, of course, but to start thinking about advantages and disadvantages.

By the end of the 2nd grade, students should know that

▶ **Information can be sent and received in many different ways. Some allow answering back and some do not. Each way has advantages and disadvantages.**

▶ **Devices can be used to send and receive messages quickly and clearly.** ∎

Grades 3 through 5

Students can start to study the internal workings of major communications systems, including those of the past. For example, they can study how the parts of the world are connected by telephone lines (many of which can be traced from a building to telephone poles and from telephone poles to the local switching office). Students can learn how telephone numbers are codes for activating switches and how these switches make a series of connections that link one location to another.

Students at this level delight in using secret codes. Their own experiences and stories about the use of codes can lead to reflections about the requirements for code use. By trying to break secret codes made by classmates, students can develop skills in finding patterns and using logic. Also, students are generally eager to use a variety of communication devices. They should gain experience using computers, audiotapes, and videotapes—as well as writing and drawing implements—to communicate information to classmates and students elsewhere.

By the end of the 5th grade, students should know that

▶ **People have always tried to communicate with one another. Signed and spoken language was one of the first inventions. Early forms of recording messages used markings on materials such as wood or stone.**

▶ **Communication involves coding and decoding information. In any language, both the sender and the receiver have to know the same code, which means that secret codes can be used to keep communication private.**

▶ People have invented devices, such as paper and ink, engraved plastic disks, and magnetic tapes, for recording information. These devices enable great amounts of information to be stored and retrieved—and be sent to one or many other people or places.

▶ Communication technologies make it possible to send and receive information more and more reliably, quickly, and cheaply over long distances. ■

Grades 6 through 8

At this level, students can understand communication systems as a series of black boxes linked together to connect people in one location with people in another location. They can recognize that each black box in the chain accepts an input signal, processes that signal, and produces and sends a new signal. Consequently, a microphone is a black box that converts sound into electricity, an amplifier is a black box that uses a weak signal and produces a stronger signal, and a speaker converts electricity into sound. Building on their experiences with electricity, students can understand how these devices need to be connected together with wire to work. Students need to experiment with simple devices such as microphones, speakers, and amplifiers before they can think about more sophisticated devices such as video cameras, cathode-ray tubes, stereo systems, and satellites.

By the end of the 8th grade, students should know that

▶ Errors can occur in coding, transmitting, or decoding information, and some means of checking for accuracy is needed. Repeating the message is a frequently used method.

▶ Information can be carried by many media, including sound, light, and objects. In this century, the ability to code information as electric currents in wires, electromagnetic waves in space, and light in glass fibers has made communication millions of times faster than is possible by mail or sound. ■

Grades 9 through 12

Students need to experience firsthand how technology helps people communicate more information to more people in less time, with greater accuracy, and fewer misunderstandings. They can begin to understand how some common communication devices transform patterns of sound or light into patterns of electricity and transmit electrical patterns across a variety of linkages and how receivers process incoming signals and convert patterns of electricity back into patterns of sound and light.

By the end of the 12th grade, students should know that

▶ Almost any information can be transformed into electrical signals. A weak electrical signal can be used to shape a stronger one, which can control other signals of light, sound, mechanical devices, or radio waves.

▶ The quality of communication is determined by the strength of the signal in relation to the noise that tends to obscure it. Communication errors can be reduced by boosting and focusing signals, shielding the signal from internal and external noise, and repeating information, but all of these increase costs. Digital coding of information (using only 1's and 0's) makes possible more reliable transmission of information.

▶ As technologies that provide privacy in communication improve, so do those for invading privacy. ■

8E INFORMATION PROCESSING

Technology has played an important role in collecting, storing, retrieving, and dealing with information as well as in transmitting it. Through experience and discussion, students should learn that writing on paper, making drawings, taking pictures with a camera, talking into a tape recorder, and entering letters and numbers into a computer are all ways of capturing and saving information. The invention of writing, moveable type, tables of data, diagrams, mathematical formulas, and filing systems have all increased the amount of information that people can handle. Large amounts of information are needed to operate modern societies, and generating, processing, and transferring information are among the most common occupations in modern countries. Students should all become comfortable using computers to manipulate information and have some idea of the processes involved. They should also explore the social consequences of increased access to information and of the fact that some people or groups have greater access than others.

Kindergarten through Grade 2

Children are often required to keep folders, notebooks, journals, and/or portfolios to organize and store their work so it can be reviewed at a later date—the essence of an information storage and retrieval system. The children can help design and use simple strategies for storing and retrieving information that is recorded in the form of words and pictures on physical media (for example, audio and video cassette tapes, paper, and photographs). Using things such as personal folders, pockets mounted on the wall, and plastic file boxes located in workstations, students can learn that things need to have places where they can be stored—and if they are stored well, they are easier to find later. Things containing the same type of information can be assigned a special color or name that make it easier to store them correctly and find them later. These experiences can provide students with the foundation they will need to address more sophisticated information-management problems in the future.

By the end of the 2nd grade, students should know that

▶ **There are different ways to store things so they can be easily found later.**

▶ **Letters and numbers can be used to put things in a useful order.**

Grades 3 through 5

Children should have the opportunity to use and investigate a range of information-handling devices such as electronic mail, audio and video recorders, and reference books. They should gather, organize, and present information in several ways, using reference books, paper files, and computers.

Students are now beginning to encounter challenging information-processing problems in their school work. These problems have one or more appropriate procedures (software) for processing data, and these procedures often can be performed more efficiently with the aid of technology (hardware). Students should be encouraged to identify the data presented in the problem, develop a procedure for processing the data, implement the procedure with the aid of technology, and evaluate the reasonableness of their results. As students encounter more sophisticated problems with more complicated data sets, the procedures and tools that they use should also become more sophisticated. Eventually, students should be gathering data, processing information, and presenting the results of their data-analysis activities.

By the end of the 5th grade, students should know that

▶ **Computers are controlled partly by how they are wired and partly by special instructions called programs that are entered into a computer's memory. Some programs stay permanently in the machine but most are coded on disks and transferred into and out of the computer to suit the user.**
continued

▶ Computers can be programmed to store, retrieve, and perform operations on information. These operations include mathematical calculations, word processing, diagram drawing, and the modeling of complex events.

▶ Mistakes can occur when people enter programs or data into a computer. Computers themselves can make errors in information processing because of defects in their hardware or software. ■

Students should use simple electronic devices for sensing, making logical decisions, counting, and storing information. It is important to put programming in perspective. Only a tiny percentage of computer users need to know how to program computers. However, working out a simple program of only a few steps can help students see the importance of logical thinking and increase their understanding of how a computer works. Programming a computer also helps students realize that all the capabilities that computers have come from human intelligence.

By the end of the 8th grade, students should know that

▶ Most computers use digital codes containing only two symbols, 0 and 1, to perform all operations. Continuous signals must be transformed into digital codes before they can be processed by a computer.

▶ What use can be made of a large collection of information depends upon how it is organized. One of the values of computers is that they are able, on command, to reorganize information in a variety of ways, thereby enabling people to make more and better uses of the collection.

▶ Computer control of mechanical systems can be much quicker than human control. In situations where events happen faster than people can react, there is little choice but to rely on computers. Most complex systems still require human oversight, however, to make certain kinds of judgments about the readiness of the parts of the system (including the computers) and the system as a whole to operate properly, to react to unexpected failures, and to evaluate how well the system is serving its intended purposes.

▶ An increasing number of people work at jobs that involve processing or distributing information. Because computers can do these tasks faster and more reliably, they have become standard tools both in the workplace and at home. ■

Students should use information devices to collect and analyze data from experiments, to simulate a variety of biological and physical phenomena, to access and organize information from databases, and to use programmable systems to control electric and mechanical devices. They should also have experience using computer models. This level is a good time to think about organisms as systems in which information is shared in genes in a code that can be interpreted by biochemical processes.

By the end of the 12th grade, students should know that

▶ Computer modeling explores the logical consequences of a set of instructions and a set of data. The instructions and data input of a computer model try to represent the real world so the computer can show what would actually happen. In this way, computers assist people in making decisions by simulating the consequences of different possible decisions.

▶ Redundancy can reduce errors in storing or processing information but increases costs.

▶ Miniaturization of information-processing hardware can increase processing speed and portability, reduce energy use, and lower cost. Miniaturization is made possible through higher-purity materials and more precise fabrication technology. ■

8F HEALTH TECHNOLOGY

Good health practices should be taught for their own sake and to provide students with an understanding of the relationship between health technology and the health of the population. Learning starts with the health of each student and the means of protecting it, then gradually moves to explanations of how the body works, what causes diseases, how they are transmitted, and how the body protects itself from disease. Along the way, students learn about the role of technology in health maintenance.

It is important not to exaggerate the importance of glamorous technologies such as dialysis machines, life-support systems, and organ-transplant surgery, because ordinary public health measures have contributed more to improving the human condition and life span. Students should learn of the advances in health and human life expectancy that have resulted from inoculations, modern waste-disposal systems, sanitary food handling, refrigeration, antibiotics, medical imaging, and other technologies now considered commonplace. Individuals and society continue to deal with difficult issues in making decisions about the use of modern medical technologies. Some of these issues are technical, some are ethical. Some of the issues, such as the worldwide population explosion, are consequences of the success of technology.

Kindergarten through Grade 2

Young children know that germs can make them sick, even though they may not know exactly what germs are. Of course, good health habits should be taught and encouraged: knowing when it's important to wash their hands, being careful about what goes into their mouths, not sneezing and coughing on others, and avoiding contact with someone who is contagiously sick. Children also know that shots and oral vaccines can help prevent certain diseases and that, if they do get sick, medicines can sometimes help them get better. This knowledge can be built upon to help students realize that science and technology contribute to good health.

By the end of the 2nd grade, students should know that

▶ Vaccinations and other scientific treatments protect people from getting certain diseases, and different kinds of medicines may help those who do become sick to recover. ■

Grades 3 through 5

Students can collect information on their own health with simple devices, such as a watch, a thermometer, and a stethoscope, and they can begin to get a sense of how such information varies. Students can even undertake projects such as designing aids for the disabled. If children visit a hospital, they can see examples of how computers and monitoring instruments are important in various aspects of health care.

By the end of the 5th grade, students should know that

▶ There are normal ranges for body measurements—including temperature, heart rate, and what is in the blood and urine—that help to tell when people are well. Tools, such as thermometers and x-ray machines, provide us clues about what is happening inside the body.

▶ Technology has made it possible to repair and sometimes replace some body parts. ■

Grades 6 through 8

Teachers can capitalize on students' interest in their changing bodies by having them monitor and assess their basic vital signs and other health-related characteristics. Using simple tools such as electronic blood-pressure devices, digital thermometers, stethoscopes, biofeedback monitors, and cardiovascular-fitness software, students can monitor their own health. The data they gather can be analyzed to show how healthy people are different.

The history of medicine and public health contains numerous accounts likely to fascinate many middle-school students. Students usually know about the marvels of modern treatments but not preventions such as sewer systems. Because the health of populations depends more on public health measures than on treatment, an effort should be made to interest students in prevention, vaccination, and other public health measures.

By the end of the 8th grade, students should know that

▶ **Sanitation measures such as the use of sewers, landfills, quarantines, and safe food handling are important in controlling the spread of organisms that cause disease. Improving sanitation to prevent disease has contributed more to saving human life than any advance in medical treatment.**

▶ **The ability to measure the level of substances in body fluids has made it possible for physicians to make comparisons with normal levels, make very sophisticated diagnoses, and monitor the effects of the treatments they prescribe.**

▶ **It is becoming increasingly possible to manufacture chemical substances such as insulin and hormones that are normally found in the body. They can be used by individuals whose own bodies cannot produce the amounts required for good health. ▪**

Grades 9 through 12

Students can understand some of the science that underlies the technology, such as genetics and molecular chemistry, which make possible genetic engineering and chemical synthesis of drugs, or radioactivity and the behavior of waves in materials, which make possible various imaging techniques. Students can routinely use information technology to store, retrieve, and analyze physiological and health information. They should also examine and discuss issues of life support and access to affordable health care.

Collection of data on their own vital signs can include response to exercise and schedule changes and be done carefully and often enough to show bodily cycles in temperature and heart rate as well as individual differences in findings that can be compared for a large group.

By the end of the 12th grade, students should know that

▶ Owing to the large amount of information that computers can process, they are playing an increasingly larger role in medicine. They are used to analyze data and to keep track of diagnostic information about individuals and statistical information on the distribution and spread of various maladies in populations.

▶ Almost all body substances and functions have daily or longer cycles. These cycles often need to be taken into account in interpreting normal ranges for body measurements, detecting disease, and planning treatment of illness. Computers aid in detecting, analyzing, and monitoring these cycles.

▶ Knowledge of genetics is opening whole new fields of health care. In diagnosis, mapping of genetic instructions in cells makes it possible to detect defective genes that may lead to poor health. In treatment, substances from genetically engineered organisms may reduce the cost and side effects of replacing missing body chemicals.

▶ Inoculations use weakened germs (or parts of them) to stimulate the body's immune system to react. This reaction prepares the body to fight subsequent invasions by actual germs of that type. Some inoculations last for life.

▶ Knowledge of molecular structure and interactions aids in synthesizing new drugs and predicting their effects.

▶ The diagnosis and treatment of mental disorders are improving but not as rapidly as for physical health. Techniques for detecting and diagnosing these disorders include observation of behavior, in-depth interviews, and measurements of body chemistry. Treatments range from discussing problems to affecting the brain directly with chemicals, electric shock, or surgery.

▶ Biotechnology has contributed to health improvement in many ways, but its cost and application have led to a variety of controversial social and ethical issues. ■

STUART DAVIS, *Rapt at Rappaport's*, 1952.

THE MATHEMATICAL WORLD

Mathematics is essentially a process of thinking that involves building and applying abstract, logically connected networks of ideas. These ideas often arise from the need to solve problems in science, technology, and everyday life—problems ranging from how to model certain aspects of a complex scientific problem to how to balance a checkbook.

SCIENCE FOR ALL AMERICANS

Mathematical knowledge is interesting in its own right and because it can contribute to the understanding of nature and human inventions. The benchmarks in this chapter deal with basic mathematical ideas, especially those with practical application. Project 2061 takes the view that adult literacy consists primarily of enhanced perception and insight. It is this knowledge that provides the basis for solving problems, making decisions, understanding the world, and learning more. Separation of knowing and doing is unusual in mathematics education, but the distinction is as important for mathematics as it is for other sciences. Consistent with the format of Chapters 1 to 11, the benchmarks in this chapter are expressed as "students should know that." The related "students should be able to" benchmarks appear in Chapter 12: Habits of Mind.

However, claiming the primary value of knowledge does not address the question of how the knowledge is acquired. As with other sciences, understanding mathematics almost always requires extensive experience using it—solving problems, communicating ideas, connecting ideas to one another. The National Council of Teachers of Mathematics' report *Curriculum and Evaluation Standards for School Mathematics* (referred to hereafter as NCTM *Standards*) is an excellent source of inspiration and of suggestions for how instruction can promote both knowledge and practice. The mathematical components of benchmarks provide a level of specificity required for the Project 2061 strategy of designing rich, interconnected curriculum blocks with explicit goals for what students will learn.

9_A NUMBERS

Numbers are everywhere and enter people's lives in many guises. School experience with numbers should foster an appreciation of the beauty and versatility of numbers and contribute to the development of *number sense*. Although difficult to define in detail, number sense is what enables literate people to judge when mathematical thinking is making sense and whether the results are reasonable. Most important, it gives people confidence to use mathematics in solving problems and communicating ideas.

If students are to gain such confidence, curricula must be designed so that (1) the roles that numbers play in different activities—sports, history, music, election communications, lotteries, coding, etc.—are made explicit and discussed; (2) time is allotted to examining some of the more fascinating mathematical ideas, such as zero, negative numbers, pi, and primes; and (3) the numbers used in problem solving are frequently those that come from actual measurements—students' whenever possible and databases when not. Inquiry and design projects provide rich opportunities for students to work on problems that interest them and that engage them in making estimates, counting, taking measurements, graphing, and otherwise using numbers in ways that contribute to the growth of number sense.

RESEARCH NOTES

page 350

Young children should have two kinds of experiences with numbers. One is simply to have fun with them. Counting and counting games in which students are challenged to count forward and backward, skip count, match numbers and things, guess how many things there are in a set and then count to see who is right, and so forth, are popular with students and help them become comfortable with numbers. These counting games should be extended to include having students compare, combine, equalize, and change numbers as well as "take away" and "add to." But counting and estimating—and of course doing sums and differences—are not the only use of numbers that students can learn in the early grades. The use of numbers for naming things, for instance, can be brought out by having students assemble a display or portfolio of all the different ways, such as car licenses and room numbers, they can discover in which numbers are for naming.

The other kind of number experience that is essential has to do with measurement (which is, after all, but a form of counting). Students should be doing things, especially in science and design projects, that require them to pose questions that can be answered only by numbers *associated with things*. In this way, they can begin to understand that answers to such questions as, say, "How big?" "How far?" or "How long?" can be, respectively, "9 pounds," "9 blocks," or "9 days"—but not "9." Although students should be encouraged to make relative physical comparisons directly whenever they can, concluding, say, that B is taller than A, C holds more than D, etc., they should

also begin to develop a preference for numerical comparisons—B is 2 inches taller than A, box C holds 14 more marbles than box D. Graphing at this level should be mostly in the form of pictographs for the purpose of relative comparisons rather than the plotting of numbers.

By the end of the 2nd grade, students should know that

▶ **Numbers can be used to count things, place them in order, or name them.**

▶ **Sometimes in sharing or measuring there is a need to use numbers *between* whole numbers.**

▶ **It is possible (and often useful) to estimate quantities without knowing them exactly.**

▶ **Simple graphs can help to tell about observations.** ■

Grades 3 through 5

At this level, students will be learning multiplication and division as necessary skills, using paper and pencil and calculators. Some of the practice needed to master these skills can be carried out using context-free numbers. But if students are to learn about the meaning of numbers and to use them properly, much of what they do must be based on solving problems in which the answers matter and the numbers used are measured quantities. A great source of number lore for students this age (and older) from which interesting problems can be crafted by the students themselves is *The Guinness Book of World Records*.

Students are now able to make more precise and varied measurements than in the earlier grades, and it is not too early to point out and discuss some of the realities of numbers based on measurement, especially that measurements are estimates that vary somewhat, that how a number is written says something about how precise the measurement was, and that specifying the unit of measurement is always necessary. These realities can be treated as general ideas and obvious examples can be given without requiring sophisticated rules.

As a practical matter, zero is important in measurement and graphing because it anchors scales, but students should have a chance now to explore it as an interesting mathematical concept. It can be made part of their introduction to the idea of a number system and place value.

By the end of the 5th grade, students should know that

▶ **The meaning of numerals in many-digit numbers depends on their positions.**

▶ **In some situations, "0" means none of something, but in others it may be just the label of some point on a scale.**

▶ **When people care about what is being counted or measured, it is important for them to say what the units are (three degrees Fahrenheit is different from three centimeters, three miles from three miles per hour).**

▶ **Measurements are always likely to give slightly different numbers, even if what is being measured stays the same.** ■

This may be the most important period of all in helping students develop an understanding of numbers. Up to now, they have dealt mostly with positive whole numbers and their manipulation, even though the numbers came from measurement. Negative numbers and fractions can now come into the picture because students will need to use them in carrying out the kinds of science and technology activities that should be on their agenda. This practical introduction to the value of fractions and negative numbers should be complemented by opportunities for students to reflect on those and other mathematical ideas, including relations of operations to one another, number systems, and abstract number patterns.

Except in instances in which the purpose is clearly practice of operations, teachers should insist that students *think about* the numbers they used in solving quantitative problems. Students should ask themselves and each other such questions as "What units should be associated with the measurements and the calculated answer?" "How many digits are enough in the answer, no matter what the calculator shows?" (Formal rules for significant figures are difficult, and most people tend to keep more digits than necessary "just to be safe," but at least they should realize that it matters and that there are ways to deal with it.) "Does the number make sense?"

By the end of the 8th grade, students should know that

▶ **There have been systems for writing numbers other than the Arabic system of place values based on tens. The very old Roman numerals are now used only for dates, clock faces, or ordering**
chapters in a book. **Numbers based on 60 are still used for describing time and angles.**

▶ **A number line can be extended on the other side of zero to represent** *negative numbers*. **Negative numbers allow subtraction of a bigger number from a smaller number to make sense, and are often used when something can be measured on either side of some reference point (time, ground level, temperature, budget).**

▶ **Numbers can be written in different forms, depending on how they are being used. How fractions or decimals based on** *measured* **quantities should be written depends on how precise the measurements are and how precise an answer is needed.**

▶ **The operations + and - are inverses of each other—one undoes what the other does; likewise x and ÷.**

▶ **The expression** *a/b* **can mean different things:** *a* **parts of size** *1/b* **each,** *a* **divided by** *b,* **or** *a* **compared to** *b.*

▶ **Numbers can be represented by using sequences of only two symbols (such as 1 and 0, on and off); computers work this way.**

▶ **Computations (as on calculators) can give more digits than make sense or are useful. ■**

Grades 9 through 12

With regard to numbers, the 9th through 12th grades are mostly an elaboration of the 6th through 8th grades. Students at this level encounter numbers and computations chiefly in the context of solving problems or learning more advanced mathematics. Through repeatedly encountering problem situations with numerical demands, students can deepen and refine their understanding of and facility with number relations, operations, ratios, estimation, measurement, graphs, and so on. Similarly, they should become more experienced in using calculators for a variety of computational tasks.

By the end of the 12th grade, students should know that

▶ Comparison of numbers of very different size can be made approximately by expressing them as nearest powers of 10.

▶ Numbers can be written with bases different from ten (which people probably use because of their 10 fingers). The simplest base, 2, uses just two symbols (1 and 0, or on and off).

▶ When calculations are made with measurements, a small error in the measurements may lead to a large error in the results.

▶ The effects of uncertainties in measurements on a computed result can be estimated. ■

HERMAN®

"Four blocks north. If it's not there, eight blocks south."

For all its popularity as a sign of academic respectability, its prominence as the gateway to advancement (to college, if not in life more generally), and its undeniable importance in science, engineering, and many other fields, algebra remains a subject about which it is not altogether clear what the average adult needs to know. The current trend toward a year of algebra for all students is often justified by citing the strong correlation between taking algebra and vocational success in life, but of course correlations do not imply cause and effect. The question that Project 2061 asked in *Science for All Americans* was what should all students learn, not how much algebra they should take—just as it did not ask how much biology or chemistry they should have.

Certainly everyone should know what algebra is, for it stands as one of the great human inventions of all time. The first step toward this understanding is for students to learn about symbols, including that the use of symbols is widespread, takes many forms, and is not the exclusive property of algebra or mathematics. Algebra, however, represents numbers, sets of numbers, quantities, and relationships with letters and signs (for operations) in a systematic way that turns out to be useful for describing relationships between variables. The second step is for students to learn what it means to manipulate symbolic statements. Students can use algebraic symbols and even make up simple symbolic statements long before they know they are "doing algebra." Later, as students encounter examples of how algebra is used in various contexts (natural and social sciences, design), they will develop a sense of what it is.

The difficult practical questions have to do with how much students should learn about the nature and uses of algebraic equations and how adept they can be expected to become in manipulating equations. Developing a sense of what equations are and how they correspond to other ways of expressing relationships among things is more important for science literacy than being able to derive or use them. Students who take a year or more of algebra often learn to manipulate symbols and solve equations (at least until exam time) but come away with little grasp of what a solution means or why anyone would want it.

Over a period of years, therefore, students should have experiences leading to the realization that symbolic equations can be used—interchangeably with graphs, tables, and words—to summarize data and to model real-world relationships (as in physics, finance, engineering, etc.). Care must be taken, however, not to go over the heads of the students. Variables should be selected for study that are interesting and observable (or measurable), and the focus should be on the simple relationships between one variable and another outlined in *Science for All Americans*.

continued

RESEARCH NOTES
page 350

◁ ALSO SEE

Science uses algebra in modeling how changes in one quantity affect changes in other quantities. Much of physics, chemistry, and engineering, and increasingly biology too, depends on algebraic representation. In Project 2061, we don't expect students to remember formulas for accelerations or parallel circuits or mass action; nor do we expect them to be able to perform algebraic manipulations or solve simultaneous equations. We do expect them to acquire an understanding of proportionality, the ability to read an algebraic formula, and to develop the ability to relate the shape of a graph to its implications for how some aspect of the world *behaves*.

The transformation of equations into graphs has been greatly simplified by calculators and computers. But before students begin to use that capability, they need to have considerable experience making data tables by numerical substitution in simple equations and then graphing the data. Then they can reverse the process by trying to find a curve, and hence a formula, that seems to fit the points on the graph. Perhaps the most practical way for them to learn about the transformation between data tables, graphs, and formulas is from using computer spreadsheet and graphics software. They can use data that interest them; make up mathematical formulas; use existing formulas (spreadsheet "functions"); carry out series calculations; and print out tables and line, bar, and circle graphs.

When algebraic equations and graphs of equations are used in studying science, the point should be made frequently that formulas and graphs are intended to describe phenomena but may not necessarily do so well. "Why doesn't it fit exactly?" is a question to which increasingly sophisticated answers should be given. Answers about errors of observation should come first, then answers about choosing the wrong formula to fit idealized data; later, answers should include uncontrolled influences and inappropriate ranges of application; finally should come the answer, "The world just doesn't seem to act as simply as the mathematics."

From the earliest grades, students should be asked to look for regularities in events, shapes, designs, and sets of numbers. Especially they should look for situations in which changes in one thing seem to be associated with changes in other things, but it would be a mistake to introduce dependence between two variables in all of its algebraic glory. A sense of function can start to be built both mathematically (as in trying the same calculator steps on different numbers) and physically (as by adjusting faucets, television-set controls, or thermostats, or observing the effects of exercise on heart and breathing rates).

By the end of the 2nd grade, students should know that

▶ **Similar patterns may show up in many places in nature and in the things people make.**

▶ **Sometimes changing one thing causes changes in something else. In some situations, changing the same thing in the same way usually has the same result.** ■

Symbols are just things that stand for other things or sets of other things or kinds of other things. They can be objects or marks, even sounds. Perhaps it is not too soon to engage students in collecting or identifying symbols—state flags, the school mascot, "happy faces," candles on birthday cakes, etc.—and making up symbols to represent things and a combination of symbols to represent relationships (specified by other students) such as "this is bigger (or faster or more expensive) than that." In this activity, students should be helped to realize that the idea of symbols is not the sole property of mathematics, and letters are not the only kind of symbol used. They should gather and compare the uses of as many different kinds of symbols as they can find in mathematics and elsewhere—hieroglyphics, numbers, icons, musical notation, etc.

The dependence of one quantity on another can be appreciated first as simply "changing x causes a change in y." That need be no more than noticing the change in y and saying whether it gets bigger or smaller. Also feasible at this level is whether a noticeable change in y requires a lot of change in x or just a little. It is probably premature to introduce dependence between two variables in formal symbols. Some foreshadowing can take place, however. The unknown box □ in equations at this level typically stands for a single value that will make the equation a true statement. Two unknown boxes (or inputs and outputs of a "function machine") allow paired sets of values to satisfy the equation. It is possible for students to figure out what y is implied by a given x – and what x would be required to give a desired y. In any case, graphs and tables, rather than equations, should be used to explore relationships between two variables.

By the end of the 5th grade, students should know that

▶ **Mathematical statements using symbols may be true only when the symbols are replaced by certain numbers.**

▶ **Tables and graphs can show how values of one quantity are related to values of another.** ∎

During these years, students can begin to see what it means to explore the relationships among different quantities by representing them as symbols and by manipulating statements that relate the symbols—and yet not themselves be ready to handle equations algebraically. That will happen if students are shown simple equations that represent some of the relationships they can extract from tables and graphs they have created and if they learn to use equations by numerical substitution. The use of substitution is suggested by the following stream-of-consciousness scenario of a student confronted with a problem that requires some algebra:

> I need to know how long would it take a dropped object to fall 10 feet. And here is the equation relating fall distance to time: $s=1/2at^2$. (I wonder why they use s—shouldn't the symbol be d for distance? Well, it doesn't matter as long as you know what it stands for.) I know s and a, but I want to find the right value for t. There is probably some way to rewrite this equation to get t by itself, but I don't have any confidence in my ability to do it right. So let's see if I can figure out what value of t would give 10 feet for s. How about 1 second? Nope, that results in a fall of 16 feet. How about 1/2 second? Nope, t is squared, so that gives just a quarter as much—4 feet. Well, try in between: 3/4 second. O.K., nine feet is pretty close. That's close enough for my purpose: a ten-foot fall takes a little over 3/4 second. (Or, if better precision were needed, "3/4 is .75, so let me try .80 . . . .")

In building on and drawing from students' experiences with patterns and regularity, emphasis shifts toward an exploration of functions—the basic notion that changes in one variable result in change in another. However, as stated in NCTM *Standards*, at this level "work with patterns should emphasize concrete situations and be informal and relatively unburdened by symbolism." More relevant than formal symbolic representation at this level is exploration of the notion of function, including maximum and minimum values, behavior at specially interesting values such as zero, approaches to limiting values, and so on.

The concept of a variable, pervasive as it is in mathematics, is difficult and often not understood. Even adult veterans of algebra may think of variables only by imagining particular numerical values for them. Letter names for variables may be taken to stand for single units (*P* to stand for a professor rather than some number of professors). Variables should not be approached through abstract definition but rather through real-world situations familiar to students in which they can understand, perhaps even be interested in, the multiple possibilities for value.

By the end of the 8th grade, students should know that

▶ **An equation containing a variable may be true for just one value of the variable.**

▶ **Mathematical statements can be used to describe how one quantity changes when another changes. Rates of change can be computed from differences in magnitudes and vice versa.**

▶ **Graphs can show a variety of possible relationships between two variables. As one variable increases uniformly, the other may do one of the following: increase or decrease steadily, increase or decrease faster and faster, get closer and closer to some limiting value, reach some intermediate maximum or minimum, alternately increase and decrease indefinitely, increase or decrease in steps, or do something different from any of these. ■**

Grades 9 through 12

Students should practice using tabular, graphical, and symbolic representations of functions and translating among them—and they should be called upon to describe their tables, graphs, and equations in clear English. With the help of calculators and computers, they should explore the effects of changing terms in an equation on the general behavior of its graph. Computing technology enables schools to provide a richer set of algebra experiences for all students than ever before. Students should spend less time plotting curves point by point but more time interpreting graphs, exploring the properties of graphs, and determining how these properties relate to the forms of the corresponding equations. Of course, students should continue to plot a few points to check the reasonableness of their graphs.

In modeling phenomena, students should encounter a variety of common *kinds* of relationships depicted in graphs (direct proportions, inverses, accelerating and saturating curves, and maximums and minimums) and therefore develop the habit of entertaining these possibilities when considering how two quantities might be related. None of these terms need be used at first, however. "It is biggest here and less on either side" or "It keeps getting bigger, but not as quickly as before" are perfectly acceptable—especially when phenomena that behave like this can be described.

In high school, students should encounter the idea that one quantity may relate, not to the amount of some other quantity, but to its rate of change—as force relates to the rate of change of velocity, or the induced electric "field" relates to the rate of change of the magnetic field. There are also many examples of the rate of change of a quantity being proportional to the quantity itself (for instance, radioactive decay, compound interest, or unhampered population growth). Prior to the availability of cheap calculators, such an ostensibly changing rate might have been treated, for purposes of general literacy, as an instance of successive multiplication.

By the end of the 12th grade, students should know that

▶ **In some cases, the more of something there is, the more rapidly it may change (as the number of births is proportional to the size of the population). In other cases, the rate of change of something depends on how much there is of something else (as the rate of change of speed is proportional to the amount of force acting).**

▶ **Symbolic statements can be manipulated by rules of mathematical logic to produce other statements of the same relationship, which may show some interesting aspect more clearly. Symbolic statements can be combined to look for values of variables that will satisfy all of them at the same time.**

▶ **Any mathematical model, graphic or algebraic, is limited in how well it can represent how the world works. The usefulness of a mathematical model for predicting may be limited by uncertainties in measurements, by neglect of some important influences, or by requiring too much computation.**

▶ Tables, graphs, and symbols are alternative ways of representing data and relationships that can be translated from one to another.

▶ When a relationship is represented in symbols, numbers can be substituted for all but one of the symbols and the possible value of the remaining symbol computed. Sometimes the relationship may be satisfied by one value, sometimes more than one, and sometimes maybe not at all.

▶ The reasonableness of the result of a computation can be estimated from what the inputs and operations are. ■

9c SHAPES

Long before they can use the language of geometry, children become aware of shape. Before entering school, they have had lots of experiences with points, lines, planes, and spaces. In school, students need to extend that knowledge, developing spatial sense and learning to see the world through the eyes of geometry. That can come from activities that require them to use geometry in constructing, drawing, measuring, visualizing, comparing, describing and transforming things. The progression of experiences should take students from recognizing shapes as wholes to recognizing explicit properties of shapes, and only then to the analysis of relationships among shapes.

RESEARCH NOTES

page 352

Because students in these grades are engaged in projects that have them collecting and building things, there are bound to be many opportunities to get them thinking about shapes. They should make drawings of the things they collect and of things they observe outside the classroom, and then discuss them from many perspectives such as color, size, and of course shape. At first, students tend to describe the shape of one thing by comparing it to something else—a marble is shaped like a basketball, a sheet of paper like a rug, a jump rope like a shoelace. As they organize different things that have sort of the same shape into groups, the need for names for the shared property will begin to become apparent to them.

Art is especially important in fostering spatial sense. Students should construct recognizable two-dimensional images (faces, people, buildings, beds, etc.) using only rectangles, triangles, and circles, and then do the reverse—that is, identify those same shapes in pictures of things. Also, a start can be made in laying the groundwork for the introduction, later, of the idea of symmetry by having students practice drawing pictures of a given object—geometrically simple ones—in which the position of the object is rotated or the observer changes position. And, in all of this, students should be given descriptive tasks that require using words such as above, below, behind, inside, outside, and upside-down.

By the end of the 2nd grade, students should know that

▶ Shapes such as circles, squares, and triangles can be used to describe many things that can be seen. ∎

The geometric description of objects includes size, orientation, symmetry, and proportions, as well as shape. Students should begin to use all these features in describing and designing things and increase substantially the number of geometric shapes and concepts they are familiar with. Concepts of area and volume should first be developed concretely, with procedures for computation following only when the concepts and some of their practical uses are well understood. Graphing can help students grasp some of the connections between quantity, shape, and position.

By the end of the 5th grade, students should know that

▶ Length can be thought of as unit lengths joined together, area as a collection of unit squares, and volume as a set of unit cubes.

▶ If 0 and 1 are located on a line, any other number can be depicted as a position on the line.

▶ Graphical display of numbers may make it possible to spot patterns that are not otherwise obvious, such as comparative size and trends.

▶ Many objects can be described in terms of simple plane figures and solids. Shapes can be compared in terms of concepts such as parallel and perpendicular, congruence and similarity, and symmetry. Symmetry can be found by reflection, turns, or slides.

▶ Areas of irregular shapes can be found by dividing them into squares and triangles.

▶ Scale drawings show shapes and compare locations of things very different in size. ∎

223

Grades 6 through 8

The expanding logical capabilities of students at this level enable them to draw inferences and make logical deductions from geometric problems. Students should investigate and use geometric ideas rather than memorizing definitions and formulas. Similarity and congruence can be explored through transformations. Figures should be oriented in various positions to aid in forming generalizations that won't be bound to standard orientations. That is made particularly convenient by computer software that performs "flips" and "stretches." Photographs, overhead projectors, and photocopying machines are other common tools for shrinking and stretching shapes.

Exploration of how linear measures, areas, and volumes change with size will strengthen the concepts themselves and help, generally, in leading students toward the ideas of scale that appear in Chapter 11: Common Themes. (Most children in this grade range expect area and volume to change in direct proportion to linear size.)

Learning to find locations in reality and on maps using rectangular and polar coordinates can contribute to an understanding of scale and illustrate one of the important connections between numbers and geometry. Shape in these grades is strongly related to spatial measurements. Students should have extensive experience in measuring and estimating perimeter, area, volume, and angles, choosing appropriate measurement units and measuring tools. As much as possible, these activities should be carried out in the context of actual projects, that is, in order to design and build something.

By the end of the 8th grade, students should know that

▶ **Some shapes have special properties: Triangular shapes tend to make structures rigid, and round shapes give the least possible boundary for a given amount of interior area. Shapes can match exactly or have the same shape in different sizes.**

▶ **Lines can be parallel, perpendicular, or oblique.**

▶ **Shapes on a sphere like the earth cannot be depicted on a flat surface without some distortion.**

▶ **The graphic display of numbers may help to show patterns such as trends, varying rates of change, gaps, or clusters. Such patterns sometimes can be used to make predictions about the phenomena being graphed.**

▶ **It takes two numbers to locate a point on a map or any other flat surface. The numbers may be two perpendicular distances from a point, or an angle and a distance from a point.**

▶ **The scale chosen for a graph or drawing makes a big difference in how useful it is. ■**

Grades 9 through 12

Deductive proofs should be encountered in discussions of proof in a larger context than just geometry. How do people know when something has been "proven" to be true? Is it the same in astronomy as in biology, chemistry as in law, geometry as in algebra? The nature of logic and evidence are topics that should come up frequently in science, history, social studies, and mathematics. Although it is not worth trying to teach all students to become good at working out Euclidean proofs, they should learn something of what such proofs entail and why they are important in mathematics.

Computers can be enormously useful at this stage. Students should use them to explore complex shapes in three dimensions, to analyze the geometry of objects of interest to them, to work out scale problems, and to graph data from their science activities. But computers cannot substitute altogether for direct experience. Thus students should, for example, have opportunities to solve problems requiring triangulation—such as the classical experience of determining the distance across a river or to the moon, which can be done with scale drawings. They should also do some mechanical drawing the old-fashioned way before using the graphic capabilities of the computer.

By the end of the 12th grade, students should know that

▶ **Distances and angles that are inconvenient to measure directly can be found from measurable distances and angles using scale drawings or formulas.**

▶ **There are formulas for calculating the surface areas and volumes of regular shapes. When the linear size of a shape changes by some factor, its area and volume change disproportionately: area in proportion to the square of the factor, and volume in proportion to its cube. Properties of an object that depend on its area or volume also change disproportionately.**

▶ **Geometric shapes and relationships can be described in terms of symbols and numbers—and vice versa. For example, the position of any point on a surface can be specified by two numbers; a graph represents all the values that satisfy an equation; and if two equations have to be satisfied at the same time, the values that satisfy them both will be found where their graphs intersect.**

▶ **Different ways to map a curved surface (like the earth's) onto a flat surface have different advantages.** ∎

225

9_D UNCERTAINTY

The sections on uncertainty, summarizing data, and sampling in Chapter 9 of *Science for All Americans* are important for learning to deal with evidence. An important distinction must be made for this goal, as for others, between what students are expected to *understand*—that is, notice, talk about, critique—and what they are expected to be able to *do themselves*—that is, plan and carry out. The principal intent is to make them informed consumers, not producers, of data. For example, students should know that people can be alert to possible bias in choosing samples that others take but may be unable to take adequate precautions against bias in designing a study of their own.

There are some very difficult ideas under this goal, ideas with which even most adults have trouble. One common misunderstanding is the belief that averages are always highly representative of a population; little or no attention is given to the range of variation around averages. This point is particularly important when comparing groups. For example, elaborating a minuscule (but believable) difference in average x for boys and girls into statements such as "Boys have high x, whereas girls have low x." So it is essential that talk about averages is *always* accompanied by some indication of the actual distribution of the data. And there is no point in introducing averages until some question arises for which the average supplies a useful answer. Some research studies suggest that learning the algorithm divorced from a meaningful context tends to block students from ever understanding what averages are for.

Another misunderstanding is the assumption that variables are always linked by cause and effect. When being told of a correlation between two variables, adults almost invariably leap quickly to imagine *a* cause or believe *the* cause that is offered to them. A correlation between A and B should always evoke four hypotheses for consideration: (1) A might cause B; (2) B might cause A; (3) A and B might have no cause between them at all, but both are caused by C; and (4) *chance alone* may have produced the apparent dependence. There may be no greater contribution of mathematics to science literacy than fostering an understanding of what a correlation is and what it is *not*.

One of the many misunderstandings of probability that teachers have to deal with is that a well-established probability will be changed by the most recent history: People tend to believe that a coin that has come up heads ten times in a row is more likely on the next flip to come up tails than heads or that the number that won the lottery last week is less likely to win this week. Those and other confusions about probability are purely mathematical and can be addressed as such, but it is also important to take up some of the questions related to how probabilities are established. Examples should come from medicine, natural catastrophes such as floods and earthquakes, weather patterns, sports events, stock-market events, elections, and other topical contexts.

Kindergarten through Grade 2

In the very earliest grades, learning can begin that will eventually lead to students' having a good grasp of everyday statistics. Children at this level can array things they collect by size and weight and then ask questions about them, such as which one is in the middle, how many are the same, and so forth. From there they can go on to make simple pictographs that show how a familiar variable is distributed and again ask questions about the distribution. They can begin to find out about sampling in the context, say, of reporting on the kinds of stones found on the school playground.

Children will be keeping track of many different phenomena, some of which they will come to see have patterns of one kind or another. From time to time they should be asked, working in small groups, to review their records to see if they can figure out if they can predict some future events. The most important part of such exercises is that the students give reasons for their predictions and for not being able to make predictions. Of course they should follow up to see if they were right or not.

By the end of the 2nd grade, students should know that

▶ **Some things are more likely to happen than others. Some events can be predicted well and some cannot. Sometimes people aren't sure what will happen because they don't know everything that might be having an effect.**

▶ **Often a person can find out about a group of things by studying just a few of them.** ■

Grades 3 through 5

The questions about data only explored in the earliest grades can now be made into formal questions. Data distributions should be made of many familiar features and quantities: heights, weights, number of siblings, or kinds of pets. The important thing to emphasize at this level is the kind of questions that can be posed and answered by a data distribution: "Where is the middle?" is a useful question; "What is the average?" probably is not. Because there is a persistent misconception, even in adults, that means are good representations of whole groups, it is especially important to draw students' attention to the additional questions, "What are the largest and smallest values?" and "How much do the data spread on both sides of the middle?" Children also should be invited to suggest some circumstances in their studies that might bias the results—for example, making measurements of student height just as a basketball team comes along or collecting only the insects that were easy to spot.

By the end of the 5th grade, students should know that

▶ **Some predictions can be based on what is known about the past, assuming that conditions are pretty much the same now.**

▶ **Statistical predictions (as for rainy days, accidents) are typically better for *how many* of a group will experience something than for which members of the group will experience it—and better for *how often* something will happen than for *exactly when*.**
continued

▶ **Summary predictions are usually more accurate for large collections of events than for just a few. Even very unlikely events may occur fairly often in very large populations.**

▶ **Spreading data out on a number line helps to see what the extremes are, where they pile up, and where the gaps are. A summary of data includes where the middle is and how much spread is around it.**

▶ **A small part of something may be special in some way and not give an accurate picture of the whole. How much a portion of something can help to estimate what the whole is like depends on how the portion is chosen. There is a danger of choosing only the data that show what is expected by the person doing the choosing.**

▶ **Events can be described in terms of being more or less likely, impossible, or certain. ■**

Building on previous experience, students can now delve into statistics in greater detail. The work should be directly related to student investigations and utilize computers. As stated in NCTM *Standards*:

Instruction in statistics should focus on the active involvement of students in the entire process: formulating key questions; collecting and organizing data; representing the data using graphs, tables, frequency distributions, and summary statistics; analyzing the data; making conjectures; and communicating information in a convincing way. Students' understanding of statistics will also be enhanced by evaluating others' arguments.

Database computer programs offer a means for students to structure, record, and investigate information; to sort it quickly by various categories; and to organize it in a variety of ways. Other computer programs can be used to construct plots and graphs to display data. Scale changes can be made to compare different views of the same information. These technological tools free students to spend more time exploring the essence of statistics: analyzing data from many viewpoints, drawing inferences, and constructing and evaluating arguments.

Students should make distributions for many data sets, their own and published sets, which have already inspired some meaningful questions. The idea of a middle to a data set should be well motivated—say, by asking for a simple way to compare two groups—and various kinds of middle should be considered. The algorithm for the mean can be learned but not without recurrent questions about what it conveys—and what it does not.

In studying data sets, questions like these should be raised: What appears most often in the data? Are there trends? Why are there outliers? How can we explain the data, and does our explanation allow a prediction of what further data would look like? What difficulties might arise when extending the explanation to similar problems? What additional data can we collect to try to verify the ideas developed from these data?

The distinction between ends and means should be kept in mind in all of this. The ultimate aim is not to turn all students into competent statisticians but to have them understand enough statistics to be able to respond intelligently to claims based on statistics; without the kind of intense effort called for here, that understanding will be elusive.

Probability, too, should be continued at this level through the use of tables of actual frequencies of events, begun in the 3rd through 5th grades. Every time, however, students should be asked to consider whether the data (necessarily collected in the past) are still applicable. How well, for example, would last year's daily temperatures apply to this year?

After they have had many occasions to count possible outcomes (such as the faces of a die) and discuss their equal probability (is each face as likely to come up as any other?), students can begin to move to generalizations about theoretical probabilities. Students' attention should consistently be drawn to the assumptions that all possible outcomes of a situation are accounted for and are all equally probable. Computers should be used to generate simulated probabilistic data for analysis, but only after students have worked on problems in which they use their own data.

By the end of the 8th grade, students should know that

▶ **How probability is estimated depends on what is known about the situation. Estimates can be based on data from similar conditions in the past or on the assumption that all the possibilities are known.**

▶ **Probabilities are ratios and can be expressed as fractions, percentages, or odds.**

▶ **The mean, median, and mode tell different things about the middle of a data set.**

▶ **Comparison of data from two groups should involve comparing both their middles and the spreads around them.**

▶ **The larger a well-chosen sample is, the more accurately it is likely to represent the whole. But there are many ways of choosing a sample that can make it unrepresentative of the whole.**■

Events can be described in terms of being more or less likely, impossible, or certain.

Grades 9 through 12

As their mathematical sophistication grows during these grades, students are able to perform and make sense of more subtleties in collecting, describing, and interpreting data. They should have multiple opportunities to plan and carry out studies of their own observations and of large databases. Their written reports should include the reasoning that went into decisions about sampling method and size, about models chosen, about the display used, and about alternative interpretations. They should look for selection bias, measurement error, and display distortion in news reports as well as in their own studies.

Important, too, is frequent discussion of reports in the news media about scientific studies. Students should identify weaknesses in the studies and offer alternative interpretations of the results—perhaps writing alternative versions of the news stories or writing letters to the editor about what the stories may have been missing.

By the end of the 12th grade, students should know that

▶ Even when there are plentiful data, it may not be obvious what mathematical model to use to make predictions from them or there may be insufficient computing power to use some models.

▶ When people estimate a statistic, they may also be able to say how far off the estimate might be.

▶ The middle of a data distribution may be misleading—when the data are not distributed symmetrically, or when there are extreme high or low values, or when the distribution is not reasonably smooth.

▶ The way data are displayed can make a big difference in how they are interpreted.

▶ Both percentages and actual numbers have to be taken into account in comparing different groups; using either category by itself could be misleading.

▶ Considering whether two variables are correlated requires inspecting their distributions, such as in two-way tables or scatterplots. A believable correlation between two variables doesn't mean that either one causes the other; perhaps some other variable causes them both or the correlation might be attributable to chance alone. A true correlation means that differences in one variable imply differences in the other when all other things are equal.

▶ For a well-chosen sample, the size of the sample is much more important than the size of the population. To avoid intentional or unintentional bias, samples are usually selected by some random system.

▶ A physical or mathematical model can be used to estimate the probability of real-world events. ■

9ᴇ Reasoning

The appearance of reasoning in this chapter in no way implies that it should be taught only in mathematics classes. Indeed, reasoning should be studied in all science courses, social-studies classes, and wherever critical thinking is being taught. Part of what is to be accomplished is for students to acquire the kind of understanding of deductive logic necessary for telling good logic from bad logic in the arguments people make. They should also become aware of why reasoning is so important in mathematics. Another part of the reasoning agenda should deal with inductive logic—making generalizations based on instances—and its uses in science and everyday life. It is important that students become clear on the limitations of such logic because of the widespread tendency of people to offer an example as a proof.

But as important as it is that students come to understand the nature of logic, it is even more important that they learn how to use logic and evidence in making valid, persuasive arguments and in judging the arguments of others. That will only happen if students have a lot of practice in formulating arguments, presenting them to classmates, responding to their criticisms, and critiquing the arguments of others. Furthermore, this experience should build over many years, becoming gradually more complex as students learn to organize evidence, and should take place in the context of interesting problems and issues raised in social-studies, history, science, and mathematics classes.

Chapter 2   THE NATURE OF MATHEMATICS
    11 A Sʏsᴛᴇᴍs (logical models)
          B Mᴏᴅᴇʟs (limits of logic for modeling)
    12   HABITS OF MIND

◁ ALSO SEE

At the beginning level, the goal is more for students to develop expectations about reasoning than for them to acquire reasoning skills. The question "How do you know?" should become routine—children should come to expect it to be asked and should feel free to ask it of others. The quality of the answer is not yet important, although there should sometimes be discussion of what is most believable in other people's answers. Science activities provide daily opportunities for students to get practice in referring to evidence.

By the end of the 2nd grade, students should know that

▶ **People are more likely to believe your ideas if you can give good reasons for them.** ∎

The quality of the answer to "How do you know?" now starts to become more important. When asked for a reason for an assertion, children are likely at this age to just repeat the assertion, add emphasis ("Just because."), or appeal to authority ("My big brother said so."). Undermining authority is not a very good idea here, but the appeal to reason can be shifted to the authority ("What do you suppose his reasons might have been?"). Supporting claims with reasons should be modeled by the teacher. In science, questions can be raised suggesting that sometimes the trouble with an argument is that the evidence offered is weak. Teachers can set the tone by asking, "Do you think it would help to collect some more samples?" "If you did the investigation over again, do you think the same thing would happen?" "What evidence might change your mind?"

At this level, students are still very concrete in their thinking, but it is probably a good time to introduce reasoning by analogy. Analogies should be simple and obvious at first, and attention should focus on how the analog is like and unlike what is being studied. Reflection on analogies should not make the students so analytical that they back away from their poetic use. Analogies should be used freely in speculation and artistic expression. But when they are used as the basis of argument, they should be challenged. ("My love is like a red, red rose; therefore. . . .")

By the end of the 5th grade, students should know that

▶ **One way to make sense of something is to think how it is like something more familiar.**

▶ **Reasoning can be distorted by strong feelings.** ∎

Many students are able to think more abstractly in the middle grades than in the prior years. Hence they can now consider the principles of reasoning in more detail and begin to appreciate the critical part that logic plays in clear thinking, whereas up to now more emphasis had been placed on the quality of the evidence being offered in support of a claim. This shift entails insisting on the careful use of particular words and phrases, such as *If. . . , then. . . , and, or, not, all,* and *some.*

Science and mathematics are obvious places for paying attention to logic, but they are not the only ones. Designing projects and troubleshooting mechanical objects and systems provide excellent opportunities for students to apply logic, and such activities have the virtue of providing concrete feedback on how good the logic was. In social studies, students should examine the use of logic in retrieving data from databases and in political and social controversies.

By the end of the 8th grade, students should know that

▶ **Some aspects of reasoning have fairly rigid rules for what makes sense; other aspects don't. If people have rules that always hold, and good information about a particular situation, then logic can help them to figure out what is true about it. This kind of reasoning requires care in the use of key words such as *if, and, not, or, all,* and *some.* Reasoning by similarities can suggest ideas but can't prove them one way or the other.**

▶ **Practical reasoning, such as diagnosing or troubleshooting almost anything, may require many-step, branching logic. Because computers can keep track of complicated logic, as well as a lot of information, they are useful in a lot of problem-solving situations.**

▶ **Sometimes people invent a general rule to explain how something works by summarizing observations. But people tend to overgeneralize, imagining general rules on the basis of only a few observations.**

▶ **People are using incorrect logic when they make a statement such as "If A is true, then B is true; but A isn't true, therefore B isn't true either."**

▶ **A single example can never prove that something is always true, but sometimes a single example can prove that something is not always true.**

▶ **An analogy has some likenesses to but also some differences from the real thing. ■**

Grades 9 through 12

Transfer of formal logic skills to real-world situations requires a great deal of practice. Claims made in print, radio, and television (including news items, editorials, letters to the editor, and advertisements) should regularly be critiqued by students for the quality of the arguments they make. Students should be able to identify the premises (whether explicit or not), logic, and evidence used, and then evaluate the claim. They should also be able to point out where something other than a sound argument is being used to convince the reader, listener, or watcher. History can provide documented cases of the uses of good and bad logic on a grand scale.

By the end of the 12th grade, students should know that

▶ **To be convincing, an argument needs to have both true statements and valid connections among them. Formal logic is mostly about connections among statements, not about whether they are true. People sometimes use poor logic even if they begin with true statements, and sometimes they use logic that begins with untrue statements.**

▶ **Logic requires a clear distinction among reasons: A reason may be *sufficient* to get a result, but perhaps is not the only way to get there; or, a reason may be *necessary* to get the result, but it may not be enough by itself; some reasons may be both sufficient and necessary.**

▶ **Wherever a general rule comes from, logic can be used in testing how well it works. Proving a generalization to be false (just one exception will do) is easier than proving it to be true (for all possible cases). Logic may be of limited help in finding solutions to problems if one isn't sure that general rules always hold or that particular information is correct; most often, one has to deal with probabilities rather than certainties.**

▶ **Once a person believes in a general rule, he or she may be more likely to notice cases that agree with it and to ignore cases that don't. To avoid biased observations, scientific studies sometimes use observers who don't know what the results are "supposed" to be.**

▶ **Very complex logical arguments can be made from a lot of small logical steps. Computers are particularly good at working with complex logic but not all logical problems can be solved by computers. High-speed computers can examine the validity of some logical propositions for a very large number of cases, although that may not be a perfect proof.** ■

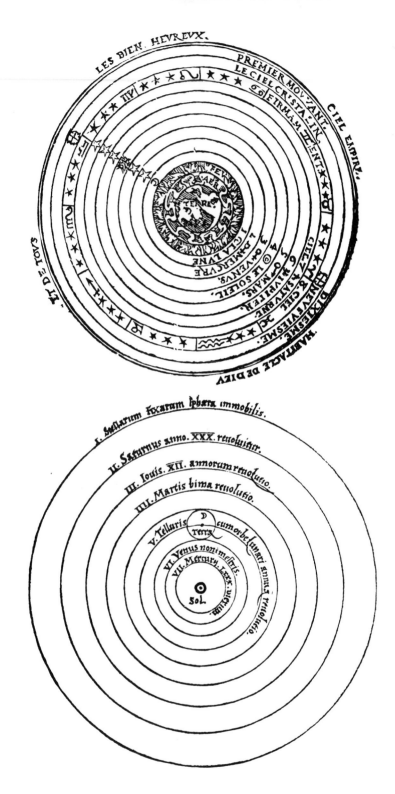

Earth-centered and sun-centered models of the solar system as depicted in the sixteenth century.

Chapter 10 HISTORICAL PERSPECTIVES

There are two principal reasons for including some knowledge of history among the recommendations. One reason is that generalizations about how the scientific enterprise operates would be empty without concrete examples. Consider, for example, the proposition that new ideas are limited by the context in which they are conceived; are often rejected by the scientific establishment; sometimes spring from unexpected findings; and usually grow slowly, through contributions from many different investigators. Without historical examples, these generalizations would be no more than slogans, however well they might be remembered.

A second reason is that some episodes in the history of the scientific endeavor are of surpassing significance to our cultural heritage. Such episodes certainly include Galileo's role in changing our perception of our place in the universe, Newton's demonstration that the same laws apply to motion in the heavens and on earth, Darwin's long observations of the variety and relatedness of life forms that led to his postulating a mechanism for how they came about, Lyell's careful documentation of the unbelievable age of the earth, and Pasteur's identification of infectious disease with tiny organisms that could be seen only with a microscope. These stories stand among the milestones of the development of all thought in Western civilization.

SCIENCE FOR ALL AMERICANS

RESEARCH NOTES
page 354

This chapter focuses on benchmarks that address the development of student understanding of selected episodes in the history of science. These benchmarks deal with history, leaving it to benchmarks in other chapters to signal when the related science and technology understandings are to be acquired. Also, it should be noted that little history of science is expected of students before they reach the 6th- to 8th-grade span, and most shows up in the 9th- to 12th-grade span. To appreciate the significance of these historical episodes, students must (1) know or at least be able to follow the science involved, and (2) be able to grasp the main features of the prevailing view at the time.

Recognizing that certain episodes in the history of science can enhance the science curriculum certainly does not imply that all science must be taught by reviewing its history or that no other history of science and technology is appropriate in the curriculum. Nor, for that matter, should it be taken to suggest that there is no need for students to study current issues related to the impact of science and technology on society.

Some educators have suggested that simple versions of these stories may help students to learn more sophisticated versions in later grades by making the main characters and story lines familiar in the early grades. It is possible, however, that simplified versions may distort both the science and the history, making learning the more sophisticated story difficult. Although this edition of *Benchmarks* does not recommended particular simplifications for students to learn, teachers and researchers could profit from collaborating in the study of the contribution that simplified stories can make to student understanding. ■

The great cosmological revolution usually associated with the name of Nicolaus Copernicus was one of the episodes in history that was truly transforming. It changed, ultimately, the sense most people have of their relation to the physical universe, and it raised difficult questions of human existence that for many people have yet to be resolved satisfactorily. The Copernican Revolution merits study by all students because it illustrates many aspects of the way science works, especially the way in which science, mathematics, and technology are intertwined and the way in which international efforts in science come together.

Prior to studying this story during the high-school years, students should become familiar with the night sky at least to the extent of having observed the moon, stars, and some of the planets with the unaided eye and with a telescope. Whether through their own observations, films, or planetarium visits, students should be helped to visualize the phenomenon at the heart of the Copernican Revolution—the seemingly irregular movement of the planets relative to the starry background.

☞ ALSO SEE

Grades 6 through 8

The scientific groundwork can now be laid to prepare students to take up in high school the issues raised by Copernicus and Galileo. Naked-eye and telescopic observations should continue, supplemented by the use of reference books, videotapes, computer programs, and planetarium visits. The emphasis should be on accurate descriptions of the appearance of the moon, stars, and planets as seen from earth and on the motion of the planets relative to the stars. Analysis of geocentric and heliocentric models can be delayed until high school.

By the end of the 8th grade, students should know that

▶ **The motion of an object is always judged with respect to some other object or point and so the idea of absolute motion or rest is misleading.**

▶ **Telescopes reveal that there are many more stars in the night sky than are evident to the unaided eye, the surface of the moon has many craters and mountains, the sun has dark spots, and Jupiter and some other planets have their own moons.** ■

Grades 9 through 12

Students now need to get the main features of the heliocentric system clearly in mind and contrast them with those of the geocentric one. People have trouble transposing frames of reference, so it is important not to rush through the story and to practice shifting frames of reference in many different physical contexts. There are films that can help show how hard it is to discern which of two objects is in motion. In studying planetary models, it is easy to bog down in making distinctions between rotating and revolving, and getting it straight may not be worth the effort it requires.

Avoid selling Copernicus' model on the basis of simplicity, for in fact it was not mathematically simpler than Ptolemy's. They were comparably complex, both using circles on circles; and both predicted comparably well where planets would be observed at any specified time. Until Kepler devised a more accurate system with elliptical orbits, choice was a matter of taste. The issue was not fully settled until Newton showed that Kepler's elliptical orbits were the natural consequence of the laws of motion.

The Copernican Revolution illustrates some of the strains that can occur between science and society when science proposes ideas that seem to violate common sense or to undermine traditional values and beliefs. This part of the story should be included but not presented as the triumph of right over wrong or of science over religion. Using selections from Galileo's *Two World Views* and Brecht's *Galileo*, along with commentaries by historians, a seminar that seeks to understand the controversy itself provides one possible capstone experience in pursuit of this goal.

By the end of the 12th grade, students should know that

▶ People perceive that the earth is large and stationary and that all other objects in the sky orbit around it. That perception was the basis for theories of how the universe is organized that prevailed for over 2,000 years.

▶ Ptolemy, an Egyptian astronomer living in the second century A.D., devised a powerful mathematical model of the universe based on constant motion in perfect circles, and circles on circles. With the model, he was able to predict the motions of the sun, moon, and stars, and even of the irregular "wandering stars" now called planets.

▶ In the 16th century, a Polish astronomer named Copernicus suggested that all those same motions could be explained by imagining that the earth was turning around once a day and orbiting around the sun once a year. This explanation was rejected by nearly everyone because it violated common sense and required the universe to be unbelievably large. Worse, it flew in the face of the belief, universally held at the time, that the earth was at the center of the universe.

▶ Johannes Kepler, a German astronomer who lived at about the same time as Galileo, showed mathematically that Copernicus' idea of a sun-centered system worked well if uniform circular motion was replaced with uneven (but predictable) motion along off-center ellipses.

▶ Using the newly invented telescope to study the sky, Galileo made many discoveries that supported the ideas of Copernicus. It was Galileo who found the moons of Jupiter, sunspots, craters and mountains on the moon, and many more stars than were visible to the unaided eye.

▶ Writing in Italian rather than in Latin (the language of scholars at the time), Galileo presented arguments for and against the two main views of the universe in a way that favored the newer view. That brought the issue to the educated people of the time and created political, religious, and scientific controversy. ■

10ʙ   UNITING THE HEAVENS AND EARTH

Students should have encountered the relevant physical concepts and laws at several levels of sophistication, at different times, and in different learning contexts prior to undertaking to learn the history associated with Newton.

Newtonian synthesis explains the observations and speculations of his time and unifies earth and sky by proposing one set of physics laws for both. This study of history provides students with an excellent opportunity to weave the strands of previous understanding into a coherent picture and to develop an appreciation of the explanatory power and logical elegance of Newton's work.

Grades 9 through 12

During the grades prior to high school, and during the early high-school years, students need to become familiar with the phenomena that the Newtonian synthesis explains and unifies, the fundamental concepts involved in the model, and the mathematics necessary to make quantitative sense out of such concepts as velocity and acceleration, the second law of motion, and the law of gravitation.

By the end of the 12th grade, students should know that

▶ **Isaac Newton created a unified view of force and motion in which motion everywhere in the universe can be explained by the same few rules. His mathematical analysis of gravitational force and motion showed that planetary orbits had to be the very ellipses that Kepler had proposed two generations earlier.**

▶ **Newton's system was based on the concepts of mass, force, and acceleration, his three laws of motion relating them, and a physical law stating that the force of gravity between any two objects in the universe depends only upon their masses and the distance between them.**

▶ **The Newtonian model made it possible to account for such diverse phenomena as tides, the orbits of planets and moons, the motion of falling objects, and the earth's equatorial bulge.**

▶ **For several centuries, Newton's science was accepted without major changes because it explained so many different phenomena, could be used to predict many physical events (such as the appearance of Halley's comet), was mathematically sound, and had many practical applications.**

▶ **Although overtaken in the 20th century by Einstein's relativity theory, Newton's ideas persist and are widely used. Moreover, his influence has extended far beyond physics and astronomy, serving as a model for other sciences and even raising philosophical questions about free will and the organization of social systems.** ■

243

10c   Relating Matter & Energy and Time & Space

Students will be very interested in the "gee whiz" aspects of relativity—the speed of light limit, time slowing down, nuclear energy release, black holes. This interest can be drawn upon to make the more important points that under extreme conditions the world may work in ways very different from our ordinary experience, and that the test of a scientific theory is not how nearly it matches common sense, but how well it accounts for known observations and predicts new ones that hadn't been expected.

One of the major difficulties is semantic rather than scientific: Einstein's concept of relativity does not declare that "everything is relative"; in fact, Galileo's concept of relative velocity comes closer to that idea. Actually, Einstein's theory holds that the speed of light is absolute. No matter how the observer is moving, his or her measurement of the speed of light always comes out the same. Einstein reformulated the laws relating to space, time, mass, and energy so that they would be valid for all observers, whatever their uniform motion might be. So "relativity theory" is as much about what is *not* relative as about what *is*.

Grades 9 through 12

Relativity is not a topic to be taken up in the elementary- and middle-school years as either history or science. To be sure, a full understanding of relativity theory is far beyond the capacity of most 17-year-olds, but it is far too important to be ignored. By treating relativity historically in high school, it is possible avoid falling into the trap of trying to teach its technical and mathematical details. The main goals should be for students to see that Einstein went beyond Newton's world view by including it as a limiting case in a more complete theory.

By the end of the 12th grade, students should know that

▶ As a young man, Albert Einstein, a German scientist, formulated the special theory of relativity, which brought about revolutionary changes in human understanding of nature. A decade later, he proposed the general theory of relativity, which, along with Newton's work, ranks as one of the greatest human accomplishments in all of history.

▶ Among the surprising ideas of special relativity is that nothing can travel faster than the speed of light, which is the same for all observers no matter how they or the light source happen to be moving.

▶ The special theory of relativity is best known for stating that any form of energy has mass, and that matter itself is a form of energy. The famous relativity equation, $E = mc^2$, holds that the transformation of even a tiny amount of matter will release an enormous amount of other forms of energy, in that the c in the equation stands for the immense speed of light.

▶ General relativity theory pictures Newton's gravitational force as a distortion of space and time.

▶ Many predictions from Einstein's theory of relativity have been confirmed on both atomic and astronomical scales. Still, the search continues for an even more powerful theory of the architecture of the universe. ■

245

10D   EXTENDING TIME

The change in the conception of the age of the earth—from a few thousand to many millions of years—proposed by scientists in the 1800s was dramatic and, for most people, beyond belief. The estimated age was unimaginably greater than the prevailing beliefs. It was also based on the assumption that the earth's features (mountains, valleys, etc.) had been formed gradually by processes still underway, not in a single, instantaneous creation.

People have difficulty imagining time spans that are vastly longer than human experience. In overturning the "sensible" notion that the earth is at most only a few thousand years old, science understandably provoked substantial opposition. The new theory was based on *indirect* evidence from fossils and rock formations and supported the even less acceptable concept of biological evolution. Thus, this episode is a good one for exploring ways in which age can be estimated and for raising questions about the lationship between science and popular beliefs.

ALSO SEE

The history of this episode can be taken up after students have studied some earth science. Their study should engage them in thinking of indirect ways to determine the age of things around them and in comparing those methods to ones used by scientists. The study of dating offers excellent opportunities to show the use and importance to science of both technology and mathematics.

By the end of the 12th grade, students should know that

▶ **Scientific evidence indicates that some rock near the earth's surface is several billion years old. But until the 19th century, most people believed that the earth was created just a few thousand years ago.**

▶ **The idea that the earth might be vastly older than most people believed made little headway in science until the publication of *Principles of Geology* by an English scientist, Charles Lyell, early in the 19th century. The impact of Lyell's book was a result of both the wealth of observations it contained on the patterns of rock layers in mountains and the locations of various kinds of fossils, and of the careful logic he used in drawing inferences from his data.**

▶ **In formulating and presenting his theory of biological evolution, Charles Darwin adopted Lyell's belief about the age of the earth and his style of buttressing his argument with vast amounts of evidence.** ■

10ᴇ Moving the Continents

The story of why science accepted the idea of moving continents only after long resistance illuminates the conservatism of the scientific enterprise. Contrary to the popular public image of scientists as radicals ready to discard their beliefs instantly in the face of contrary "facts," the plate-tectonics episode shows that it sometimes takes a large accumulation of evidence over an extended period of time to provoke a dramatic shift in what most scientists in a discipline accept as true.

The history of the rise of the theory of plate tectonics shows that the acceptance of a theory depends on its explanatory power as well as on the evidence that supports it. As it has turned out, the modern theory of plate tectonics makes sense out of such a large and diverse array of phenomena related to the earth's surface that it now serves as a unifying principle in geology. In a sense, plate tectonics does for geology what evolution does for biology.

Chapter **1**   THE NATURE OF SCIENCE
4 ᴄ Processes that Shape The Earth

◁ **ALSO SEE**

Grades 9 through 12

This bit of history should probably be taken up in high school, after students have acquired descriptive knowledge about the earth's surface—the shapes and locations of the continents and ocean basins, the nature of earthquakes and volcanoes and their distribution on a world map, etc.

By the end of the 12th grade, students should know that

▶ The idea of continental drift was suggested by the matching shapes of the Atlantic coasts of Africa and South America, but rejected for lack of other evidence. It just seemed absurd that anything as massive as a continent could move around.

▶ Early in the 20th century, Alfred Wegener, a German scientist, reintroduced the idea of moving continents, adding such evidence as the underwater shapes of the continents, the similarity of life forms and land forms in corresponding parts of Africa and South America, and the increasing separation of Greenland and Europe. Still, very few contemporary scientists adopted his theory.

▶ The theory of plate tectonics was finally accepted by the scientific community in the 1960s, when further evidence had accumulated in support of it. The theory was seen to provide an explanation for a diverse array of seemingly unrelated phenomena, and there was a scientifically sound physical explanation of how such movement could occur. ▪

10F UNDERSTANDING FIRE

Apart from the story of Lavoisier—who he was, when and where he lived, what he did that was so important—this episode illustrates several aspects of the scientific endeavor. These are (1) the power of concepts, in this case the conservation of matter; (2) the importance of careful measurement, here that of combustion products; (3) the way in which different lines of investigation sometimes converge, in this instance, those of Lavoisier and Dalton; and (4) the role of communication in advancing science, here illustrated by Lavoisier's system for naming substances and describing reactions.

Lavoisier and Dalton were not, of course, solely responsible for the development of the science of chemistry. In the actual study of chemistry and its origins, many of the other strands will need to be brought into the story. Lavoisier and the controversy over the nature of burning provide a dramatic focus for the story.

Chapter **1** THE NATURE OF SCIENCE

4 D STRUCTURE OF MATTER

E ENERGY TRANSFORMATIONS

ALSO SEE

Grades 6 through 8

Students should have opportunities to become familiar with many kinds of (safe) chemical reactions and with the ways things behave or change in the process, and to gain experience doing elementary qualitative analysis. That will provide a background for developing the Lavoisier/Dalton story, parts of which can be told as students are introduced to atomic theory and the conservation of matter. During this time, students should also gain practice in describing chemical reactions in general, and burning in particular, in terms of elements and compounds, atoms and molecules. They cannot be expected to become knowledgeable about details of atomic structure or bonding.

By the end of the 8th grade, students should know that

▶ **From the earliest times until now, people have believed that even though millions of different kinds of material seem to exist in the world, most things must be made up of combinations of just a few basic kinds of things. There has not always been agreement, however, on what those basic kinds of things are. One theory long ago was that the basic substances were earth, water, air, and fire. Scientists now know that these are not the basic substances. But the old theory seemed to explain many observations about the world.**

▶ **Today, scientists are still working out the details of what the basic kinds of matter are and of how they combine, or can be made to combine, to make other substances.**

▶ **Experimental and theoretical work done by French scientist Antoine Lavoisier in the decade between the American and French revolutions led to the modern science of chemistry.**

▶ **Lavoisier's work was based on the idea that when materials react with each other many changes can take place but that in every case the total amount of matter afterward is the same as before. He successfully tested the concept of conservation of matter by conducting a series of experiments in which he carefully measured all the substances involved in burning, including the gases used and those given off.**

▶ **Alchemy was chiefly an effort to change base metals like lead into gold and to produce an elixir that would enable people to live forever. It failed to do that or to create much knowledge of how substances react with each other. The more scientific study of chemistry that began in Lavoisier's time has gone far beyond alchemy in understanding reactions and producing new materials.** ■

This is the time to enrich the Lavoisier story by bringing in Dalton and by emphasizing the importance of the consistent use of language, scientific classification, and symbols in establishing the modern science of chemistry. For some students, the study of chemical bonds, equations, and structures will reinforce the usefulness of symbolic representations.

By the end of the 12th grade, students should know that

▶ Lavoisier invented a whole new field of science based on a theory of materials, physical laws, and quantitative methods, with the conservation of matter at its core. He persuaded a generation of scientists that his approach accounted for the experimental results better than other chemical systems.

▶ Lavoisier's system for naming substances and describing their reactions contributed to the rapid growth of chemistry by enabling scientists everywhere to share their findings about chemical reactions with one another without ambiguity.

▶ John Dalton's modernization of the ancient Greek ideas of element, atom, compound, and molecule strengthened the new chemistry by providing a physical explanation for reactions that could be expressed in quantitative terms.

▶ While the basic ideas of Lavoisier and Dalton have survived, the advancement of chemistry since their time now makes possible an explanation of the bonding that takes place between atoms during chemical reactions in terms of the inner workings of atoms. ■

10G SPLITTING THE ATOM

The story of the discovery of radioactivity and the structure of the nucleus of the atom, along with the incredible results that followed in this century, is drama of the highest order. It also illuminates several features of the scientific enterprise: the role of accidental discovery, the interdependence of disciplines, the ability of women to do outstanding work in both empirical and theoretical science, and the impact of science on world affairs.

Before students can appreciate the significance of the work of the Curies and the others, they must have some understanding of the mass/energy relationship and the physics of nuclear fission and fusion, and they should be familiar with the general history of World War II and the postwar uses of nuclear energy. Prior to the middle grades, nothing is to be gained by delving into the history of radioactivity and nuclear energy, for the science is too abstract for students to grasp and the history too remote for them to care about. Perhaps the earliest introduction should be in the form of the story of Madame Curie, many features of which will capture the imagination of boys and girls—as long as the technical details of her work are not the main focus.

By the end of the 8th grade, students should know that

▶ **The accidental discovery that minerals containing uranium darken photographic film, as light does, led to the idea of radioactivity.**

▶ **In their laboratory in France, Marie Curie and her husband, Pierre Curie, isolated two new elements that caused most of the radioactivity of the uranium mineral. They named one radium because it gave off powerful, invisible rays, and the other polonium in honor of Madame Curie's country of birth. Marie Curie was the first scientist ever to win the Nobel prize in two different fields—in physics, shared with her husband, and later in chemistry.** ■

Grades 9 through 12

The focus of the "splitting-the-atom" story should be on the discovery of nuclear fission and its impact on world affairs. It is important not to overlook the science in this episode when considering the ethical and national-security considerations associated with fission and fusion. It is a measure of its significance that books for the general reader continue to emerge on this subject.

By the end of the 12th grade, students should know that

▶ The Curies made radium available to researchers all over the world, increasing the study of radioactivity and leading to the realization that one kind of atom may change into another kind, and so must be made up of smaller parts. These parts were demonstrated by other scientists to be a small, dense nucleus that contains protons and neutrons and is surrounded by a cloud of electrons.

▶ Ernest Rutherford of New Zealand and his colleagues discovered that the heavy radio active element uranium spontaneously splits itself into a slightly lighter nucleus and a very light helium nucleus.

▶ Later, Austrian and German scientists showed that when uranium is struck by neutrons, it splits into two nearly equal parts plus one or two extra neutrons. Lise Meitner, an Austrian physicist, was the first to point out that if these fragments added up to less mass than the original uranium nucleus, then Einstein's special relativity theory predicted that a large amount of energy would be released. Enrico Fermi, an Italian working with colleagues in the United States, showed that the extra neutrons trigger more fissions and so create a sustained chain reaction in which a prodigious amount of energy is given off.

▶ A massive effort went into developing the technology for the two nuclear fission bombs used on Japan in World War II, nuclear fusion weapons that followed, and reactors for the controlled conversion of nuclear energy into electric energy. Nuclear weapons and energy remain matters of public concern and controversy.

▶ Radioactivity has many uses other than generating energy, including in medicine, industry, and scientific research in many different fields.■

253

10ᴴ   EXPLAINING THE DIVERSITY OF LIFE

The educational goal should be for all children to understand the concept of evolution by natural selection, the evidence and arguments that support it, and its importance in biology. The study of this history provides a good opportunity to feature the importance in science of careful observation and description and to illustrate that not all scientific advances depend on experimentation. Also, the history of this episode should be brought up to current times with regard to its acceptance and rejection by people.

Grades 9 through 12

Because of the complexity of the evidence and the arguments that must be examined, a clear understanding of species evolution probably cannot be achieved earlier than high school. So the full-blown episode will await or accompany students' studying the science. But students in earlier grades will certainly be developing the evidence base for which the theory attempts to account. Darwin's voyage aboard the *Beagle* provided the main observations that led him on his intellectual journey toward evolution by natural selection. The *Voyage of the Beagle* is a great story that can first illustrate the source of Darwin's puzzlement and later provide a picture of the many complex processes of science ideas in the making.

By the end of the 12th grade, students should know that

▶ **The scientific problem that led to the theory of natural selection was how to explain similarities within the great diversity of existing and fossil organisms.**

▶ **Prior to Charles Darwin, the most widespread belief was that all known species were created at the same time and remained unchanged throughout history. Some scientists at the time believed that features an individual acquired during its lifetime could be passed on to its offspring, and the species could thereby gradually change to fit its environment better.**

▶ Darwin argued that only biologically inherited characteristics could be passed on to offspring. Some of these characteristics were advantageous in surviving and reproducing. The offspring would also inherit and pass on those advantages, and over generations the aggregation of these inherited advantages would lead to a new species.

▶ The quick success of Darwin's book *Origin of Species*, published in the mid-1800s, came from the clear and understandable argument it made, including the comparison of natural selection to the selective breeding of animals in wide use at the time, and from the massive array of biological and fossil evidence it assembled to support the argument.

▶ After the publication of *Origin of Species*, biological evolution was supported by the rediscovery of the genetics experiments of an Austrian monk, Gregor Mendel, by the identification of genes and how they are sorted in reproduction, and by the discovery that the genetic code found in DNA is the same for almost all organisms.

▶ By the 20th century, most scientists had accepted Darwin's basic idea. Today that still holds true, although differences exist concerning the details of the process and how rapidly evolution of species takes place. People usually do not reject evolution for scientific reasons but because they dislike its implications, such as the relation of human beings to other animals, or because they prefer a biblical account of creation. ∎

255

10ı DISCOVERING GERMS

Students believe that germs exist, having had the idea drummed into them from infancy. But in fact the existence of microorganisms was not easy to establish, and their connection to particular diseases even harder. Studying the development of germ theory provides a good opportunity to highlight several important attributes of science, including that it depends on technology, that sometimes investigations designed to solve a practical problem lead to fundamental scientific discoveries, and that a major breakthrough frequently requires the work of different scientists working independently.

Early microbiologists

Grades 6 through 8

In contrast to many of the other historical episodes, the study of germ theory is one worth trying in middle school. The science needed can be developed during these grades. The story of Pasteur's discovery that microbes can cause disease is straightforward, the role played by microscopes in making germs in diseased tissues visible follows nicely from students' own microscopic observations, and the implications for sanitary practice and disease prevention are things that students routinely encounter.

By the end of the 8th grade, students should know that

▶ **Throughout history, people have created explanations for disease. Some have held that disease has spiritual causes, but the most persistent biological theory over the centuries was that illness resulted from an imbalance in the body fluids. The introduction of germ theory by Louis Pasteur and others in the 19th century led to the modern belief that many diseases are caused by microorganisms—bacteria, viruses, yeasts, and parasites.**

▶ **Pasteur wanted to find out what causes milk and wine to spoil. He demonstrated that spoilage and fermentation occur when microorganisms enter from the air, multiply rapidly, and produce waste products. After showing that spoilage could be avoided by keeping germs out or by destroying them with heat, he investigated animal diseases and showed that microorganisms were involved. Other investigators later showed that specific kinds of germs caused specific diseases.**

▶ **Pasteur found that infection by disease organisms—germs—caused the body to build up an immunity against subsequent infection by the same organisms. He then demonstrated that it was possible to produce vaccines that would induce the body to build immunity to a disease without actually causing the disease itself.**

▶ **Changes in health practices have resulted from the acceptance of the germ theory of disease. Before germ theory, illness was treated by appeals to supernatural powers or by trying to adjust body fluids through induced vomiting, bleeding, or purging. The modern approach emphasizes sanitation, the safe handling of food and water, the pasteurization of milk, quarantine, and aseptic surgical techniques to keep germs out of the body; vaccinations to strengthen the body's immune system against subsequent infection by the same kind of microorganisms; and antibiotics and other chemicals and processes to destroy microorganisms.**

▶ **In medicine, as in other fields of science, discoveries are sometimes made unexpectedly, even by accident. But knowledge and creative insight are usually required to recognize the meaning of the unexpected.** ∎

257

10ⱼ HARNESSING POWER

Students need to learn about the nature of tools, the geographical distribution of material and energy resources, and how people lived and worked in the 18th century and earlier in order to grasp the nature and impact of the Industrial Revolution. Students study the Industrial Revolution to some degree as a part of world history, although the scientific and technical aspects of it are often shortchanged.

It is such a central episode in human experience that students should encounter aspects of it in the elementary and middle grades and in geography, government, literature, and science, as well as in history and technology classes. Once they are familiar with the 18th century's Industrial Revolution, students can compare the 20th century's Information Revolution to it.

THE BORN LOSER ART SANSOM

Grades 6 through 8

Students should acquire some knowledge of the Industrial Revolution in social studies, and from science and technology they should acquire a grasp of how steam engines and pumps work.

By the end of the 8th grade, students should know that

▶ Until the 1800s, most manufacturing was done in homes, using small, handmade machines that were powered by muscle, wind, or running water. New machinery and steam engines to drive them made it possible to replace craftsmanship with factories, using fuels as a source of energy. In the factory system, workers, materials, and energy could be brought together efficiently.

▶ The invention of the steam engine was at the center of the Industrial Revolution. It converted the chemical energy stored in wood and coal, which were plentiful, into mechanical work. The steam engine was invented to solve the urgent problem of pumping water out of coal mines. As improved by James Watt, it was soon used to move coal, drive manufacturing machinery, and power locomotives, ships, and even the first automobiles. ■

Grades 9 through 12

An important goal in teaching about the Industrial Revolution is to help students understand it in both its contemporary and modern contexts. Is the computer the steam engine of our times? Or is it the electrical generator? Will information become as important politically and economically as energy? What impact has the chemical, and now the biochemical, revolution had on how people live and work? These and other such questions are raised here to suggest some of the issues students should examine in the context of the impact of technology on society.

By the end of the 12th grade, students should know that

▶ The Industrial Revolution happened first in Great Britain because that country made practical use of science, had access by sea to world resources and markets, and had an excess of farm workers willing to become factory workers.

▶ The Industrial Revolution increased the productivity of each worker but it also increased child labor and unhealthy working conditions, and it gradually destroyed the craft tradition. The economic imbalances of the Industrial Revolution led to a growing conflict between factory owners and workers and contributed to the main political ideologies of the 20th century.

▶ The Industrial Revolution is still underway as electric, electronic, and computer technologies change patterns of work and bring with them economic and social consequences. ■

259

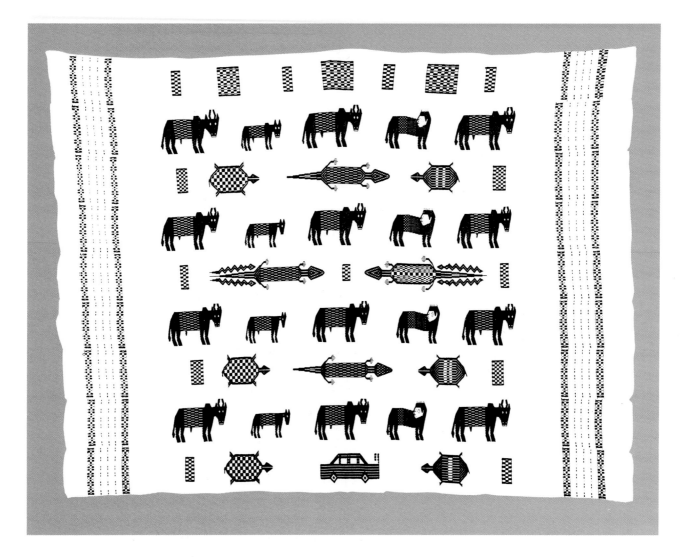

Blanket, Niger, 1960.

Chapter 11 COMMON THEMES

A   SYSTEMS
B   MODELS
C   CONSTANCY AND CHANGE
D   SCALE

Some important themes pervade science, mathematics, and technology and appear over and over again, whether we are looking at an ancient civilization, the human body, or a comet. They are ideas that transcend disciplinary boundaries and prove fruitful in explanation, in theory, in observation, and in design.

SCIENCE FOR ALL AMERICANS

Some powerful ideas often used by mathematicians, scientists, and engineers are not the intellectual property of any one field or discipline. Indeed, notions of system, scale, change and constancy, and models have important applications in business and finance, education, law, government and politics, and other domains, as well as in mathematics, science, and technology. These common themes are really ways of thinking rather than theories or discoveries. (Energy also represents a prominent tool for thinking in science and technology, but because it is part of the *content* of science, it is not included here as a theme.) *Science for All Americans* recommends what all students should know about those themes, and the benchmarks in the four sections below suggest how student understanding of them should grow over the school years. Although the context of both *Science for All Americans* and *Benchmarks* is mainly science, mathematics, and technology, other contexts are identified here to emphasize the general usefulness of these themes. ■

11ᴀ   Sᴇᴛᴇᴍs

One of the essential components of higher-order thinking is the ability to think about a whole in terms of its parts and, alternatively, about parts in terms of how they relate to one another and to the whole. People are accustomed to speak of political systems, sewage systems, transportation systems, the respiratory system, the solar system, and so on. If pressed, most people would probably say that a system is a collection of things and processes (and often people) that interact to perform some function. The scientific idea of a system implies detailed attention to inputs and outputs and to interactions among the system components. If these can be specified quantitatively, a computer simulation of the

RESEARCH NOTES

page 355

system might be run to study its theoretical behavior, and so provide a way to define problems and investigate complex phenomena. But a system need not have a "purpose" (e.g., an ecosystem or the solar system) and what a system includes can be imagined in any way that is interesting or useful. Students in the elementary grades study many different kinds of systems in the normal course of things, but they should not be rushed into explicit talk about systems. That can and should come in middle and high school.

Children tend to think of the properties of a system as belonging to individual parts of it rather than as arising from the interaction of the parts. A system property that arises from interaction of parts is therefore a difficult idea. Also, children often think of a system only as something that is made and therefore as obviously defined. This notion contrasts with the scientific view of systems as being defined with particular purposes in mind. The solar system, for example, can be defined in terms of the sun and planets only, or defined to include also the planetary moons and solar comets. Similarly, not only is an automobile a system, but one can think of an automotive system that includes service stations, oil wells, rubber plantations, insurance, traffic laws, junk yards, and so on.

The main goal of having students learn about systems is not to have them talk about systems in abstract terms, but to enhance their ability (and inclination) to attend to various aspects of particular systems in attempting to understand or deal with the whole system. Does the student troubleshoot a malfunctioning device by considering connections and switches—whether using the terms *input*, *output*, or *controls* or not? Does the student try to account for what becomes of all of the input to the water cycle—whether using the term *conservation* or not? The vocabulary will be helpful for students once they have had a wide variety of experiences with systems thinking, but otherwise it may mistakenly give the impression of understanding. Learning about systems in some situations may not transfer well to other situations, so systems should be encountered through a variety of approaches, including designing and troubleshooting. Simple systems (a pencil or mousetrap), of course, should be encountered before more complex ones (a stereo system, a plant, the continuous manufacture of goods, ecosystems, or school government).

A persistent student misconception is that the properties of an assembly are the same as the properties of its parts (for example, that soft materials are made of soft molecules). Sometimes it is true. For example, a politically conservative organization may be made up entirely of conservative individuals. But some features of systems are unlike any of their parts. Sugar is sweet, but its component atoms (carbon, oxygen, and hydrogen) are not. The system property may result from what its parts are like, but the parts themselves may not have that property. A grand example is life as an emergent property of the complex interaction of complex molecules.

Curious Avenue   **by Tom Toles**

Students in the elementary grades acquire the experiences that they will use in the middle grades and beyond to develop an understanding of systems concepts and their applications. They also can begin to attend to what affects what. Frequent discussion of how one thing affects another lays the ground for recognizing interactions. Another tack for focusing on interaction is to raise the question of when things work and when they do not—owing, say, to missing or broken parts or the absence of a source of power (batteries, gasoline).

Students should practice identifying the parts of things and how one part connects to and affects another. Classrooms can have available a variety of dissectable and rearrangeable objects, such as gear trains and toy vehicles and animals, as well as conventional blocks, dolls, and doll houses. Students should predict the effects of removing or changing parts.

By the end of the 2nd grade, students should know that

▶ **Most things are made of parts.**

▶ **Something may not work if some of its parts are missing.**

▶ **When parts are put together, they can do things that they couldn't do by themselves.** ■

Hands-on experience with a variety of mechanical systems should increase. Classrooms can have "take-apart" stations where a variety of familiar hardware devices can be taken apart (and perhaps put back together) with hand tools. Devices that are commonly purchased disassembled can be provided, along with assembly instructions, to emphasize the importance of the proper arrangement of parts (and incidentally, the importance of language-arts skills, which are needed to read and follow instructions).

By the end of the 5th grade, students should know that

▶ **In something that consists of many parts, the parts usually influence one another.**

▶ **Something may not work as well (or at all) if a part of it is missing, broken, worn out, mismatched, or misconnected.** ■

Grades 6 through 8

Systems thinking can now be made explicit—suggesting analysis of parts, subsystems, interactions, and matching. But descriptions of parts and their interaction are more important than just calling everything a system.

Student projects should now entail analyzing, designing, assembling, and troubleshooting systems—mechanical, electrical, and biological—with easily discernable components. Students can take apart and reassemble such things as bicycles, clocks, and mechanical toys and build battery-driven electrical circuits that actually operate something. They can assemble a sound system and then judge how changing different components affects the system's output, or observe aquariums and gardens while changing some parts of the system or adding new parts. The idea of system should be expanded to include connections among systems. For example, a can opener and a can may each be thought of as a system, but they both—together with the person using them—form a larger system without which neither can be put to its intended use.

By the end of the 8th grade, students should know that

▶ **A system can include processes as well as things.**

▶ **Thinking about things as systems means looking for how every part relates to others. The output from one part of a system (which can include material, energy, or information) can become the input to other parts. Such feedback can serve to control what goes on in the system as a whole.**

▶ **Any system is usually connected to other systems, both internally and externally. Thus a system may be thought of as containing subsystems and as being a subsystem of a larger system.** ■

265

Grades 9 through 12

Students should have opportunities—in seminars, projects, readings, and experiments—to reflect on the value of thinking in terms of systems and to apply the concept in diverse situations. They should often discuss what properties of a system are the same as the properties of its parts and what properties arise from interactions of its parts or from the sheer number of parts. They should learn to see feedback as a standard aspect of systems. The definitions of *negative* and *positive* feedback may be too subtle, but students can understand that feedback may oppose changes that do occur (and lead to stability), or may encourage more change (and so drive the system toward one extreme or another). Eventually, they can see how some delay in feedback can produce cycles in a system's behavior.

By the end of the 12th grade, students should know that

▶ A system usually has some properties that are different from those of its parts, but appear because of the interaction of those parts.

▶ Understanding how things work and designing solutions to problems of almost any kind can be facilitated by systems analysis. In defining a system, it is important to specify its boundaries and subsystems, indicate its relation to other systems, and identify what its input and its output are expected to be.

▶ The successful operation of a designed system usually involves feedback. The feedback of output from some parts of a system to input for other parts can be used to encourage what is going on in a system, discourage it, or reduce its discrepancy from some desired value. The stability of a system can be greater when it includes appropriate feedback mechanisms.

▶ Even in some very simple systems, it may not always be possible to predict accurately the result of changing some part or connection. ■

11ʙ Models

Physical, mathematical, and conceptual models are tools for learning about the things they are meant to resemble. Physical models are by far the most obvious to young children, so they should be used to introduce the idea of models. Dolls, stuffed animals, toy cars and airplanes, and other everyday objects can stimulate discussions about how those things are like and unlike the real things. The term *model* should probably be used to refer only to physical models in the early grades, but the notion of likeness will be the central issue in using any kind of model.

The usefulness of conceptual models depends on the ability of people to imagine that something they do not understand is in some way like something that they do understand. Imagery, metaphor, and analogy are every bit as much a part of science as deductive logic, and as much at home in science as in the arts and humanities. Students cannot be expected to become adept in the use of conceptual models, however, until they get to know quite a bit about materials, things, and processes in the accessible world around them through direct, hands-on experience. The curriculum emphasis, therefore, should be on a rich variety of experiences, not on generalizations about conceptual models. Moreover, students need to acquire images and understandings that come from drawing, painting, sculpting, playing music, acting in plays, listening to and telling stories, reading, participating in games and sports, doing work, and living life.

By their nature, mathematical models are usually more abstract than physical and conceptual models. The connection of mathematics to concrete matters, and hence its value for modeling, could be substantially stronger if mathematics were often taught as part of science, social studies, technology, health, gym, music, and other subjects, rather than only during "mathematics time." One of the drawbacks of teaching mathematics entirely as a separate subject is that mathematics is taught before real-world problems are identified, so the related exercises may have mostly to do with learning the procedures rather than with solving interesting problems.

RESEARCH NOTES
page 357

◁ ALSO SEE

267

Kindergarten through Grade 2

Every opportunity should be taken to get students to talk about how the things they play with relate to real things in the world. The more imaginative the conversation the better, for insisting upon accuracy at this level may hinder other important developments.

By the end of the 2nd grade, students should know that

▶ **Many of the toys children play with are like real things only in some ways. They are not the same size, are missing many details, or are not able to do all of the same things.**

▶ **A model of something is different from the real thing but can be used to learn something about the real thing.**

▶ **One way to describe something is to say how it is like something else.** ∎

Grades 3 through 5

As students develop beyond their natural play with models, they should begin to modify them and discuss their limitations. What happens if wheels are taken off, or weight is added, if different materials are used, or if the model gets wet? Is that what would happen to the real things? Students also can begin to compare their objects, drawings, and constructions to the things they portray or resemble (real bears, houses, airplanes, etc.). Since students are being introduced to geometry, graphs, and other mathematical concepts, they should at the same time reflect on how these representations relate to nature. Similarly, what they are learning in the arts and humanities can supply analogies. Students can begin to formulate their own models to explain things they cannot observe directly. By testing their models and changing them as more information is acquired, they begin to understand how science works.

By the end of the 5th grade, students should know that

▶ **Seeing how a model works after changes are made to it may suggest how the real thing would work if the same were done to it.**

▶ **Geometric figures, number sequences, graphs, diagrams, sketches, number lines, maps, and stories can be used to represent objects, events, and processes in the real world, although such representations can never be exact in every detail.** ∎

Now models and their use can be dealt with much more explicitly than before because students have a greater general knowledge of mathematics, literature, art, and the objects and processes around them. Also, student use of computers should have progressed beyond word processing to graphing and simulations that compute and display the results of changing factors in the model. All of these things can give students a grasp of what models are and how they can be compared by considering their consequences. Students should have many opportunities to learn how conceptual models can be used to suggest interesting questions, such as "What would the atmosphere be like if its molecules were to act like tiny, high-speed marshmallows instead of tiny, high-speed steel balls?"

The use of physical models also can increase in sophistication. Students should discover that physical models on a reduced scale may be inadequate because of scaling effects: With change in scale, some factors change more than others so things no longer work the same way. The drag effects of water flow past a model boat, for example, are very different from the effects on a full-sized boat.

By the end of the 8th grade, students should know that

▶ **Models are often used to think about processes that happen too slowly, too quickly, or on too small a scale to observe directly, or that are too vast to be changed deliberately, or that are potentially dangerous.**

▶ **Mathematical models can be displayed on a computer and then modified to see what happens.**

▶ **Different models can be used to represent the same thing. What kind of a model to use and how complex it should be depends on its purpose. The usefulness of a model may be limited if it is too simple or if it is needlessly complicated. Choosing a useful model is one of the instances in which intuition and creativity come into play in science, mathematics, and engineering.** ■

Grades 9 through 12

In the upper grades, considerable emphasis should be placed on mathematical modeling because it epitomizes the nature and power of models and provides a context for integrating knowledge from many different domains. The main goal should be getting students to learn how to create and use models in many different contexts, not simply to recite generalizations about models. They can acquire such generalizations too, but that will occur through discussions of models already studied. Research in developmental psychology implies that high-school students may understand that the best model isn't found yet, or that different people prefer different models while waiting for more evidence, but not that there may be no "true" model at all.

By the end of the 12th grade, students should know that

▶ **The basic idea of mathematical modeling is to find a mathematical relationship that behaves in the same ways as the objects or processes under investigation. A mathematical model may give insight about how something really works or may fit observations very well without any intuitive meaning.**

▶ **Computers have greatly improved the power and use of mathematical models by performing computations that are very long, very complicated, or repetitive. Therefore computers can show the consequences of applying complex rules or of changing the rules. The graphic capabilities of computers make them useful in the design and testing of devices and structures and in the simulation of complicated processes.**

▶ **The usefulness of a model can be tested by comparing its predictions to actual observations in the real world. But a close match does not necessarily mean that the model is the only "true" model or the only one that would work. ∎**

Much of science and mathematics has to do with understanding how change occurs in nature and in social and technological systems, and much of technology has to do with creating and controlling change. Constancy, often in the midst of change, is also the subject of intense study in science. The simplest account to be given of anything is that it does not change. Because scientists are always looking for the simplest possible accounts (that are true), they are always delighted by any aspect of anything that doesn't change even when many other aspects do. Indeed, many historians and philosophers regard conservation laws in physics (such as for mass, energy, or electric charge) to be among the greatest discoveries in science. Somewhat different aspects of constancy are described by the terms *stability, conservation, equilibrium, steady state,* and *symmetry.* These various ideas are interrelated in some subtle ways. But memorizing the distinct meanings for these terms is not a high priority. More important is being able to think about what is happening.

Symmetry is another kind of constancy—or more generally, invariance—in the midst of change. Equilibrium, steady states, and conservation might all be thought of as showing symmetry. But more typically, symmetry implies a pattern whose appearance stays the same when it undergoes a change such as rotation, reflection, stretching, or displacement. The symmetry can be geometrical or more general, as in a social order, set of computer operations, or classification of atomic particles.

When change occurs in a variable, a major issue is the rate at which change occurs. Clearly students have to make sense of a constant rate of change before they can consider increasing or decreasing rates. Yet understanding a constant rate of change is not as simple as it might seem, because of the difficulty of the idea of rate. Graphs would seem to be an immense help for semiquantitative descriptions of change—such as whether the rate is constant, increasing, saturating, etc. But the research results are that, unless the graph is of literal altitude, graph heights and slopes are puzzling to most children. The goal for all Americans should be modest: to understand a graph of any familiar variable against time in terms of reading it and interpreting its ups and downs in a story about what is going on. Eventually, steepness as well as direction of change can become part of the story.

Considering the pattern of change usually involves a scale of observations and a scale of analysis. The rock may appear to sit there on the ground unchanging, but at a distance scale 10^8 times smaller its atoms are chaotically restless, and at a scale of 10^8 times larger its planet is turning and orbiting. An ecological system may seem stable over a few centuries, but over days

RESEARCH NOTES
page 357

☞ ALSO SEE

individuals come and go, and over millions of years it is greatly transformed.

Very, very small differences in what a system is like now may produce very large differences in what it is like later. That's not a difficult idea even in the middle school. What is harder to understand is that no matter how small the initial uncertainty may be, the behavior is eventually unpredictable. At the finest level, that of individual atoms, uncertainty is unavoidable. So the future is not determined by the present. For example, long-range weather forecasting now seems to be impossible—in principle, not just because of the limits of observation and analysis.

For the most part, change should not be taught as a separate subject. At every opportunity throughout the school years, the theme of change should be brought up in the context of the science, mathematics, or technology being studied. The first step is to encourage children to attend to change and describe it. Only after they have a storehouse of experience with change of different kinds are they ready to start thinking about patterns of change in the abstract. When students have such a background, a short capstone course on the subject of change could help them integrate their knowledge of patterns of change in physical, biological, social, and technological systems.

When collecting and observing the things around them, students can look for what changes and what does not and question where things come from and where things go. They may note, for instance, that most animals move from place to place but most plants stay in place, that water left in an open container gradually disappears but sand does not, and so forth. Such activities can sharpen students' observation and communication skills and instill in them a growing sense that many different kinds of change go on all the time. Students should be encouraged to take, record, and display counts and simple measurements of things over time. This activity can provide them with many opportunities to learn and use elementary mathematics. To begin to work toward ideas of conservation, mathematics exercises in which the sum stays the same may be helpful—e.g., "How many ways can you add whole numbers to get 13?"

By the end of the 2nd grade, students should know that

▶ **Things change in some ways and stay the same in some ways.**

▶ **People can keep track of some things, seeing where they come from and where they go.**

▶ **Things can change in different ways, such as in size, weight, color, and movement. Some small changes can be detected by taking measurements.**

▶ **Some changes are so slow or so fast that they are hard to see.** ■

With greater emphasis than before on measurement, graphing, and data analysis, students can make progress toward understanding some very important notions about change. At this stage, becoming familiar with a large and varied set of actual examples of change is more important than being able to recite the generalizations set out in the benchmarks.

Notions of symmetry can begin with identifying patterns whose appearance stays the same when they undergo some change (such as rotation, reflection, stretching, or displacement). Children generally are interested in exploring the shapes of things (plants and animals, themselves, buildings, vehicles, toys, etc.) and looking for regularities of shape. Students should have many experiences in discussing and depicting all sorts of change: continuing in the same direction, reaching a high or low value, repeatedly reversing direction, and so on.

By the end of the 5th grade, students should know that

▶ **Some features of things may stay the same even when other features change. Some patterns look the same when they are shifted over, or turned, or reflected, or seen from different directions.**

▶ **Things change in steady, repetitive, or irregular ways—or sometimes in more than one way at the same time. Often the best way to tell which kinds of change are happening is to make a table or graph of measurements. ∎**

Constancy in a system can be represented in two ways: as a constant sum or as compensating changes. When the quantity being considered is a *count* (as of students or airplanes), then constancy of the total is obvious. When the quantity being considered is a *measure* on a continuous scale, rather than a packaged unit, then "it has to come from somewhere and go somewhere" may be a more directly appreciable principle. For example, it seems easier to see that heat lost from one part of a system has to show up somewhere else than to say that the total measure for the whole system has to stay the same. This may be particularly true when the quantity can take various, interconvertible forms—say, forms of energy or monetary value.

In these grades, students can look for more sophisticated patterns, including rates of change and cyclic patterns. Invariance may be found in change itself: The water in a river changes, but the rate of flow may be constant; or the rate of flow may change seasonally, but the cycle may have a constant cycle length.

The idea of a series of repeating events is not difficult for students—that is what their day-by-day and week-by-week lives are like. Cyclic variation in a magnitude is more difficult. The cycle length is its simplest feature, whereas the range of variation has little interest unless students are familiar with and care about the variable. (A variation of one degree in body temperature—because of its relevance to whether they have to stay home from school—may be more interesting to students than a tenfold variation in the number of cases of measles.)

273

By the end of the 8th grade, students should know that

▶ **Physical and biological systems tend to change until they become stable and then remain that way unless their surroundings change.**

▶ **A system may stay the same because nothing is happening or because things are happening but exactly counterbalance one another.**

▶ **Many systems contain feedback mechanisms that serve to keep changes within specified limits.**

▶ **Symbolic equations can be used to summarize how the quantity of something changes over time or in response to other changes.**

▶ **Symmetry (or the lack of it) may determine properties of many objects, from molecules and crystals to organisms and designed structures.**

▶ **Cycles, such as the seasons or body temperature, can be described by their cycle length or frequency, what their highest and lowest values are, and when these values occur. Different cycles range from many thousands of years down to less than a billionth of a second.** ■

Most of what is appropriate to study about constancy and change in the high-school years has at least been touched upon in the earlier years, though mostly in a qualitative or semiquantitative way. Although it is still not necessary to become intensely quantitative, many of the applications of the ideas take on more concrete meaning when calculations are made.

Stability, like many concepts in science, has to be considered in some context of scale. On a familiar scale of space and time, a mountain may appear stable for centuries. Yet on the atomic scale, the mountain is a continuous hubbub of restless motion and absorption and radiation of energy. On the scale of millions of years, mountains rise up from plains and erode away. In a practical sense, stability of some object or system means only that for present purposes one does not notice or have to worry about changes in it.

Perhaps the most important ideas to be dealt with are the conservation laws, rates of change, and the general notion of evolutionary change. The emphasis on conservation laws should probably be practical— that is, should show how those concepts led, and continue to lead, to advances in science. The historical cases studied can contribute to this understanding. Rates of change that are approximately constant (or averageable) make possible a variety of practical calculations. Changing rates need not be calculated but can be identified in graphs and sketched. Especially important is the case in which change rate is proportional to how much there already is (as in population growth or radioactive decay).

Evolutionary change is a general concept, of which biological evolution is only one instance. Another point is more philosophical: Although evolution is the kind of change that emerges from and is influenced by the past, the past appears not to completely determine the future.

Two major arguments for indeterminism are included in *Benchmarks*—the principle of uncertainty at the submicroscopic level and the additional uncertainty owing to the complexity of systems and their sensitivity to vanishingly small differences in conditions. These arguments are not easy to grasp but at least students should be given a chance to debate them. Many students may be reassured to learn that scientists do not claim to be able to predict the future in every detail, nor do they claim that nature is a mechanical system in which every occurrence is already determined.

By the end of the 12th grade, students should know that

▶ **A system in equilibrium may return to the same state of equilibrium if the disturbances it experiences are small. But large disturbances may cause it to escape that equilibrium and eventually settle into some other state of equilibrium.**

▶ **Along with the theory of atoms, the concept of the conservation of matter led to revolutionary advances in chemical science. The concept of conservation of energy is at the heart of advances in fields as diverse as the study of nuclear particles and the study of the origin of the universe.**

▶ **Things can change in detail but remain the same in general (the players change, but the team remains; cells are replaced, but the organism remains). Sometimes counterbalancing changes are necessary for a thing to retain its essential constancy in the presence of changing conditions.**

▶ **Graphs and equations are useful (and often equivalent) ways for depicting and analyzing patterns of change.**

▶ **In many physical, biological, and social systems, changes in one direction tend to produce opposing (but somewhat delayed) influences, leading to repetitive cycles of behavior.**

▶ **In evolutionary change, the present arises from the materials and forms of the past, more or less gradually, and in ways that can be explained.**

▶ **Most systems above the molecular level involve so many parts and forces and are so sensitive to tiny differences in conditions that their precise behavior is unpredictable, even if all the rules for change are known. Predictable or not, the precise future of a system is not completely determined by its present state and circumstances but also depends on the fundamentally uncertain outcomes of events on the atomic scale. ■**

11D SCALE

Most variables in nature—size, distance, weight, temperature, and so on—show immense differences in magnitude. As their sophistication increases, students should encounter increasingly larger ratios of upper and lower limits of these variables. But that is only the starting point for the idea of changes of scale. The larger idea is that the way in which things work may change with scale. Different aspects of nature change at different rates with changes in scale, and so the relationships among them change, too. Probably the most easily demonstrated example is that as something changes size, its volume changes out of proportion to its area. So properties that depend on volume (such as mass and heat capacity) increase faster than properties that depend on area (such as bone strength and cooling surface). Therefore a large container of hot water cools off more slowly than a small container, and a large animal must have proportionally thicker legs than a small animal of otherwise similar shape.

As another consequence of disproportional change of properties, some "laws" of science (such as how friction depends on speed) are valid only within a certain range of circumstances. New and sometimes surprising kinds of phenomena can appear at extremely large or small values of a variable. For example, a star many times more massive than the sun can eventually collapse under its own gravity to become a black hole from which not even light can escape.

Looking at how things change with scale requires familiarity with the range of values and with how to express the range in numbers that make some sense. So children should start by noticing extremes of familiar variables and how things may be different at those extremes. There is no problem here, in that most children are entranced by "biggest," "littlest," "fastest," and "slowest"—giants and superlatives in general. In any case, scale should be introduced explicitly only when students already have a rich ground of experiences having to do with magnitudes and the effects of changing them.

The range of numbers that people can grasp increases with age. No benefit comes from trying to foist exponential notation on children who can't grasp its meaning at all. It has been argued that people really can't comprehend a range of more than about 1,000 to 1 at any one moment. One can think of a meter being a thousand millimeters (they are there to be seen in a quick look at a meter stick) and that a kilometer is a thousand meters (it can be run off in a few minutes)—but one may not be able to think of a kilometer as a million millimeters. A million becomes meaningful, however, as a thousand thousands, once a thousand becomes comprehensible. Particularly important senses of scale to develop for science literacy are the immense size of the cosmos, the minute size of molecules, and the enormous age of the earth (and the life on it).

Kindergarten through Grade 2

Children at this level are not yet comfortable enough with numbers to succeed much in comparing magnitudes. Their attention should be drawn repeatedly to simple comparisons in observations: What is smaller or larger, what might be still smaller or larger, what is the smallest or largest they could imagine, and do such things exist? A sense of changes in scale can be encouraged by perspective-taking games that challenge imagination (for example, "What would other people look like to you if you were as tall as a house or as small as an ant?").

By the end of the 2nd grade, students should know that

► **Things in nature and things people make have very different sizes, weights, ages, and speeds.** ■

Grades 3 through 5

Children at this level tend to be fascinated by extremes. That interest should be exploited to develop student math skills as well as a sense of scale. Students may not have the mathematical sophistication to deal confidently with ratios and with differences among ratios but the observational groundwork and familiarity with talking about them can begin. At the very least, students can compare speeds, sizes, distances, etc., as fractions and multiples of one another.

Students should now be building structures and other things in their technology projects. Through such experience, they can begin to understand both the mathematical and engineering relationships of length, area, and volume. They can be challenged to measure things that are hard to measure on account of being very small or very large, very light or very heavy.

By the end of the 5th grade, students should know that

► **Almost anything has limits on how big or small it can be.**

► **Finding out what the biggest and the smallest possible values of something are is often as revealing as knowing what the usual value is.** ■

Grades 6 through 8

As students' familiarity with very large and very small numbers, ratios, and powers of ten improves, extremes of scale become more meaningful. The use of ratios can now be explicit and comparisons of extremes that exceed 10^{10} may make some sense to students. Alternative representations of great scale differences should be used—such as Charles Eames' *Powers of Ten* film and Haldane's classic essay, "On Being the Right Size." Indeed, this essay might very well serve as the centerpiece for a seminar or short course dealing with the importance of size in nature and in construction.

The topic of scale also lends itself to the use of computer simulation, in which the user can change scales at will, and to the use of elementary statistics—large collections of things may have to be represented by summaries such as averages or typical examples. Approximate powers of ten (*orders of magnitude*) can be learned if students have become comfortable with estimates and approximations. This use of exponents for comparisons does not justify teaching the full apparatus of exponential notation to all students.

Understanding the notion that things necessarily work differently on different scales is more difficult than recognizing extremes, hence students should study a variety of different examples (for instance, cooling rates of different-sized containers of water, strength of different-sized constructions from the same material, flight characteristics of different-sized model airplanes).

By the end of the 8th grade, students should know that

▶ **Properties of systems that depend on volume, such as capacity and weight, change out of proportion to properties that depend on area, such as strength or surface processes.**

▶ **As the complexity of any system increases, gaining an understanding of it depends increasingly on summaries, such as averages and ranges, and on descriptions of typical examples of that system.** ■

Grades 9 through 12

Facility with powers of ten can make it easier to describe great differences of scale, but not necessarily to make them comprehensible. Students can bootstrap their comprehension of magnitude only by a few factors of ten at a time, perhaps grasping each new level only in terms of the previous one. For instance, once students have come to terms with a million, then they may have a better sense of what it means to say there are over a billion galaxies, each with over a billion stars.

Mathematical sophistication can now also include abstract, algebraic representation of the effects of powers; properties that increase by the square of linear size or the cube; and the relation between those increases. Still, the most important point is not the precise ratio of x^3 to x^2, but the more approximate idea that one changes out of proportion to the other—therefore relationships change. Things, systems, and models that work well on one scale may work less well, or not at all, if greatly expanded or shrunk. A meter-wide amoeba, for example, would never be able to get enough nutrients and oxygen through its surface to survive; a meter-long bird built like a sparrow could not fly.

By the end of the 12th grade, students should know that

▶ **Representing large numbers in terms of powers of ten makes it easier to think about them and to compare things that are greatly different.**

▶ **Because different properties are not affected to the same degree by changes in scale, large changes in scale typically change the way that things work in physical, biological, or social systems.**

▶ **As the number of parts of a system increases, the number of possible interactions between pairs of parts increases much more rapidly.** ■

Chapter 12 HABITS OF MIND

A   VALUES AND ATTITUDES

B   COMPUTATION AND ESTIMATION

C   MANIPULATION AND OBSERVATION

D   COMMUNICATION SKILLS

E   CRITICAL-RESPONSE SKILLS

Throughout history, people have concerned themselves with the transmission of shared values, attitudes, and skills from one generation to the next. All three were taught long before formal schooling was invented. Even today, it is evident that family, religion, peers, books, news and entertainment media, and general life experiences are the chief influences in shaping people's views of knowledge, learning, and other aspects of life. Science, mathematics, and technology—in the context of schooling—can also play a key role in the process, for they are built upon a distinctive set of values, they reflect and respond to the values of society generally, and they are increasingly influential in shaping shared cultural values. Thus, to the degree that schooling concerns itself with values and attitudes—a matter of great sensitivity in a society that prizes cultural diversity and individuality and is wary of ideology—it must take scientific values and attitudes into account when preparing young people for life beyond school.

Similarly, there are certain thinking skills associated with science, mathematics, and technology that young people need to develop during their school years. These are mostly, but not exclusively, mathematical and logical skills that are essential tools for both formal and informal learning and for a lifetime of participation in society as a whole.

Taken together, these values, attitudes, and skills can be thought of as habits of mind because they all relate directly to a person's outlook on knowledge and learning and ways of thinking and acting.

SCIENCE FOR ALL AMERICANS

Page from a LEONARDO DA VINCI *notebook, fifteenth-sixteenth centuries.*

In *Science for All Americans*, Project 2061 expresses the view that education has multiple purposes and that those purposes should serve as criteria for specifying what students need to know and be able to do. The criteria are philosophical and utilitarian, individual and social. While they speak to the intrinsic value of knowing for its own sake, they emphasize also the need for education to prepare students to make their way in the real world, a world in which problems abound—in the home, in the workplace, in the community, on the planet.

Hence, preparing students to become effective problem solvers, alone and in concert with others, is a major purpose of schooling. Science, mathematics, and technology can contribute significantly to that end because in their different ways they are enterprises in the business of searching for solutions to problems ranging from the highly theoretical to the entirely concrete. Moreover, in their interactions with society, science and technology create the context for many personal and community issues.

There is a large and growing literature on problem solving. Aside from exhortation, a staple of most educational writing (including this document, to be sure), the problem-solving literature deals mostly with what skills need to be learned, why skills should be expressed behaviorally, and how to teach the desired skills. After a study of that literature, wide consultation with experts, and intense discussion, Project 2061 has reached conclusions that are reflected in the content and language of this chapter. Chief among them are the following:

Students' ability and inclination to solve problems effectively depend on their having certain knowledge, skills, and attitudes.

Quantitative, communication, manual, and critical-response skills are essential for problem solving, but they are also part of what constitutes science literacy more generally. That is why they are brought together here as scientific habits of mind rather than more narrowly as problem-solving skills or more generally as thinking skills.

Learning to solve problems in a variety of subject-matter contexts, if supplemented on occasion by explicit reflection on that experience, may result in the development of a generalized problem-solving ability that can be applied in new contexts; such transfer is unlikely to happen if either varied problem-solving experiences or reflection on problem solving is missing.

The problem of rote learning is primarily a pedagogical one that applies to skills as well as knowledge, and it is not solved simply by stating learning goals in one way instead of another.

In the light of those conclusions, it is useful to explain why the skill goals in this chapter are separated from the knowledge goals in Chapters 1 through 11. One reason is that the knowledge called for in the previous 11 chapters responds to all of the science literacy criteria mentioned earlier, not solely those having to do with problem solving. Another reason is that the skills advocated in this chapter need to be learned in the context of all of the knowledge chapters and thus would have to be repeated chapter after chapter if we tried to present knowledge and skill goals in tandem. Finally, the skills are significant in their own right as part of what it means to be science-literate, and presenting them together should make it easier to consider them as such.

It is widely argued that listing intended learning goals in specific detail is unwise because teachers will simply have their students memorize the individual entries as isolated facts. The same danger applies to stating skills in detail—procedures also can be memorized without comprehension, as veterans of "the scientific method" and mathematics algorithms can attest. Project 2061's response is the same in both cases, namely that there are better ways to deal with the problem of rote learning than by remaining vague on what knowledge and skills we want students to acquire.

The phrase "Students should know that . . ." used in benchmarks in the preceding chapters means that students should be able to connect one idea to other ideas and use it in thinking about new situations and in problem solving. But we surely want students to be likely to make such connections, not merely be able to do so. Similarly, with respect to this chapter, we want students not only to acquire certain skills but also to be inclined to use them in new situations, outside as well as inside school. Thus when the benchmarks specify that "Students should be able to" do something, we take that to mean they will in fact do so when appropriate circumstances present themselves.

One manifestation of such inclination is what someone thinks about when reading news articles. For example, on reading that trees were being logged for an important new drug found in their bark, the science-literate person might wonder about the yield from a single tree, the amount of drug needed, and how long a new tree would take to grow; or about the possibility of synthesizing the drug instead; or about what species in the forest might suffer from the loss of those particular trees; or about how complex ecological interactions are and the need for computer software to track the implications; or about possible bias in whoever was responsible for considering those various possibilities. ■

12A VALUES AND ATTITUDES

Honesty is a desirable habit of mind not unique to people who practice science, mathematics, and technology. It is highly prized in the scientific community and essential to the scientific way of thinking and doing. The importance of honesty is urged on children from every quarter, and most children are able to say what the general principle is. What honesty means in practice, however, probably comes from their seeing firsthand how it is applied in many different situations. In school science, mathematics, and technology, there are numerous opportunities to show what honesty means and how it is valued. Science: Always report and record what you observe, not what you think it ought to be or what you think the teacher wants it to be, and do not erase your notes. Math: Do not change an answer from a calculation because it is different from what others get. Technology: If your design has limitations, say so.

Children are curious about things from birth. Curiosity does not have to be taught. The problem is the reverse: how to avoid squelching curiosity while helping students focus it productively. By fostering student curiosity about scientific, mathematical, and technological phenomena, teachers can reinforce the trait of curiosity generally and show that there are ways to go about finding answers to questions about how the world works. Students will gradually come to see that some ways of satisfying one's curiosity are better than others and that finding good answers and solutions is as much fun as raising good questions.

Balancing open-mindedness with skepticism may be difficult for students. These two virtues pull in opposite directions. Even in science itself, there is tension between an openness to new theories and an unwillingness to discard current ones. As students come up with explanations for what they observe or wonder about, teachers should insist that other students pay serious attention to them. Students hearing an explanation of how something works proposed by another student or by teachers and other authorities should learn that one can admire a proposal but remain skeptical until good evidence is offered for it.

Highest priority should be given to encouraging the curiosity about the world that children bring to school. Natural phenomena easily capture the attention of these youngsters, but they should be encouraged to wonder about mathematical and technological phenomena as well. Questions about numbers, shapes, and artifacts, for example, should be treated with the same interest as those about rocks and birds. Typically, children raise questions that are hard to answer. But some of their questions are possible to deal with, and some of the impossible questions can be transformed.

As students learn to write, they should start keeping a class list of things they wonder about, without regard to how easy it might be to answer their own questions. Teachers should then help them learn to pick from the list the questions they can find answers to by doing something such as collecting, sorting, counting, drawing, taking something apart, or making something. At this level, questions that can be answered descriptively are to be preferred over those requiring abstract explanations. Students are more likely to come up with reasonable answers as to "how" and "what" than as to "why."

Still, students should not be expected to confine themselves to empirical questions only. Some questions requiring an explanation for an answer can be taken up to foster scientific habits of thought. Thus, to the question, "Why don't plants grow in the dark?" students should learn that scientists would respond by asking, "Is it true that plants don't grow in the dark?" and "How do you know?" or "How can we find out if it is true?" If the facts are correct, then reasons can be offered. Presumably children, like scientists, will propose different explanations, and some children may have a need to establish whose ideas are good or best. Comparisons will come in time, when students are able to imagine ways to make judgments. Everyone's ideas should be valued, and differing opinions should be regarded as interesting and food for thought.

By the end of the 2nd grade, students should

▶ **Raise questions about the world around them and be willing to seek answers to some of them by making careful observations and trying things out.** ■

285

Grades 3 through 5

Sustaining curiosity and giving it a scientific cast is still a high priority. Students should advance in their ability to frame their questions about the world in ways that lead to their finding answers by conducting investigations, building and testing things, and consulting reference works. In doing so, whether working alone or in teams, students should be required to keep written records in bound notebooks of what they did, what data they collected, and what they think the data mean. Emphasis should be placed on honesty in record-keeping rather than on reaching correct conclusions. To the extent that a judgment is made by one group of students about another's conclusions, it should be on the basis of its correspondence to the evidence presented, not on what a book says is true.

The thrust of the science experience is still to learn how to answer interesting questions about the world that can be answered empirically. But now students should also sometimes think up and propose explanations for their findings. In this introduction to the world of theory, the main point to stress is that for any given collection of evidence, it is usually possible to invent different explanations, and it is not always easy to tell which will prove to be best. That is one reason that scientists pay attention to ideas that may differ from what they personally believe.

By the end of the 5th grade, students should

▶ **Keep records of their investigations and observations and not change the records later.**

▶ **Offer reasons for their findings and consider reasons suggested by others.** ■

Grades 6 through 8

The scientific values and attitudes that are the focus of this section have all been introduced in the previous grades. Now they can be reinforced and developed further. Care should be taken, in an effort to cover content, not to stop fostering curiosity. Time needs to be found to enable students to pursue scientific questions that truly interest them. Inquiry projects, individual and group, provide that opportunity. Such projects also establish realistic contexts in which to emphasize the importance of scientific honesty in describing procedures, recording data, drawing conclusions, and reporting conclusions.

Consideration of the nature and uses of hypotheses and theory in science can give operational substance to the scientific habits of openness and skepticism. Hypotheses and explanations serve somewhat different purposes, but they both are judged, ultimately, by reference to evidence. Students can come to see that a hypothesis does not have to be correct—one can believe it or not—but that to be taken seriously, it should indicate what evidence would be needed to decide whether or not it is true, thus incorporating the notions of both openness and skepticism.

In this same vein, a start can be made toward legitimizing the notion that there are often several different ways of making sense out of a body of existing information. Having teams invent two or more explanations for a set of observations, or having different teams independently come up with explanations for the same set of observations, can lead to discussions of the nature of scientific explanation that are grounded in reality. Developmental psychologists doubt that alternative explanations are seriously examined by most students at this level, but at least the possibility of alternatives can be planted, not as an abstract notion but as something stemming from students' own experience.

By the end of the 8th grade, students should

▶ **Know why it is important in science to keep honest, clear, and accurate records.**

▶ **Know that hypotheses are valuable, even if they turn out not to be true, if they lead to fruitful investigations.**

▶ **Know that often different explanations can be given for the same evidence, and it is not always possible to tell which one is correct. ■**

Grades 9 through 12

Skepticism is not just a matter of willingness to challenge authority, though that is an aspect of it. It is a determination to suspend judgment in the absence of credible evidence and logical arguments. Students can learn its value in science, and that is important. Given that most of them will not be scientists as adults, the educational challenge is to help students internalize the scientific critical attitude so they can apply it in everyday life, particularly in relation to the health, political, commercial, and technological claims they encounter.

Openness to new and unusual ideas about how the world works can now be developed in the study of historical cases as well as in the context of continuing inquiry projects. The Copernican Revolution, for example, illustrates the eventual success of ideas that were initially considered outrageous by nearly everyone. This and other cases also illustrate that ideas in science are not easily or quickly accepted. Some such mixture of openness and conservatism will serve most people and societies well.

By the end of the 12th grade, students should

▶ **Know why curiosity, honesty, openness, and skepticism are so highly regarded in science and how they are incorporated into the way science is carried out; exhibit those traits in their own lives and value them in others.**

▶ **View science and technology thoughtfully, being neither categorically antagonistic nor uncritically positive. ■**

12ʙ Computation and Estimation

The scientific way of thinking is neither mysterious nor exclusive. The skills involved can be learned by everyone, and once acquired they can serve a lifetime regardless of one's occupation and personal circumstances. That is certainly true of the ability to think quantitatively, simply because so many matters in everyday life, as in science and many other fields, involve quantities and numerical relationships.

Computation is the process of determining something by mathematical means. Its value is acknowledged by the prominence accorded mathematics in school systems everywhere. Unfortunately, that preferred status has not been matched by results. It turns out that being able to get correct answers to the problems at the end of the chapter or on a work sheet or test is no guarantee of problem-solving ability in real situations. That ought not to be surprising, given that in traditional mathematics teaching, problems lack interesting real-world contexts; that memorization of algorithms by drill is not matched by learning when to use them; that numbers are used without units or attention to significance; and that students receive little, if any, help in learning how to judge how good their answers are.

In the real world, there is no need for people to make a calculation if the answer to their question is already known and easily available; they just need to know how to look it up— which is, of course, something that scientists and engineers do frequently. But in most situations, answers are not known and so making judgments about answers is as much a part of computation as the calculation itself. That is why the benchmarks in this section emphasize the need for students to develop estimation skills and the habit of checking answers against reality.

Estimation skills can be learned, but only if teachers make sure that students have lots of practice estimating (which happens if estimation is routinely treated as a standard part of problem solving). But there is no fixed set of all-purpose steps for students to memorize. If students are frequently called upon to explain how they intend to calculate an answer before carrying it out, they find that making step-by-step estimations is not hard and contributes to thinking through the problem at hand. They also gain confidence in their ability to figure out ahead of time approximately what the answer will be—bigger than this and smaller than that—if they do the calculation properly.

But a computationally "correct" answer is not necessarily a sensible one. If a computation leads to the result that an adult elephant weights 1.2 pounds, most people know that something is wrong, because elephants are enormous animals and a pound isn't much weight. Reality tells them to check their computation. Did they use appropriate mathematics? Were the numerical inputs correct? Is the decimal point in the right place? What about the unit the answer is expressed in?

Developing good quantitative thinking skills and learning about the world go together. It is not sufficient for students to learn how to perform mathematical operations in the abstract if they are to become effective

RESEARCH NOTES

page 358

problem solvers and to be able to express their arguments quantitatively whenever appropriate. Hence, at every level, the teaching of science, technology, social studies, health, physical education, and perhaps other subjects should include problem solving that requires students to make calculations and check their answers against their estimates and their knowledge of whatever the problem pertains to. As much as possible, the problems should emerge from student activities— surveys, laboratory investigations, building projects, physical-education performance data, etc.—and the content being studied rather than from prepackaged word problems. Computational skills can be learned in contexts outside of mathematics courses.

Where do calculators and computers come into the picture? The answer is, nearly everywhere. Computers imbedded in cash registers, self-help gasoline pumps, automatic teller machines, and the like do much of the arithmetic that adults formerly had to do by paper and pencil. The inexpensive, hand-held calculator makes it possible for people to apply their knowledge of basic mathematics to the quantitative matters they encounter throughout the day instantly and on the spot. And computers, with their easy-to-use spreadsheet, graphing, and database capabilities, have become tools that everyone can use, at home and at work, to carry out extensive quantitative tasks.

Undoubtedly calculators and computers can vastly extend the mathematical capabilities of everyone, for they offer a precision and speed that few people can match. But their power can be of no avail, or even detrimental, unless they are used skillfully and with understanding. These instruments do not compensate for human reasoning errors or for poor mathematics, often deliver answers with misleading precision, and are prone to operator error.

Science literacy includes being able to use electronic tools thoughtfully and with confidence. This skill calls for students to be able to select appropriate algorithms, carry out basic mathematical operations on paper, judge the reasonableness of the results of a calculation, and round off insignificant numbers. Students should start using calculators and computers early and use them in as many different contexts as possible. That will increase the likelihood that students will learn to use them effectively, including learning when it is sufficient to make a mental estimate, when to use paper and pencil, and when to draw on the help of a calculator or computer. This early, continuing, and broadly based experience has another advantage: Properly used over time, calculators and computers can actually help students learn mathematics and acquire quantitative thinking skills.

In this section and those that follow, there are no grade-level commentaries. According to reviewers, skill benchmarks are less likely to be misunderstood than knowledge or attitude benchmarks, and hence, section essays are sufficient to cover all grades.

Kindergarten through Grade 2

By the end of the 2nd grade, students should be able to

▶ Use whole numbers and simple, everyday fractions in ordering, counting, identifying, measuring, and describing things and experiences.

▶ Readily give the sums and differences of single-digit numbers in familiar contexts where the operation makes sense to them and they can judge the reasonableness of the answer.

▶ Give rough estimates of numerical answers to problems before doing them formally.

▶ Explain to other students how they go about solving numerical problems.

▶ Make quantitative estimates of familiar lengths, weights, and time intervals and check them by measurements. ■

Grades 3 through 5

By the end of the 5th grade, students should be able to

▶ Add, subtract, multiply, and divide whole numbers mentally, on paper, and with a calculator.

▶ Use fractions and decimals, translating when necessary between decimals and commonly encountered fractions—halves, thirds, fourths, fifths, tenths, and hundredths (but not sixths, sevenths, etc.).

▶ Judge whether measurements and computations of quantities such as length, area, volume, weight, or time are reasonable in a familiar context by comparing them to typical values.

▶ State the purpose of each step in a calculation.

▶ Read and follow step-by-step instructions in a calculator or computer manual when learning new procedures. ■

Grades 6 through 8

By the end of the 8th grade, students should be able to

▶ Find what percentage one number is of another and figure any percentage of any number.

▶ Use, interpret, and compare numbers in several equivalent forms such integers, fractions, decimals, and percents.

▶ Calculate the circumferences and areas of rectangles, triangles, and circles, and the volumes of rectangular solids.

▶ Find the mean and median of a set of data.

▶ Estimate distances and travel times from maps and the actual size of objects from scale drawings.

▶ Insert instructions into computer spreadsheet cells to program arithmetic calculations.

▶ Determine what unit (such as seconds, square inches, or dollars per tankful) an answer should be expressed in from the units of the inputs to the calculation, and be able to convert compound units (such as yen per dollar into dollar per yen, or miles per hour into feet per second).

▶ Decide what degree of precision is adequate and round off the result of calculator operations to enough significant figures to reasonably reflect those of the inputs.

▶ Express numbers like 100, 1,000, and 1,000,000 as powers of 10.

▶ Estimate probabilities of outcomes in familiar situations, on the basis of history or the number of possible outcomes. ■

Grades 9 through 12

By the end of the 12th grade, students should be able to

▶ Use ratios and proportions, including constant rates, in appropriate problems.

▶ Find answers to problems by substituting numerical values in simple algebraic formulas and judge whether the answer is reasonable by reviewing the process and checking against typical values.

▶ Make up and write out simple algorithms for solving problems that take several steps.

▶ Use computer spreadsheet, graphing, and database programs to assist in quantitative analysis.

▶ Compare data for two groups by representing their averages and spreads graphically.

▶ Express and compare very small and very large numbers using powers-of-ten notation.

▶ Trace the source of any large disparity between an estimate and the calculated answer.

▶ Recall immediately the relations among 10, 100, 1000, 1 million, and 1 billion (knowing, for example, that 1 million is a thousand thousands).

▶ Consider the possible effects of measurement errors on calculations. ■

12c   MANIPULATION AND OBSERVATION

Construing habits of mind to include manipulation and observation skills raises no eyebrows in science. Scientists know that finding answers to questions about nature means using one's hands and senses as well as one's head. The same is true in medicine, engineering, business, and many other fields, and so it should be in everyday life.

Tools, from hammers and drawing boards to cameras and computers, extend human capabilities. They make it possible for people to move things beyond their strength, move faster and farther than their legs can carry them, detect sounds too faint to be heard and objects too small or too far away to be seen, project their voices around the world, store and analyze more information than their brains can cope with, and so forth. In daily living, people have little need to use telescopes, microscopes, and the sophisticated instruments used by scientists and engineers in their work. But the array of mechanical, electrical, electronic, and optical tools that people can use is no less than awesome.

What people use tools for and how thoughtfully they use them is another matter, however. Tools can of course be used for banal or noble, even ignoble, purposes, and used with or without much regard for consequences. Education for science literacy implies that students be helped to develop the habit of using tools, along with scientific and mathematical ideas and computation skills, to solve practical problems and to increase their understanding, throughout life, of how the world works. A very common problem people encounter is that things don't work right. In many instances, the problem can be diagnosed and the malfunctioning device fixed using ordinary troubleshooting techniques and tools.

Kindergarten through Grade 2

Grades 3 through 5

By the end of the 2nd grade, students should be able to

▶ Use hammers, screwdrivers, clamps, rulers, scissors, and hand lenses, and operate ordinary audio equipment.

▶ Assemble, describe, take apart and reassemble constructions using interlocking blocks, erector sets, and the like.

▶ Make something out of paper, cardboard, wood, plastic, metal, or existing objects that can actually be used to perform a task.

▶ Measure the length in whole units of objects having straight edges. ■

By the end of the 5th grade, students should be able to

▶ Choose appropriate common materials for making simple mechanical constructions and repairing things.

▶ Measure and mix dry and liquid materials (in the kitchen, garage, or laboratory) in prescribed amounts, exercising reasonable safety.

▶ Keep a notebook that describes observations made, carefully distinguishes actual observations from ideas and speculations about what was observed, and is understandable weeks or months later.

▶ Use calculators to determine area and volume from linear dimensions, aggregate amounts of area, volume, weight, time, and cost, and find the difference between two quantities of anything.

▶ Make safe electrical connections with various plugs, sockets, and terminals. ■

By the end of the 8th grade, students should be able to

▶ Use calculators to compare amounts proportionally.

▶ Use computers to store and retrieve information in topical, alphabetical, numerical, and key-word files, and create simple files of their own devising.

▶ Read analog and digital meters on instruments used to make direct measurements of length, volume, weight, elapsed time, rates, and temperature, and choose appropriate units for reporting various magnitudes.

▶ Use cameras and tape recorders for capturing information.

▶ Inspect, disassemble, and reassemble simple mechanical devices and describe what the various parts are for; estimate what the effect that making a change in one part of a system is likely to have on the system as a whole. ▪

By the end of the 12th grade, students should be able to

▶ Learn quickly the proper use of new instruments by following instructions in manuals or by taking instructions from an experienced user.

▶ Use computers for producing tables and graphs and for making spreadsheet calculations.

▶ Troubleshoot common mechanical and electrical systems, checking for possible causes of malfunction, and decide on that basis whether to make a change or get advice from an expert before proceeding.

▶ Use power tools safely to shape, smooth, and join wood, plastic, and soft metal. ▪

12D COMMUNICATION SKILLS

Good communication is a two-way street. It is as important to receive information as to disseminate it, to understand other's ideas as to have one's own understood. In the scientific professions, tradition places a high priority on accurate communication, and there are mechanisms, such as refereed journals and scientific meetings, to facilitate the sharing of new information and ideas within various disciplines and subdisciplines. Science-literate adults share this respect for clear, accurate communication, and they possess many of the communication skills characteristic of the scientific enterprise.

Accurate communication within a science discipline results in part from the use of technical language. An unintentional side effect of reliance on specialized terms, however effective it may be within a discipline, is that it impedes communication between specialists and between the specialists and the general public. For the general public, science writers for newspapers, magazines, and television undertake to translate the highly technical language of each discipline into language accessible to the educated adult. In doing that, they assume that an educated reader is familiar with some of the central ideas of science and is able to read material that uses the basic language and logic of mathematics. *Science for All Americans* describes the knowledge base for such educated readers, and *Benchmarks* points the way to the development of such adults. The communication skills below are intended to complement that knowledge base.

There is an aspect of quantitative thinking that may be as much a matter of inclination as skill. It is the habit of framing arguments in quantitative terms whenever possible. Instead of saying that something is big or fast or happens a lot, a better approach is often to use numbers and units to say how big, fast, or often, and instead of claiming that one thing is larger or faster or colder than another, it is better to use either absolute or relative terms to say how much so. Communication becomes more focused when "big" is replaced with "3 feet" or "250 pounds" (very different notions of what constitutes bigness) and "happens a lot" with "17 times this year compared to 2 or 3 times in each of the previous 10 years" or "90 to 95% of the time." And just as students should develop this way of thinking, they should demand it of others and not be satisfied with vague claims when quantitative ones are possible and relevant.

Chapter 8 D COMMUNICATION
E INFORMATION PROCESSING
9 THE MATHEMATICAL WORLD

ALSO SEE

Kindergarten through Grade 2

By the end of the 2nd grade, students should be able to

▶ Describe and compare things in terms of number, shape, texture, size, weight, color, and motion.

▶ Draw pictures that correctly portray at least some features of the thing being described. ■

Grades 3 through 5

By the end of the 5th grade, students should be able to

▶ Write instructions that others can follow in carrying out a procedure.

▶ Make sketches to aid in explaining procedures or ideas.

▶ Use numerical data in describing and comparing objects and events. ■

Grades 6 through 8

By the end of the 8th grade, students should be able to

▶ Organize information in simple tables and graphs and identify relationships they reveal.

▶ Read simple tables and graphs produced by others and describe in words what they show.

▶ Locate information in reference books, back issues of newspapers and magazines, compact disks, and computer databases.

▶ Understand writing that incorporates circle charts, bar and line graphs, two-way data tables, diagrams, and symbols.

▶ Find and describe locations on maps with rectangular and polar coordinates. ∎

Grades 9 through 12

By the end of the 12th grade, students should be able to

▶ Make and interpret scale drawings.

▶ Write clear, step-by-step instructions for conducting investigations, operating something, or following a procedure.

▶ Choose appropriate summary statistics to describe group differences, always indicating the spread of the data as well as the data's central tendencies.

▶ Describe spatial relationships in geometric terms such as perpendicular, parallel, tangent, similar, congruent, and symmetrical.

▶ Use and correctly interpret relational terms such as *if . . . then . . .* , *and, or, sufficient, necessary, some, every, not, correlates with*, and *causes.*

▶ Participate in group discussions on scientific topics by restating or summarizing accurately what others have said, asking for clarification or elaboration, and expressing alternative positions.

▶ Use tables, charts, and graphs in making arguments and claims in oral and written presentations. ∎

12E   CRITICAL-RESPONSE SKILLS

In everyday life, people are bombarded with claims—claims about products, about how nature or social systems or devices work, about their health and welfare, about what happened in the past and what will occur in the future. These claims are put forth by experts (including scientists) and nonexperts (including scientists), by honest people and charlatans. In responding to this barrage, trying to separate sense from nonsense, knowledge helps.

But apart from what they know about the substance of an assertion, individuals who are science literate can make some judgments based on its character. The use or misuse of supporting evidence, the language used, and the logic of the argument presented are important considerations in judging how seriously to take some claim or proposition. These critical response skills can be learned and with practice can become a lifelong habit of mind.

By the end of the 2nd grade, students should

▶ Ask "How do you know?" in appropriate situations and attempt reasonable answers when others ask them the same question. ■

RESEARCH NOTES

page 360

Grades 3 through 5

By the end of the 5th grade, students should

▶ Buttress their statements with facts found in books, articles, and databases, and identify the sources used and expect others to do the same.

▶ Recognize when comparisons might not be fair because some conditions are not kept the same.

▶ Seek better reasons for believing something than "Everybody knows that . . ." or "I just know" and discount such reasons when given by others. ∎

Grades 6 through 8

By the end of the 8th grade, students should

▶ Question claims based on vague attributions (such as "Leading doctors say...") or on statements made by celebrities or others outside the area of their particular expertise.

▶ Compare consumer products and consider reasonable personal trade-offs among them on the basis of features, performance, durability, and cost.

▶ Be skeptical of arguments based on very small samples of data, biased samples, or samples for which there was no control sample.

▶ Be aware that there may be more than one good way to interpret a given set of findings.

▶ Notice and criticize the reasoning in arguments in which (1) fact and opinion are intermingled or the conclusions do not follow logically from the evidence given, (2) an analogy is not apt, (3) no mention is made of whether the control groups are very much like the experimental group, or (4) all members of a group (such as teenagers or chemists) are implied to have nearly identical characteristics that differ from those of other groups. ∎

Grades 9 through 12

By the end of the 12th grade, students should

▶ Notice and criticize arguments based on the faulty, incomplete, or misleading use of numbers, such as in instances when (1) average results are reported, but not the amount of variation around the average, (2) a percentage or fraction is given, but not the total sample size (as in "9 out of 10 dentists recommend..."), (3) absolute and proportional quantities are mixed (as in "3,400 more robberies in our city last year, whereas other cities had an increase of less than 1%), or (4) results are reported with overstated precision (as in representing 13 out of 19 students as 68.42%).

▶ Check graphs to see that they do not misrepresent results by using inappropriate scales or by failing to specify the axes clearly.

▶ Wonder how likely it is that some event of interest might have occurred just by chance.

▶ Insist that the critical assumptions behind any line of reasoning be made explicit so that the validity of the position being taken—whether one's own or that of others—can be judged.

▶ Be aware, when considering claims, that when people try to prove a point, they may select only the data that support it and ignore any that would contradict it.

▶ Suggest alternative ways of explaining data and criticize arguments in which data, explanations, or conclusions are represented as the only ones worth consideration, with no mention of other possibilities. Similarly, suggest alternative trade-offs in decisions and designs and criticize those in which major trade-offs are not acknowledged. ■

BACKGROUND

Chapter 13 THE ORIGIN OF BENCHMARKS

We believe that users of *Benchmarks for Science Literacy* want to know how the benchmarks came into being. Who developed them? How did they go about it? What kinds of help did they have? Why is *Benchmarks* in its present form? How thoroughly was it reviewed? The purpose of this chapter is to answer such questions.

School-Based Research and Development Teams

Project 2061's *Science for All Americans (SFAA)*, published in 1989 after study and debate by scientists, mathematicians, engineers, and educators, specified literacy goals in science, mathematics, and technology for all high-school graduates. But setting these adult literacy goals was just the first step toward reforming science education. A needed second step was to create a set of tools for educators to use in designing K-12 curricula that would meet the content standards of *SFAA*. Chief among those tools, it was thought at the beginning, would be curriculum models that could serve as examples of alternative ways to configure the K-12 experience so as to obtain the desired science-literacy outcomes.

Project staff considered who would be best equipped to take on such a demanding assignment. Scientists and engineers know their subjects but are at a distance from the classroom. Learning and education researchers understand the difficulties children have, but only in a rather narrow range of topics. Classroom teachers, despite their keen sense of what interests children and what they learn under current conditions, often lack a full kindergarten-through-graduation perspective on education as well as needed resources, training, and time to envision radical departures from the current curriculum. We decided that if traditional constraints were removed and adequate time and resources were provided, school teachers and administrators, advised by education specialists and backed by scientists, would be most likely to develop intellectually sound curriculum models and other curriculum-design tools that would prove credible to other teachers.

The Project 2061 teachers would have to be exceptional individuals—leaders in their districts and willing to take risks. They would have to be well versed in the major ideas of science, mathematics, and technology; have a broad educational perspective, encompassing several disciplines and spanning the entire K-12 learning process; and have expertise in designing curricula. This array of qualities was unlikely to be found in even the most exceptional team of teachers in a chosen district but was certainly worth cultivating.

Inter-team work on curriculum models at Project 2061 summer conference in Seattle, 1991.

We also decided to have district teams rather than an assembled national group. Members of each district team would then be close enough to discuss issues at length and often. The various teams would also be very different from one another by virtue of locale, demographics, and available resources, so that they might collectively represent the nation. The Project recruited teams of school teachers and administrators from six sites around the country—in rural Georgia; in suburban McFarland, Wisconsin; and in urban Philadelphia, San Antonio, San Diego, and San Francisco.

So that the teams could plan for 13 years of schooling in science, mathematics, and technology, team members were chosen to bring all grades and subjects to the process. Each team had 5 elementary teachers, 5 middle-school teachers, 10 high-school teachers, 1 principal from each level, and 2 curriculum specialists. The teachers had taught the life and physical sciences, social studies, mathematics, technology, and also other disciplines.

Each 25-member team received clerical support, computers and computer training, office space, reference materials, travel funds, and other resources they needed for research and development. The school districts involved agreed to release team members an average of four days per month from their classrooms to work on Project 2061 tasks. Faculty from local universities provided consultation and technical assistance to the teams as requested throughout the year. Consultants from around the country offered their expertise at annual summer conferences, where staff and teams met to advance mutual tasks.

We asked each team to design a curriculum model that could be used by school districts to plan curricula that serve local needs and meet the goals in *SFAA*. The teams were encouraged to be as imaginative as possible, and not let local barriers limit their vision of what a K-12 curriculum *could* look like. Instead, they were asked to keep track of obstacles encountered or envisioned and alert the Project 2061 staff to them. Special groups were commissioned to write papers (called *Blueprint* papers) to consider how all aspects of the American system of schooling would need to respond to support these new models.

To foster continuity of ideas, the teams were asked to discern useful connections within and among the typically separate disciplines (the natural and social sciences, mathematics, technology—and the humanities, for that matter) and to consider all 13 grades when specifying desirable and reasonable progress toward the goals in *SFAA*.

The Nature of the Work

As it turned out, the work of the teams included many different tasks having to do with curriculum design and the implementation of reform. Two that undergirded all others were to gain a deep understanding of the substance and nature of science literacy and to identify the antecedent ideas that would be needed for students to make conceptual and psychological sense of the ideas in *SFAA*. During four joint summer conferences and three academic years at their own sites, the teams immersed themselves in this work. One product of that effort is *Benchmarks*.

Understanding SFAA

SFAA deliberately omits much of the traditional content of science, mathematics, and technology found in today's curricula and textbooks and yet contains material with which most teachers are not altogether familiar. Few teachers have had the opportunity to become familiar with how science really works, study the history of science, explore themes that cut across disciplines, or learn engineering concepts. Moreover, there is little cross-over in the education, training, and experience secondary-school teachers have of science, mathematics, and technology, and the background of most high-school science teachers is limited to the biological, earth and space, or physical sciences. And it is not unusual for elementary teachers to have had very little of any of those subjects in college.

To deal with this, several steps were taken. Especially during their first joint summer meeting, but also to some extent during the following summers and academic years, team members

- met with scientists, engineers, mathematicians, historians, architects, and physicians to learn about their current work and raise questions about its applications and relation to environmental and social issues;

- attended lectures and read widely, especially in the history and philosophy of science;

- participated in inquiry and design projects;

- prepared matrices to highlight the connections among ideas in various chapters of *SFAA* and analyzed the mathematics content of *SFAA* in relation to the NCTM *Standards;* and

- worked in cross-discipline, cross-grade discussion groups to share understandings and help one another.

Considering Antecedents to Each SFAA Outcome

The team members had to imagine what progress students could make toward each *SFAA* goal, a process that came to be called mapping because it required groups to link more sophisticated ideas in later grades to the more primitive ones suitable in the earlier years. Consider, for example, the concept of the structure of matter. What must students be taught first in order to understand the basic ideas about atoms and molecules and their structure and how their arrangements and activities underlie all phenomena? When should students learn that everything is made of invisibly small atoms linked together in many different patterns? What other ideas must they have before they can understand this one?

A partial map for the structure-of-matter literacy goal identified in Chapter 4 of *SFAA* is shown below. Progression of student understanding is organized around four story lines: *properties, common ingredients, invisibly small pieces, and conservation of matter.* Each of the story lines can be followed as one looks at the sequence of boxes. Eventually, all four story lines converge upon the idea of atoms and molecules (see page 306).

Other maps highlighted the importance of connections among disciplines. For example, understanding the scientific explanation for the evolution of life, a goal in Chapter 5 of *SFAA*, depends on some precursor knowledge of the physical sciences, mathematics, some common themes that cut across disciplines, and the nature of scientific inquiry. Four story lines are involved: *evidence from existing organisms, fossil evidence, mechanism of selection, and origin of new traits.* Also involved are notions of proportion and summary characteristics of a population.

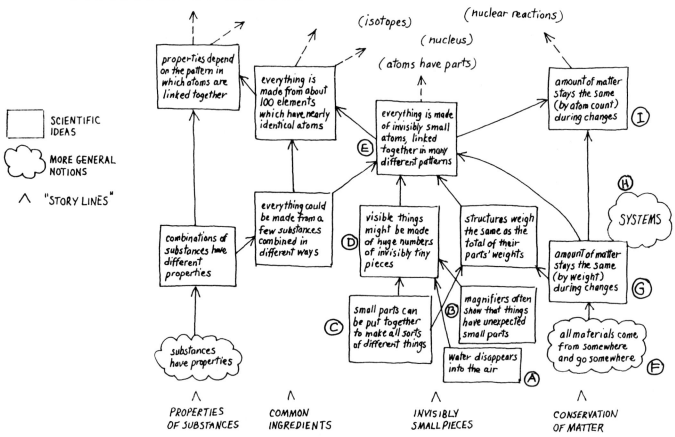

STRUCTURE OF MATTER

(isotopes)

(nuclear reactions)

(nucleus)

(atoms have parts)

properties depend on the pattern in which atoms are linked together

everything is made from about 100 elements which have nearly identical atoms

everything is made of invisibly small atoms, linked together in many different patterns (E)

amount of matter stays the same (by atom count) during changes (I)

SCIENTIFIC IDEAS

MORE GENERAL NOTIONS

∧ "STORY LINES"

combinations of substances have different properties

everything could be made from a few substances combined in different ways

visible things might be made of huge numbers of invisibly tiny pieces (D)

structures weigh the same as the total of their parts' weights

(H)

SYSTEMS

amount of matter stays the same (by weight) during changes (G)

substances have properties

small parts can be put together to make all sorts of different things (C)

magnifiers often show that things have unexpected small parts (B)

all materials come from somewhere and go somewhere (F)

water disappears into the air (A)

∧
PROPERTIES OF SUBSTANCES

∧
COMMON INGREDIENTS

∧
INVISIBLY SMALL PIECES

∧
CONSERVATION OF MATTER

The cluster of boxes in the *invisibly small pieces* story line begins with three ideas. One is that water seems to disappear, a phenomenon young children can observe (A). A second idea is that magnifiers often show that things have unexpectedly small parts, which young children can learn from experience (B). Third, there is the idea that small parts can be put together to make all sorts of different things, which children can learn by working with interlocking blocks (C).

All three ideas contribute to understanding the more advanced idea that things are made of huge numbers of invisibly small pieces (D). This story line, along with several others, prepares students to learn a central idea in understanding the structure of matter–that everything is made of invisibly small atoms, linked together in many different patterns (E). At this

box, all four story lines converge for the first time.

In the *conservation of matter* story line, students must first acquire the notion that all materials come from somewhere and go somewhere—and do not just appear or dissappear (F). That helps students understand the idea that the amount of matter stays the same (by weight) during changes (G). This idea is supported by the notion of systems (H)—that all systems have inputs and outputs that can be accounted for.

Once these ideas are in place, students can begin to gain an even more sophisticated understanding—that the amount of matter stays the same during changes, not just experimentally by weight, but conceptually and inevitably—because the same collection of atoms is just rearranged (I).

The team members looked to published research findings for help. From the very first summer, the teams were introduced to the research on children's learning in science, mathematics, and technology through the work of several prominent researchers in the field. We began to collect and distribute to the teams research articles on children's ideas about various *SFAA* topics (see Chapter 15: The Research Base). Where research was lacking, the teams relied on their own experience with students. Indeed, part of the rationale for having school-based teams was to enable team members to probe the understanding of their own students. When experience with a topic was lacking, help was sought from more experienced colleagues.

As a result of their experience in mapping, the team members came to appreciate the value of working in K-12, cross-discipline groups. Secondary-school teachers could provide some background about the science; elementary-school teachers brought insights about the ideas of young children. When someone would assert that a concept was possible for all 5th- or 8th-graders to know, others learned to ask for evidence. All the teams began to think about what would count as evidence of student understanding.

A Common Set of Benchmarks

For two years, the teams worked independently to map many *SFAA* sections, representing their thinking in a variety of forms. Some used commercial graphics packages to display connections among topics; some used Hypertext outlines that displayed only a few connections at once; others used charts or lists of "student statements" that tried to capture how students might express their knowledge. But whatever form they used, the teams were convinced that mapping raised basic questions about sequencing of instruction that they would have to answer in designing their K-12 curriculum models.

By 1991, three major forces had converged to encourage us to write a *common* set of expectations for students. First, the teams' progress on curriculum models was being held up by the shortage of maps. Their need for the antecedent goals contained in maps or charts (on which their curriculum plans depended) was outpacing their rate of production of maps. The analytical nature of the work was "mind frying" as one team member described it, so groups could spend only a few hours at a time at the task. And district pressures were mounting for teams to share their curriculum models, so that Project 2061 ideas could find their way as soon as possible into classrooms. Second, there was a growing demand for the maps themselves. The teams and staff noted that audiences at Project 2061 presentations were intrigued by the thinking that went into making maps and wanted to use them for their own curriculum-reform efforts. Finally, the teams were finding a need for more polished statements for colleagues at their own sites to use while developing curriculum. The draft maps and charts, while stimulating the thinking of map makers during a session, weren't nearly as helpful to those who had not participated.

307

The question arose whether teams needed to agree on a common set of student expectations to be refined for public use. Analysis of each other's work convinced teams that it was both possible and desirable.

In what form could the thinking of over two years of work (plus the prior three years in developing *SFAA*) be captured helpfully for others? As the teams, consultants, and staff struggled to agree on a format that would do justice to the substance and thought, several issues arose—such as whether to use maps or lists or essays or some combination to represent the progression of student understanding and which grade levels to use as checkpoints (see Chapter 14: Issues and Language). Suffice it to say that discussions were long and intense. Although arguments were made for both sides of issues, decisions leading to the existing benchmarks were based on the staff's desire to maintain the "less is more" focus of goals while fostering creativity and variety of means. Based on the maps and charts of the six teams, a common set of benchmarks was drafted by spring 1992, reviewed by the teams and consultants, and then revised accordingly. Use of the benchmarks led to further revisions in form and substance.

In the months prior to the summer 1992 meeting, "Science Education Standards" got on the nation's education-reform agenda, and the National Academy of Sciences—through its National Research Council (NRC)—was asked to develop standards for education in science. In spite of different notions as to what standards might mean in the context of science education, all advocates seemed to want, above all, recommendations for "what students should know and be able to do at certain grade levels"—precisely what Project 2061 had been formulating as benchmarks.

The national interest in standards presented Project 2061 with a welcome opportunity: By providing the NRC working group with the June 1992 draft of *Benchmarks*, we were able to contribute to their thinking and help shape national standards.

Nationwide Review

In early 1993, the draft benchmarks were ready for the scrutiny of groups and individuals knowledgeable and concerned about science education. The Project convened its advisory board, the National Council on Science and Technology Education, in November 1992 to get its recommendations on conducting a thorough and meaningful review. The Council suggested important constituencies whose opinions should be solicited, and suggested questions important for reviewers to consider. The draft version of *Benchmarks for Science Literacy* was sent out in January and February to thousands of potential reviewers, who were asked to appraise the technical accuracy of the science in the document, the necessity and sufficiency of precursors provided to anticipate later concepts, the appropriateness of grade-level placement, the acceptability of the language, and the overall usefulness of the benchmarks.

The Project was particularly interested in group review, through which *Benchmarks* would be subject to discussion. Members from each of the Project 2061 teams conducted review sessions in which participants could debate the strengths and weaknesses of the draft. In addition, over 75 volunteer groups from around the country responded to an invitation in the Project newsletter to review the document. Many of these volunteers, as it turned out, were educators who were involved in revamping their district curricula and hence were eager to get ideas for their own schools as well as influence a national project.

Reviews were solicited from scientific and educational organizations, which were encouraged to assemble review teams to debate and discuss the draft. Other reviews came from state curriculum groups, research and development centers, regional educational laboratories, the American Association for the Advancement of Science sections, Project consultants and affiliates from all phases of the Project, publishers, and individual scientists and educators in science, mathematics, and technology. Several states accepted the Project's invitation to set up teams to review the draft document. The Project received over 1,300 responses from groups and individuals throughout the country. Those responses came from 46 states; urban and rural school districts; elementary-, middle-, and high-school teachers; scientists and engineers from diverse disciplines; and groups traditionally under-represented in mathematics, science, and technology.

Often the reviewers pointed out issues that had been debated during development of the draft—refreshing the debate yet again. Sometimes the reviewers had policy differences with the Project (such as suggestions that we should not include mathematics, the social sciences, or technology). Some of their comments made us realize that certain sections should be rewritten more clearly to help readers understand our intent.

After a rewrite based on all the comments, leaders of the Project teams' review groups and consultants from mathematics, science, social-studies, and technology education convened to discuss still-unresolved issues. The National Council on Science and Technology Education likewise met to discuss the rewrite before publication.

These benchmarks have been written and rewritten, appraised and reappraised, and argued about at length by the Project teams, staff, and consultants. After all of this sustained, hard work, we could yield to the temptation to fold our arms and say, "We did our best; job done." But because we want the document to be as helpful as possible, we expect to update it periodically. Field use of the benchmarks over the next few years and continuing research on how children learn will no doubt tell us more about which benchmarks should be shifted, eliminated, elaborated . . . or even left alone. ∎

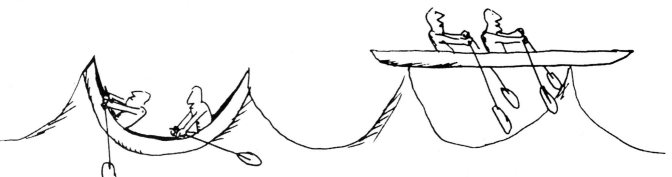

PAUL KLEE, *Mural from the Temple of Longing Thither, 1922/30.*

Published reports typically imply that their recommendations were arrived at quickly and without confusion or dissent. But smooth sailing does not describe Project 2061 as it wrote *Benchmarks for Science Literacy* (or, for that matter, its predecessor, *Science for All Americans—SFAA).* There was much debate on substance and style, on meaning and language, on organization, and on what was missing and what was superfluous. The first part of this chapter describes some of the issues debated by the Project 2061 team members, consultants, and staff during the three years it took them to formulate the benchmarks. The second part has to do with our desire to be clear on what meaning we attach to the language we use.

Essays or Lists

As pointed out in Chapter 13: The Origin of Benchmarks, the early work that led eventually to the benchmarks started out as "progression-of-understanding" mapping. It became convenient later to collect the entries from the maps into lists and then enlarge the lists without always writing out maps in full detail. As the lists grew, so did the conviction among members of the six Project teams that the lists by themselves were too stark, too isolated from instructional issues, and not clear enough about what "knowing" means in different cases.

The teams developed three kinds of materials—maps, lists, and essays—suitable for their task but not necessarily for publication. Maps were not the best choice for publication, for although they were central to developing the benchmarks, they were far too

specific and complex and would require much written explanation anyway. The lists and essays presented two important problems, one substantive, one organizational.

The substantive problem was this: Essays might give the publication the appearance of a curriculum guide, rather than a statement of goals supported by commentary. But readers wanting more than what a brief essay provides might then think the report disappointing and inadequate as a guide. We hope our introduction and our efforts to clarify what is and is not a benchmark will lessen this danger.

The organizational difficulty was deciding what constitutes a benchmark in *Benchmarks for Science Literacy.* Are they individual listed items or groups of items with a companion essay? In the first view, for instance, we would say that 5C: Cells is composed of 17 benchmarks distributed across the grades. In the second view, we would say there are four benchmarks, one for each grade span, each one a set of items-plus-essay. The conceptual difference here is not trivial. It seriously affects how *Benchmarks* will be used. The individual items are not to be taught or tested. They help guide the development of a curriculum. In developing learning experiences (curricula), designers will select benchmark items from different sections of a chapter and from different chapters. This is difficult if one must move an entire list and its essay from place to place. Moreover, it is awkward to talk or write about benchmarks when the term refers to a report rather than referring to the parts that make it up. We are putting *Benchmarks* on computer disk to make that sort of moving, talking, and writing easier.

311

Characterizing Knowledge

In writing *Benchmarks*, we found that language issues emerged in two different but related contexts, one scientific, the other educational. The first has to do with the terminology used in the benchmarks to indicate what is to be learned by students, the second with how best to indicate the sense in which students are to acquire the benchmarks.

SFAA uses only those technical terms that scientists believed ought to be part of every adult's vocabulary. The clear purpose was to free teachers from spending most of their time and energy teaching science vocabulary and let them concentrate on teaching meaningful science. The pressure to cover the curriculum and test the students often leads people—teachers, administrators, test makers, and parents—to be willing to accept the glib use of technical terms as evidence of understanding. Students will soon forget all of those technical words anyway. Few adults can confidently distinguish between revolve and rotate, reflect and refract, meiosis and mitosis, mass and weight, orders and families, igneous and metamorphic rocks, nimbus and cumulus clouds, mitochondria and ribosomes.

In that spirit, *Benchmarks* was faced with a more difficult language problem in trying to convey accurately what children in the lower grades should learn. It wouldn't do just to match the children's language exactly—*Benchmarks* is for educators, not students—yet to use the adult technical language of *SFAA* could encourage its premature teaching. The solution was to try to say in plain English what the quality of the learning should be and use technical terms only when it was time to make them part of a student's permanent vocabulary. For example, with regard to Section 4B: The Earth, a K-2 benchmark is given as "Water left in an open container disappears, but water in a closed container does not disappear,"

rather than as "Water in an open container evaporates".

On the education side, the issue is somewhat different. As noted elsewhere, *Benchmarks* uses "know" and "know how" to lead into each set of benchmarks. The alternative would have been to use a finely graded series of verbs, including "recognize, be familiar with, appreciate, grasp, know, comprehend, understand," and others, each implying a somewhat greater degree of sophistication and completeness than the one before. The problem with the graded series is that different readers have different opinions of what the proper order is. We decided, after trying many possibilities, that the best approach is to use one verb universally—"know"—and then specify more accurately just what the student "knows." The figure to the right illustrates the proposition.

Another alternative would have been to use a potentially large set of action verbs that describe different *observable behaviors*—for example: "The student should be able to describe (or explain, give examples of)." An advantage of using action verbs is that they persistently imply that mere recitation of memorized facts will not be acceptable. Because of the strong tendency for any set of goals to regress into catechism, this advantage was seen as highly important by team members. They wanted to constrain the methods of instruction and assessment used by teachers to reach the benchmarks. Indeed, action-verb statements may be tantamount to assessment items. If the major use of benchmarks had been to construct tests, action verbs would have provided the best chance of making the test fit the goal. But that was not the intent.

A disadvantage of using action verbs is that the choice among them is arbitrary. Although sometimes there is a clear difference in intent between, say, "give examples of" and "explain," often they would be equally reasonable ways to exhibit the same understanding.

Option 1   GRADED WORDS, SAME TARGET

     BE AWARE that   atoms are made of electrons, protons, and neutrons.

     KNOW that   atoms are made of electrons, protons, and neutrons.

     UNDERSTAND that   atoms are made of electrons, protons, and neutrons.

Option 2   SAME WORD, GRADED TARGETS

     KNOW that   **atoms have parts.**

     KNOW that   **atoms are made of electrons, protons, and neutrons.**

     KNOW that   **atoms have a nucleus of neutrons and positive protons, surrounded by a cloud of negative electrons.**

Using any particular action verb would be limiting and might imply a unique performance that was not intended. Moreover, unless the specific action is exactly the outcome desired, use of the verb results in ambiguity about just what underlying knowledge was intended. We avoided action verbs for this reason.

For example, consider a benchmark: "The students should know that scientific problems have sometimes led to development of new mathematics." For those who prefer behavioral language, a reasonable substitute would be "The students should be able to give examples of cases in which a scientific problem led to new mathematics." But is that performance the real goal? A "reflective literacy" form of the goal would be more like, "When reading in the paper about a new scientific breakthrough, the student should wonder whether new mathematics had to be developed to solve the problem," although it is of course difficult to ascertain what is in a student's mind.

Perhaps somewhere along the line students would have to have been able to give examples to show that they had acquired understanding of and belief in the generalization. Once learned, however, the generalization might come to mind even if all the examples had been forgotten. The active availability of the generalization might better be assessed by an essay question, such as "What can lead to the development of new mathematics?" for which a satisfactory answer would have to include mention of new scientific developments. Unfortunately, such an answer could be learned by rote—but then, just as undesirable, so could a couple of relevant examples.

"Grain Size"

How big should a benchmark be? At one extreme, it might be a simple declarative sentence. At the other, it might be a paragraph of description and explanation. *SFAA* was mostly written in the latter form, but the general sentiment for *Benchmarks* went the other way, especially for the lower grades. Simplicity favors precision; complexity allows coherence.

For a while, two principles were used in making case-by-case decisions on benchmark grain size. One principle was that putting small ideas back together into bigger ideas is easier for most users than trying to grasp all the complexity of a bigger idea. The second principle was that an idea should be separated into parts if it is likely to be learned in parts. The parts would not be learned in isolation, of course, but in rich and meaningful contexts.

For example, "Atoms are made of electrons, protons, and neutrons" might be divisible into two parts: "An atom has a center, or nucleus, surrounded by electrons" and "An atomic nucleus consists of protons and neutrons." These two aspects of atomic structure are learnable separately in that sequence, as history shows. If a teacher wishes to bypass the historical sequence and teach or assess the structure of the atom as a unity, then adding the separate statements together is no problem. It does not make sense, however, to divide the second statement again, into "An atomic nucleus contains protons" and "An atomic nucleus may contain neutrons." These are more easily learned together.

The example also shows how a small grain size makes intent more specific. A benchmark stating summarily that the student should "know the structure of atoms" gives no clue about how elaborate that knowledge should be. If the statement is divided into components as above, the reader better understands the writer's intended level of sophistication.

Authors of the early *Benchmarks* drafts attempted maximum separation, only to find that listing all components separately does little to advance the coherence of the ideas. *SFAA* and *Benchmarks* both claim to describe a coherent fabric of understanding, not a collage of bits and pieces. So there was also some justification for emphasizing the coherence of benchmarks by grouping them into related sets. This was done. Unfortunately, grouping benchmarks to increase cogency can also result in distorted, diluted, or lost meanings, and where this appeared to be a danger, we went back to greater division of the ideas.

For example, consider the K-2 statement

> People can often learn about things around them by just observing them carefully, but sometimes they can learn more by doing something to them and noting what happens.

This was meant mostly to encourage exploration and to distinguish between passive observation and experimental intervention. Some team members wanted to increase the statement's cogency by connecting it to other ideas to make a larger grain size, so that the statement would become

> People can often learn about things around them by just observing them carefully, but sometimes they can learn more by doing something to them and noting what happens. People try to explain things as well as tell what they see. "Seeing what happens" can provide evidence for their explanation.

There is clearly a shift from exploration to explanation and hypothesis testing. But will the exploration get lost? Will poking at things be seen as legitimate only when it serves to test an explanation? And doesn't the addition make the benchmark more sophisticated and no longer appropriate for the K-2 span? In this instance, the decision went against the larger aggregation, but in others the decision went in the other direction. In general, this first edition of *Benchmarks* employs a moderate amount of grouping of fine grains, hopefully giving an optimal balance of grouping and separation.

Grade Levels

When Project 2061 began formulating benchmarks, the question of which benchmark fits which grade level naturally arose. One school of thought argued that there should be no grade or age specifications at all, but rather attainment levels. Different students progress at different rates and so to set up any checkpoints would do a disservice to very slow and very fast students. This position was especially attractive to the teams exploring the possibility of ungraded curricula. Another school of thought opted for benchmarks at grades 4, 8, and 12 on the grounds that that was the way the nation was headed, and Project 2061 would risk being ignored if it did not get in step. Variations included starting with preschool, instead of kindergarten, or ending at grade 10 or 14, instead of grade 12.

As work progressed on curriculum models, the need to sort the benchmarks into some few bands related to cognitive and psychological child development became clear to all six of the Project teams. Independently, the teams concluded that grades K-2, 3-5, 6-8, and 9-12 were acceptable proxies for early childhood, childhood, early adolescence, and adolescence and young adulthood—as long as the boundaries were not rigid. This agreement allowed work to go forward apace, but the comments of some reviewers later suggested that there is no consensus among educators on the issue.

Connections

A central Project 2061 premise is that the useful knowledge people possess is richly interconnected. *SFAA* speaks of the physical setting and the living environment, not chemistry, physics, geology, astronomy, or biology, and it asks for students to become aware of the similarities between the natural and social sciences and to learn about some of the interdependencies of science, mathematics, and technology. The curriculum need not be totally integrated to achieve those ends, but we are aware of the division of opinion on how strongly the Project should press for integrated or interdisciplinary studies, if at all. This issue will be explored in detail in *Designs for Science Literacy*.

There is another sense in which the project is concerned with connections—namely those from one grade span to the next. On any given topic, the benchmarks in grades 3-5 usually build on benchmarks already stated for K-2 and will contribute to the benchmarks stated later for grades 6-8. Such connections have been called "strands" or "story lines" in the curriculum. A concept may increase in sophistication as it proceeds linearly from one grade span to the next; two or more concepts at one level may converge at the next level to form a more complex idea; or a concept may, at the next level, connect to two or more others. For example:

> The understanding that *different plants and animals have external features that help them thrive in different kinds of places* can later be extended conceptually to *individuals of the same kind differ in their characteristics and sometimes the differences give individuals an advantage in surviving and reproducing.*

Knowing that magnifiers *often show that things have unexpectedly small parts* and that *water disappears invisibly into the air* helps to understand, at the next level, that *visible things might be made of huge numbers of invisibly tiny pieces.*

The idea that *all substances are made of invisibly small atoms that can recombine in various ways* contributes, at the next level, to understanding both why *the total amount of matter stays the same during changes in substances* and how *everything can be made from a small number of elements recombined in different ways.*

When benchmarks were represented as progression-of-understanding maps, the connections were explicit and obvious. In one early version of *Benchmarks*, small codes were used to identify the story-line relevance of each listed statement, but some users found the coding device to be more distracting than helpful. The codes were dropped because readers found them to be confusing and because they did not identify cross-chapter connections, but the problem of how best to highlight the vertical connections among the benchmarks remains. At issue is priorities: Some team members and consultants believe that in principle (at least for Project 2061) emphasizing connections is so important that a high priority should be given to finding a satisfactory solution, whereas others regard it as of less importance, believing that the curriculum developers themselves should undertake the task of creating strands of connected benchmarks. The solution adapted—at least for this first edition of *Benchmarks*, was to avoid complicated coding and to indicate some of the main interconnections by cross-references in the margins. Feedback from actual use will be considered in deciding how subsequent editions will deal with the connection issue.

Vocabulary

Education lacks a true technical vocabulary. There are few terms that all practitioners in the field agree upon, and hence precision is difficult to achieve in education writing. Jargon, on the sunny side called "technical language," exists in education, as in most fields, and borrowing terms from others or coining new ones is an honorable, if often abused, practice not restricted to education.

The intent of this Project 2061 report is to say something important about education and say it clearly, not to expand literary horizons. Since education is everybody's business, it is important that *Benchmarks for Science Literacy* be understandable by the general public as well as by teachers and other educators—an admirable aim not easily attained. The difficulty, of course, is that many education-related words in everyday use convey different meanings to different people, that different educators assign different meanings to the same term (even in their journals), and that many education words have different meanings for educators and the public. Consider some typical examples of very popular terms for which there is little agreement on precise meaning in or out of the education community: "choice," "know," "basics," "literacy," "standards," "reform."

The difficulty goes beyond the imprecision of the education vocabulary. Part of the trouble stems from the widespread use of catch phrases in education discourse. The popularity of such phrases is not surprising, for like slogans, they often condense a whole argument or point of view and have political wattage. "World-Class Standards," "Less Is More," "Hands-on Science," "Authentic Testing," "Outcomes-Based Education, " and "Systemic Reform" are current examples; "The Whole Child" and "Schools Without Walls" not very distant ones. They do have meanings and can sometimes be used, but unless carefully defined, their meaning is vague and tends to stifle critical discussion. Besides, they go out of favor about as fast as they come in.

Project 2061 responds to the language problem in two ways. We have tried to convey each idea with as few terms as possible and then use the terms consistently. For instance, *Benchmarks* uses "know" throughout and does not try to select a term having just the right shade of meaning from among "recognize," "be aware of," "be familiar with," "grasp," "understand," and other possible synonyms. Also, we have tried to make clear what we mean by the terms we do use—especially those we believe might otherwise be misconstrued or that are particularly important to our story.

The terms and phrases discussed below are mostly those that have a prominent place in the titles and content of our publications or in our descriptions of Project 2061. We have included a few catch phrases that we use from time to time, and a few that we do not use, sometimes to the dismay of their advocates.

All. The "All" in *SFAA* is intended to emphasize inclusiveness as strongly as possible. No individuals or groups are to be excluded from an opportunity to become science literate, nor are any to be presumed unable to become science literate. We believe that the science, mathematics, and technology understandings and skills spelled out in *SFAA* and *Benchmarks* are within the reach of all but the most severely mentally and emotionally handicapped individuals. To realize that goal, however, it will be necessary to redesign the basic curriculum, change teaching practices, and reform many other parts of the school system.

Still, in the real and imperfect world, "all" cannot possibly be absolute. When pressed for an operational definition, we have settled for "at least 90% of all future adults will have acquired at least 90% of the knowledge and skills recommended in *SFAA*." However, the main reason for adopting an optimistic goal—all—is to ensure that no students are preemptively deprived of the chance to receive a basic education in science, mathematics, and technology.

Back to Basics. Insofar as it means that the focus of schooling should be on those ideas and skills that most enable students to live interesting and productive lives, "back to basics" is an excellent slogan. Agreement, however, on what those basics are has not been easy to come by. Clearly, what was good enough for most of this century is no longer good enough and will surely be less so in the next century. Science and technology have changed what can be taken to be basic. If "basic" can include scientific reasoning, elements of the picture that scientists paint of how the world works, notions of what mathematicians do, and how the technological world relates to science and to society, then let us go "Forward to Basics."

Benchmarks. According to *Websters's New World Dictionary* (1984): "**bench mark 1.** a surveyor's mark made on a permanent landmark of known position and altitude; it is used as a reference point in determining other altitudes **2.** a standard or point of reference in measuring or judging quality, value, etc. . . . " Given the second meaning, the bench mark metaphor seems apt enough to justify its use as a name for the goal statements set out in this report. The Project 2061 benchmarks (many dictionaries now use the single-word form) are offered as reference points for analyzing existing or proposed curricula in the light of science-literacy goals.

Blueprints. For any good-sized building these days, there is a virtual library of elaborately detailed blueprints for the foundation, the structural supports,

317

the electrical system, ventilation, and so on. The blueprints we have in mind, however, are more like architects' sketches (the ones with the little trees and people) of what new buildings would look like all together and how the various parts relate to each other. For each component of the education system, *Blueprints for Reform* will sketch out what the major issues are, identify what needs to be done to support the proposed Project 2061 curriculum reforms, and suggest some strategies for effecting those changes.

Core Curriculum. The idea of a core of essential studies for all students is an ancient one, the fragmentation of the curriculum and the appearance of different expectations for different classes of students being rather recent developments historically. In part, that change was a response to the need in increasingly technological and increasingly democratic societies for more people to become educated. As education expanded, it could no longer be assumed that what was best for an elite class, ruling or otherwise, was also best for everyone else. In today's world, nations struggle to create school systems that balance the needs of society with those of the individual, academic considerations with vocational ones, and basic education with education focused on special needs and talents.

Project 2061 believes that a decent balance of those interests can in fact be achieved. It has set out, however, to focus primarily on that part of the curriculum that concerns itself with the science, mathematics, and technology learning expected of all students—the science-literacy core. In *Designs for Science Literacy*, Project 2061 will discuss the relationship between the science literacy core and the arts and humanities core and the part of the curriculum that goes beyond the core.

Curriculum. "Curriculum" is an everyday word that is not easy to tie down. For some educators, it refers to a general sketch of what should happen in schools. For others, it refers to the day-by-day experiences that students actually have. Papers in the journals speak of a "planned" curriculum, which is different from the "taught" curriculum (the instruction actually delivered to students), which is different from the "learned" curriculum (what students actually learn). Project 2061 makes no such distinctions, using the term in all of these senses—curriculum as planned by teachers and administrators, as delivered by teachers, and as experienced by students; curriculum as an overview of the scope and sequence of student experiences; curriculum as a detailed delineation of learning experiences. Context makes it clear enough which specific meaning is intended.

No matter how curriculum may be defined, curriculum reform is a key element in the overall Project 2061 reform strategy. Is Project 2061 therefore a curriculum project? Not in the usual sense of designing a single curriculum (K-12, in our case), or producing materials to support the curriculum, or setting up implementation sites. But Project 2061 is formulating a variety of tenable K-12 curriculum plans (called models in the architectural and scientific sense of the word) that local school districts can use, along with other Project 2061 tools, in designing a curriculum that takes into account local circumstances and state requirements and meets the science literacy goals set out in *SFAA*.

The argument is frequently heard that reforming the curriculum will not result in significant and lasting K-12 reform. That is true, just as it is true for *any other single change*. That is why Project 2061, through its *Blueprints for Reform*, is looking at the other aspects of the education system in conjunction with its efforts to promote curriculum reform. These are described in Chapter 16: Beyond Benchmarks.

Curriculum Blocks. Today's curriculum, as planned and delivered, is composed of building blocks that are nearly standard in duration. Subjects in the lower grades and courses usually involve daily classes and run for a semester or year. Traditional "units" are typically about an hour a day for two to four weeks, and one after another in a fixed sequence. Project 2061 believes that greater time flexibility is desirable and so will design an illustrative variety of units that would range from a few minutes a day for several years (such as recording weather data) to all day for several weeks. We use the term "curriculum block" for this variable chunk of curriculum time.

Furthermore, traditional curriculum blocks are all much alike in form, essentially didactic. But many things can be learned better through projects (inquiry and design), seminars, and independent study. Project 2061 curriculum blocks will therefore vary in form as well as in duration and allow some variations in sequence. We intend each block to provide students with some coherence of purpose beyond the usual "topics," "units," and courses. For more on their nature and use, see Chapter 16: Beyond Benchmarks.

Curriculum Models. There are many plausible ways to schedule learning over 13 years of school that result in all students reaching the science-literacy goals recommended in *SFAA* and *Benchmarks*. To illustrate the variety of possibilities, Project 2061 is working out a few alternative models of curricula. Each will be a sketch of a K-12 curriculum with a rationale for how the various curriculum blocks were selected and organized. Project 2061 will describe several models in *Designs for Science Literacy* to illustrate what the possibilities are. In education circles, a "model curriculum" is exemplary practice that can be visited and observed.

A Project 2061 curriculum model is more an architectural rendition or a scientific model than a reality to be copied.

Habits of Mind. The education literature speaks of "scientific attitude," "thinking skills," "higher-order thinking skills," "quantitative reasoning," "number sense," "estimation skills," "calculator and computer skills," "problem-solving skills," and "decision-making skills." These all have to do with bringing the intellect to bear on practical matters in one way or another. Although each has its own nuances, Project 2061 has elected to use the term "habits of mind" to cover that whole territory and a little more in addition. The additions include manipulation skills, communications skills, and critical-response skills.

Project 2061 takes the position that skills have limited problem-solving value in the absence of the ability to use them knowledgeably, just as knowledge has limited problem-solving value in the absence of the ability to apply it skillfully. Knowledge and skills are both essential and can be learned together. In fact, they should be learned together most of the time. In *SFAA* and *Benchmarks* they are in separate chapters, but *Designs for Science Literacy* will describe how to merge them, and almost all the individual Project 2061 curriculum blocks will engage students in gaining knowledge and skills simultaneously.

Hands-On. In their understandable eagerness to move away from the dull bookishness of so much science and mathematics teaching, educators and others have pressed for more "hands-on" learning, sometimes implying, however, that students will learn well only by manipulating physical objects. Hands-on experience is important but does not guarantee meaningfulness. It is possible to have rooms full of students doing interesting and enjoyable hands-on work that leads nowhere conceptually. Learning also requires reflection that helps students to make sense of their activities. Hands-on activities contribute most to learning when they are part of a well-thought-out plan for how students will learn over time.

Interdisciplinary. With relation to curriculum, this term is used to refer to many different possibilities, from mere "coordination" among separate disciplines, to courses made up of still-identifiable chunks of two or more disciplines (as in many general science courses), to courses that are integrated around topics and issues that cut across many disciplines. For many educators, the relationship is usually confined to disciplines within a domain (chemistry and biology, for example); for some it is more adventurous (say, physics and geometry, or geology and economics); and for still others it is even more sweeping, reaching into the arts and humanities. As far as Project 2061 is concerned, students' actual learning experiences can occur in totally integrated contexts, or in segregated subjects, or (more likely) in a great variety of possible mixes of both—as long as they result in the achievement by all students of the science-literacy goals of *SFAA* and *Benchmarks*.

But with regard to curriculum, there are three senses in which Project 2061 is solidly behind integration. First, *integrated planning*—the curriculum in science, mathematics, and technology (and perhaps more) should be the result of the collaboration of teachers from all the relevant subjects and all grade levels, not a parceling out to grade and subject specialists. Second, *interconnected knowledge*—the students' experiences should be designed to help them see the relationships among science, mathematics, and technology and between them and other human endeavors. Third, *coherence*—the students' experiences need to add up to more than a collection of miscellaneous topics, whether under themes (everything about, say, salmon), discipli-

nary subject headings (Principles of Chemistry), or activities ("neat things for kids to do").

Know. As noted earlier in this chapter, *Benchmarks* uses "know" throughout where others might use "learn," "recognize," "be aware of," "be familiar with," "grasp," and "understand" to indicate the particular state of knowing they have in mind for each recommendation. We urge readers to substitute at will. The rationale for this can be found in the Characterizing Knowledge section of this chapter.

Such a cavalier approach does not, unfortunately, quite get Project 2061 out of the woods. Does Project 2061 have something special in mind, it is fair to ask, by "know" (and its cognate, "know how," when referring to skill acquisition), since that is its term of choice? The answer is mostly, no. The standard dictionary definitions will do: To know is to apprehend something with clarity, or to possess specified knowledge; to know how is to be capable of doing something, to possess certain skills. Project 2061 adds one condition, namely that what counts is lasting knowledge and skills, not just what one would know or be able to do *at the moment* of completing school or reaching any particular grade level. Such lasting outcomes may be difficult to test for but that is not sufficient reason to be satisfied with transient learning.

Less Is More. This slogan asserts that there is more profit in learning fewer things better than in learning more things poorly. The hundreds of Project 2061 benchmarks led some reviewers of the draft manuscript to ask how that could be "less." *Benchmarks* demands more of students than is now customary—more depth, more connectedness, more relevance. But it demands of them far less memorization of isolated facts and concepts than the great compendium of miscellaneous topics confronting them today in the required science and mathematics curriculum. Learning important ideas in any useful way simply takes more time than has usually been assumed, at least in part because many ideas in science and mathematics are abstract and not in accord with everyday experience.

Long Term. Project 2061 bills itself as long term, by which it means to signal that it recognizes that significant reform takes time and defies quick fixes, and that the Project itself will participate in the reform movement as long as it is needed. One positive consequence of this approach is to enable the work of the Project to be approached thoughtfully and systematically with the continuous involvement of educators, scientists, and interested citizens; a less-positive one is that its contributions stretch out over a considerable length of time, even though the need for them is right now. But long term does not mean that the project's work will not be finished until 2061. See **2061**, below.

Mathematics. Mathematics is the oldest branch of science, yet it is just as modern and lively as any other field of science. It is an enterprise in its own right and contributes to advances in nearly every domain from the arts and humanities to business, engineering, and the other sciences. But it is first a science—the science of patterns is how the Mathematical Sciences Education Board of the National Academy of Sciences has put it—and it has undeniably intimate connections with the rest of science. Whatever else it is, mathematics is part of the scientific enterprise and hence part of what constitutes science literacy.

Reform of mathematics education is generally being led by the National Council of Teachers of Mathematics (NCTM), which published *Curriculum and Evaluation Standards for School Mathematics* shortly after *SFAA* was published. The two studies were conducted independently, so it is encouraging that they largely agree. Project 2061 has endorsed the NCTM report.

Precursor. Project 2061 uses the term "precursor" in its ordinary dictionary meaning, that is, as something that precedes and is the source of something else. In *Benchmarks*, a precursor is a statement of an idea that leads directly or indirectly to one of the adult-level understandings in *SFAA*. Precursors for the earlier grades use simple, nontechnical language to indicate how young children might express their understanding; later precursors use technical terms when it is believed that students can use them with understanding.

Some precursors are simply less-sophisticated versions of later goal statements. For example, understanding that in a chemical reaction the total amount of material stays the same *by weight* comes before knowing that in such reactions the total amount of matter remains the same *by atom count*. Some precursors contribute to one or more adult goals but may be more or less distinct from them. For example, knowing what fossils are and how they are formed precedes understanding that fossils provide evidence for descent from common ancestors.

Reform. Revolution, reform, reorganization, restructuring, revision, revitalization . . . . There seem to be many words for reform, but in one way or another they all suggest radical change—well, more or less radical. The Project 2061 view was expressed in the introduction to *SFAA*: Sweeping changes in the entire educational system from kindergarten through 12th grade will have to be made if the United States is to become a nation of science-literate citizens.

Science. By "science," Project 2061 means basic and applied natural and social science, basic and applied mathematics, and engineering and technology, and their interconnections—which is to say the scientific enterprise as a whole. The basic point is that the ideas and practice of science, mathematics, and technology are so closely intertwined that we do not see how

321

education in any one of them can be undertaken well in isolation from the others. The title *Science for All Americans* was selected for reasons of economy: the alternative, "Social and Natural Science, Mathematics, and Technology for All Americans," not to mention plugging in "basic and applied" frequently, seemed altogether too clumsy and, if used in the text, needlessly tedious for readers.

Science Literacy. A literate person is an educated person, one having certain knowledge or competencies. But of course the rules keep changing with regard to precisely which knowledge and competencies define literacy—the ability to write one's name and read a simple prose passage long since having been replaced by more demanding requirements. In today's world, adult literacy has come to include knowledge and competencies associated with science, mathematics, and technology. Project 2061 has undertaken, in *SFAA*, to identify the knowledge and habits of mind that people need if they are to live interesting, responsible, and productive lives in a culture in which science, mathematics, and technology are central—that is, to describe what constitutes the substance of science literacy.

People who are literate in science are not necessarily able to *do* science, mathematics, or engineering in a professional sense, any more than a music-literate person needs be able to compose music or play an instrument. Such people *are* able, however, to use the habits of mind and knowledge of science, mathematics, and technology they have acquired to think about and make sense of many of the ideas, claims, and events that they encounter in everyday life. Accordingly, science literacy enhances the ability of a person to observe events perceptively, reflect on them thoughtfully, and comprehend explanations offered for them. In addition, those internal perceptions and reflections

can provide the person with a basis for making decisions and taking action.

Sites. There are six school-district sites in the United States that collaborate with Project 2061 in fostering educational reform that will eventually lead to nation-wide science literacy. They function as centers where Project 2061 team members explore new curriculum design possibilities, create tools for curriculum redesign, identify high-quality learning and teaching materials related to science literacy, generate ideas for curriculum blocks, test innovative techniques and technologies, and develop and implement strategies for system reform. But just as Project 2061 is not a curriculum project in the traditional sense, neither are the sites *demonstration* sites where visitors can go to see a Project 2061 curriculum in action, since no such thing exists. See Chapter 13: The Origin of Benchmarks for information on the location of the Project 2061 sites.

Standards. A standard, in its broadest sense, is something against which other things can be compared for the purpose of determining accuracy, estimating quantity, or judging quality. In practice, standards may take the form of *requirements* established by authority, *indicators* such as test scores, or *operating norms* approved of and fostered by a profession.

But that bypasses more interesting and important issues: For what aspects of science education do fully spelled-out national standards make sense? Are there some aspects for which setting national standards is unnecessary or undesirable? What do we have now in the way of standards that educators, scientists, and the public should support? What is already in the works? How can standards best be expressed? Who will monitor national standards? How, and for what purposes?

Extracting a consensus is precisely what is *not* needed, because it would reflect where we are rather than where we should be headed. Moreover, to be of much use, standards must be limited in number and lasting in significance. In that way, standards will free educators to concentrate on the quality of student learning rather than on its sheer quantity.

Systemic Reform. Presumably this term, now widely used in education, is intended to refer to reform initiatives that pertain to whole systems in contrast to parts of systems. That much is easy. The hard part comes in trying to define the boundaries and the parts of the particular systems that are to be reformed. More than merely an arbitrary collection of elements, a system is a set of interacting, interrelated, or interdependent parts that can be regarded as a collective entity. Furthermore, the individual parts may be objects, processes, ideas, rules, or yet other systems. A local school district is a system, but is the state it is in a part of its system (or vice versa)? And are the universities from around the country that supply its teachers part of the system even though the school district has no control over them? Is a project engaged in systemic reform simply by virtue of its trying to change one or a few parts of the system?

Because systems tend to restore themselves when only one or a few of its parts are disturbed, the Project 2061 position is that for nationwide reform to occur in science, mathematics, and technology education, the curriculum must undergo major changes in content and organization, and everything that interacts with the curriculum must be changed to accommodate the curriculum changes. This includes policy (local, state, and federal), teacher education (wherever it takes place), the design of learning materials, assessment practices, and much more. It was in response to that

belief that Project 2061 has initiated twelve studies that will lead to *Blueprints for Reform.*

Technology. Technology is even older than mathematics and science. Indeed, the latter may both have developed at first in response to the need to build things and solve practical problems, although discoveries in science and mathematics today often precede practical uses. In any case, although technology still has a life of its own, it is becoming much more closely tied to mathematics and science and hence is an essential part of the scientific enterprise. Understanding technology and its connections to science and mathematics is therefore necessary for science literacy. Unfortunately, technology does not have a place in the general curriculum, so academic students fail to learn about technology or develop engineering problem-solving skills. Furthermore, the technology taught in technology-education classes (formerly industrial arts, and before that, "shop") is often so singlemindedly vocational that teachers fail to teach about technology in social or scientific contexts. Project 2061 is trying to help adjust both sides of that equation.

Themes. As used in both *SFAA* and *Benchmarks*, themes identify some ways of thinking that cut across many fields of science, mathematics, and technology. The curriculum should be designed to help students gradually understand and be able to use these themes.

It is not unusual for teachers, curriculum writers, and developers of instructional materials to use themes as conceptual organizers. Some educators have used the topics in Chapter 11 of *SFAA* for that purpose, some going so far as to try to organize all other topics around them. Although we have no objection to such efforts, that was not the purpose we had in mind. In any case, themes in Project 2061 are not meant to be more important than other ideas, but are expected to arise naturally out of the content, rather than be imposed as a contrivance.

Threshold. *Benchmarks* defines a series of attainment thresholds for students as they progress through school. "Threshold" implies more than a partial understanding. It requires that students' understanding be sufficient for them to make sense of what they have already learned and sufficient for them to be able to learn more. Less than the attainment threshold may still be partial understanding, but it is really too little to build on reliably. Project 2061 proposes that all students reach or surpass the thresholds in *Benchmarks.* As a practical matter, the education in science, mathematics, and technology that most students receive today is far below the science-literacy expectations of Project 2061. Until major changes occur in the curriculum and in the way students are taught, any worry that the floor will become the ceiling is premature.

In deciding to express science-literacy outcomes in terms of thresholds, Project 2061 takes a calculated risk. We are as concerned about students who are below average (in any particular ability relevant to science) as we are about average and above-average students. We have set goals that they can all reach. This implies that we expect most students to learn considerably more than the floor recommended in Project 2061. Knowledgeable educators have estimated that no more than a third of students ever exceed any goals that are set for everyone. If so, we might expect two-thirds to fall short of the Project 2061 minimums. One popular strategy is to ask for more than you think you can get and then settle for less. But our strategy is to play it straight: *Benchmarks* specifies what we intend at least 90% of all students (see All, above) to achieve, average students to surpass, and outstanding students to leave far behind.

Tools. In referring to Project 2061 products, the tool metaphor is used to emphasize their instrumental nature. If it is true that prepackaged curricula cannot usually be imported successfully into local school districts and that consequently curricula must be formulated by their users, it is no less true that they need sophisticated resources to do so. Project 2061 is trying to serve their need.

Topic. We use the term "topic" sparingly in Project 2061 because it is open-ended. It allows a wide variety of content to be included under it. We have deliberately made the benchmarks as specific as they are in order to limit the choice of core topics to those ideas that are centrally important. A long listing of topics might encourage retention of material from the present, overstuffed curriculum. For example, if "atomic structure" were listed as a topic for the middle school, some readers might feel justified in including everything that might come under that heading—such as electron shells and electro-negativities—long before they would likely make sense to most students.

2061. Why 2061? The answer involves a couple of coincidences, some imagery, the purposes of the project, and a bit of romance. Coincidence one: The project happened to get under way in 1985, a year in which Comet Halley was in the earth's neighborhood. Coincidence two: The period of the comet, about 76 years, closely approximates the average human lifespan in developed nations. Thus, we can expect about half of the people born in 1985 to live to see the next appearance of Comet Halley—a dramatic reminder that education is for a lifetime. We wanted to keep our purpose clearly before us—namely to help transform the schools of America to enable them to prepare all their graduates to live full, interesting, and responsible lives.

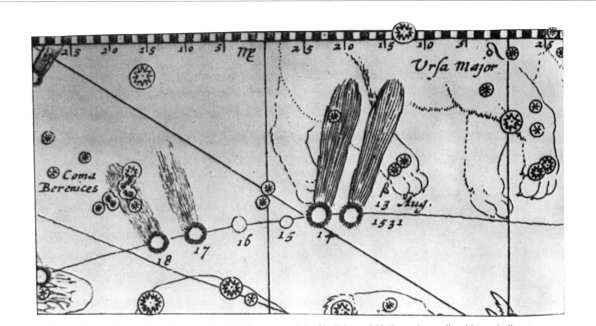

Peter Apian's observation of a comet in 1532 later named for Sir Edmund Halley, who predicted its periodic return.

Now imagine that you could have traveled with Comet Halley, visiting earth at lifespan intervals, starting with the year Edmund Halley first observed it: 1682, 1758/59, 1835, 1910, 1985/86. You would, of course, have seen incredible changes each time: changes on the face of the planet; changes in the life on it; above all, changes in the human population, its size and distribution, its behavior. From Newton to Einstein and beyond in a few lifespans.

At least since Newton's time, most of the dramatic changes in the human situation have occurred primarily because of advances in science, mathematics, and technology. Moreover, most of those changes were unforeseen at the time (and, to be fair, were probably unforeseeable). We have assumed that the same will be true between now and 2061: Science and technology will shape the future, but we cannot predict what the results will be. The comet will return in 2061, no doubt about it—but what will human existence be like?

Project 2061 is predicated on the belief that the quality of life in 2061 will depend above all on the education received by this generation of children and the next. Those young people need to leave school with a solid education in science, mathematics, and technology—one that will enable them to participate intellectually and emotionally in science, the great adventure of our times, and to become responsible and productive members of society. Education must prepare them for an uncertain future and it must include understandings and habits of mind that can serve as tools for thinking throughout life. That is why our full name is "Project 2061: Science Literacy for a Changing Future." ∎

GEORGE SEURAT, *Child In White (Study for a Summer Sunday on the Grande Jatte), 1884.*

Chapter 15 THE RESEARCH BASE

The Role of Research
The Nature of the Research Literature
Research Findings by chapter and section
References

The Role of Research

The task in preparing this report has been to create clear and specific benchmarks and place them at a reasonable grade level. To allow for differences among students, grade spans of 3 or 4 years are used rather than single years, although even then it is difficult to ascertain when students are able to learn particular concepts and skills effectively. Overestimation of what students can learn at a given age results in student frustration, lack of confidence, and unproductive learning strategies, such as memorization without understanding. Underestimating what students can learn results in boredom, overconfidence, poor study habits, and a needlessly diluted education. So it is important to make decisions about what to expect of students and when on the basis of as much good information as possible.

The presence of a topic at a grade level in current textbooks or curriculum guides is not reliable evidence that it can be learned meaningfully at that grade. For example, atoms and molecules sometimes appear in a 4th-grade science reader. Yet extensive research on how children learn about these ideas suggests postponement until at least 6th grade and perhaps until 8th grade for most students.

As noted in Chapter 13: The Origin of Benchmarks, the benchmark decisions have been based on inputs from three sources: the Project 2061 school-district teams, learning specialists, and published research. Although each of those sources has its own particular limitations, each contributed significantly to the final product.

The single most important source of knowledge on student learning comes from thoughtful teachers. They have firsthand experience in helping students acquire science, mathematics, and technology knowledge and skills. Their input is limited, however, by the realities of the usual teaching situation. Teachers have little time to conduct careful assessments of student learning, lack instruments for assessing richly connected learnings and higher-order thinking skills, and rarely have opportunities to compare their experiences with others who teach the same concepts and skills. The Project 2061 teachers were able to overcome some of these limitations as a result of having ready access to learning specialists and time for exploring questions of benchmark placement with other teachers. Whenever benchmark-writing conferences were held with Project 2061 teams, several research consultants participated as well. They helped by elaborating on their knowledge of important topics on which they had already done research, extrapolating that knowledge to other topics and helping the teachers extend and interpret their own experience.

Researchers have the advantage of being able to work out a careful design, having time and other resources (including special training in research methods) that teachers seldom have, and undergoing systematic peer review. Research studies into whether students at a certain age can understand a certain idea fall into four categories:

(1) Students do understand the idea after traditional instruction—or even without any instruction at all. So placement at the given grade level is all right and might possibly be earlier.

(2) Students do not understand the idea after traditional instruction. (This is the most common category of published research.) So the idea may be intrinsically too sophisticated for the grade level or may not have been adequately prepared for. The research may provide clues about how to design more effective instruction.

(3) Students do understand the idea after special instruction. So placement at the given grade level is possible, depending on whether adequate additional resources are likely to be available.

(4) Students still do not understand even after special instruction. (This is fairly common.) The idea has to be simplified, better prepared for, or postponed until students are more ready.

As those categories imply, research findings have led to several different kinds of adjustments in writing benchmarks:

(1) Stating less-sophisticated precursors of an idea. For example, research suggests that the notion of a "fair comparison" can be understood in lower grades as a preliminary form of the later concept of a controlled experiment.

(2) Adding prerequisite components for learning outcomes. For example, research draws attention to the need for understanding how people see things by reflected light as a prerequisite to a benchmark for understanding the phases of the moon.

(3) Changing benchmarks to different grade levels. For example, research shows that natural selection is still a difficult idea for many college students—even after special instruction. So the benchmark for natural selection was moved from 8th grade (where some teachers thought it could be taught) to 12th grade.

But research, too, has its limitations. For decades, the chief tool for investigations of what children know was multiple-choice testing, but researchers eventually found such tests often inflate estimates of students' understanding and disguise their misunderstandings. Studies of students' general cognitive development over the decades have usually proven to be too general to be of much help in making decisions about particular concepts and grade placements. Nonetheless, developmental psychology does provide some guidance where there are no specific research findings. (For example, the late acquisition of the concept of proportionality implies that it is very unlikely that students could understand probability in a quantitative way before middle school.)

Beginning in the late 1970s, cognitive studies in many different countries began to focus on the learning of particular science and mathematics concepts, often using in-depth interviews of students in place of conventional testing. Interviews have revealed that students usually have ideas about how the world works even before instruction. Some of the students' ideas work fairly well in familiar contexts and are highly resistant to change. Moreover, students' lack of understanding or misunderstanding of ideas in science is often masked by their ability to memorize the right words. In-depth interviews with students have often shown consistent deficits or peculiarities in student understanding that had not been identified by conventional testing. Still, the total number of concepts or skills investigated is very small and is unevenly distributed across fields.

Evidence on learning from both teacher experience and research ought to be interpreted cautiously because it necessarily refers to today's students taught in today's schools by today's teachers. There are so many variables operating in the learning process—teacher and parent expectations, the learning environment, the methods and materials used, the previous knowledge and experience of individual learners, and more—that the failure of students to learn something currently leaves open the question of whether they could have done so if they had had ideal learning conditions from the beginning.

Some learning difficulties students have may not be modifiable at all until brain maturation allows them to use higher levels of abstract thinking, whereas some current learning difficulties might be readily ameliorated by improved resources such as better books, more hands-on work, or more time for teachers to plan.

So benchmark statements and placements, including estimates of what might reasonably be possible in the future, have been attempted with the evidence from various sources—teachers, general principles of developmental psychology, research on specific topics, and prominent researchers.

There is a vast literature on education, a fraction of which reports research findings. That research literature is spread across different fields of study, grade levels, kinds of schools, and aspects of education. Studies dealing with the learning of specific science, mathematics, or technology content, while small compared to the whole body of educational literature, are growing in number and sophistication.

Research progress in a field can be inferred from an examination of review articles, published bibliographies, and handbooks. They are also good starting places for beginning a literature search. With regard to the topics covered in this report, the following were found to be very helpful:

Elementary School Social Studies: Research as a Guide to Practice, edited by Virginia Atwood. Washington, DC: National Council for the Social Studies, 1986. 176 pp.

This book examines the research base on when and how elementary-school students develop concepts, skills, and attitudes associated with social sciences. Chapters discuss and summarize research findings related to topics in citizenship and law-related education, intercultural and multicultural education, geography, history, economics, anthropology, and sociology.

Children's Ideas in Science, edited by Rosalind Driver, Edith Guesne, and Andrée Tiberghien. Milton Keynes, UK: Open University Press, 1985. 208 pp.

This book documents and explores conceptions that students aged between 10 and 16 hold about topics related to light, heat and temperature, force and motion, the structure of matter, and the earth as a cosmic body. It also examines how students' conceptions develop with teaching.

Handbook of Research on Mathematics Teaching and Learning, edited by Douglas Grouws. A project of the National Council of Teachers of Mathematics. New York: Macmillan Publishing Company, 1992. 771 pp.

The *Handbook* presents a comprehensive view and analysis of available research on mathematics teachers, teaching mathematics, and the learning process for mathematics content. Areas for further research needed are identified, and, where appropriate, implications of research for classroom practice are provided. Several chapters concern students' learning in content areas that have received considerable research attention: addition and subtraction, multiplication and division, rational numbers, estimation, geometry and spacial reasoning, and probability and statistics.

Bibliography. Students' Alternative Frameworks and Science Education, by Helga Pfundt and Reinders Duit. Kiel, Germany: Institute for Science Education at the University of Kiel, 1991. 270 pp.

The bibliography documents and categorizes research into students' conceptions in science. It contains about 2000 citations to journal articles, research reports, conference papers and whole books on students' learning in physics, biology, chemistry, and earth science. Entries are classified by type of issue. Issues include general considerations concerning research on students' conceptions, relations between students' conceptions and scientific conceptions; relations between the development of student conceptions and the development of notions in the history of science, relations between everyday language and students' conceptions, methods of investigation, investigations of students' conceptions, instruction taking students' conceptions into account, investigations of teachers' conceptions, and consequences of students' conceptions research on teacher training. Investigations of students' and teachers' conceptions are further classified by content area.

Handbook of Research on Social Studies Teaching and Learning, edited by James Shaver. A project of the National Council for the Social Studies. New York: Macmillan Publishing Company, 1991. 661 pp.

The *Handbook* presents a comprehensive view and analysis of available research on social-studies teachers, teaching social studies, and the learning process for social-studies content. Needed future research efforts are identified and methodological issues for research on social studies education are raised. Several chapters in the *Handbook* examine student characteristics and development as relevant to social studies. Research findings are summarized related to topics in government, civics, and law; multicultural education; history; economics; geography; and anthropology, sociology, and psychology.

In searching for papers relevant to our benchmarks, Project 2061 also turned to the research literature from other countries and to proceedings from international research conferences. The references below cite the work of researchers from 10 countries in addition to the United States. With few exceptions, English-language versions were found and cited. The bibliography currently contains citations to over 40 professional journals in several categories. Major journals carrying research papers in science, mathematics, and technology education include

Journal of Research in Science Teaching
International Journal of Science Education (former European Journal of Science Education)
Science Education
School Science and Mathematics
Physics Education
Journal of Biological Education
Journal of Chemical Education

Research in Science and Technological Education

Journal for Research in Mathematics Education

Educational Studies in Mathematics

Theory and Research in Social Studies Education

There are also research articles relevant to learning concepts in science and mathematics in a variety of psychology journals including

Cognition and Instruction

Child Development

Developmental Psychology

Journal of Educational Psychology

There are a number of journals that occasionally carry research articles and provide helpful commentary on general principles of when and how students learn ideas, including *The Science Teacher, Science and Children, The Physics Teacher, The American Biology Teacher, Mathematics Teacher,* and *The Technology Teacher.* Also, education associations sometimes publish monographs on research, an example being the series "What Research Says to the Teacher," published by the National Science Teachers Association from 1978 (edited by Mary Budd Rowe) to the present (edited by Robert Yager). Generally informing Project 2061's interpretation of research findings was long familiarity with the theoretical work of educational psychologists such as Jerome Bruner, Robert Gagné, and David Ausubel and with the successes and failures of the national curriculum projects that were undertaken from the late 1950s to the early 1980s.

The references that follow are organized to match chapters and sections of *Benchmarks*, which in turn mostly match those of *Science for All Americans.* The list is very selective and includes only those references that met two criteria. One was relevance—some excellent papers were not included because they did not bear on one of the *Benchmarks* topics. The other criterion was quality—papers, however relevant, were bypassed if they were seen to have design flaws or their evidence or argument was weak. Even then, however, not all relevant and good papers are included. In many cases, a single paper has been used as representative of a number of similar reports.

It will immediately be clear that mathematics and the physical sciences have had the benefit of many more studies than have other fields. Perhaps that is because the subject matter lends itself to research more easily; in the next few years, though, perhaps the attention to cognitive research will increase in all fields.

1 THE NATURE OF SCIENCE

Research on students' understanding of the nature of science has been conducted for more than 30 years. The earlier part of the research investigated students' understanding about scientists and the scientific enterprise and about the general methods and aims of science (Cooley & Klopfer, 1961; Klopfer & Cooley, 1963; Mackey, 1971; Mead & Metraux, 1957; Welch & Pella, 1967). More recent studies have added students' understanding of the notion of "experimentation," the development of students' experimentation skills, students' understanding of the notions of "theory" and "evidence," and their conceptions of the nature of knowledge. The available research is reviewed in Lederman (1992).

Research on the nature of science focuses mainly on the middle-school and high-school grades. There are few studies that investigate what elementary-school learning experiences are effective for developing an understanding of the nature of science, although Susan Carey's and Joan Solomon's work is a beginning in that direction (Carey, Evans, Honda, Jay, & Unger, 1989; Solomon, Duveen, Scot, McCarthy, 1992).

Research in the 1960s and 70s used multiple-choice questionnaires. Recent studies using clinical interviews reveal discrepancies between researchers' and students' understanding of the questions and the proposed answers in those questionnaires. This finding raises doubt about the earlier studies' findings because almost none of them used the clinical interview to corroborate the questionnaires. Therefore, the following remarks draw mainly upon the results of the relatively recent interview studies.

1A THE SCIENTIFIC WORLD VIEW

Although most students believe that scientific knowledge changes, they typically think changes occur mainly in facts and mostly through the invention of improved technology for observation and measurement. They do not recognize that changed theories sometimes suggest new observations or reinterpretation of previous observations (Aikenhead, 1987; Lederman & O'Malley, 1990; Waterman, 1983). Some research indicates that it is difficult for middle-school students to understand the development of scientific knowledge through the interaction of theory and observation (Carey et al., 1989), but the lack of long-term teaching interventions to investigate this issue makes it difficult to conclude that students can or cannot gain that understanding at this grade level.

1B SCIENTIFIC INQUIRY

Experimentation. Upper elementary- and middle-school students may not understand experimentation as a method of testing ideas, but rather as a method of trying things out or producing a desired outcome (Carey et al., 1989; Schauble et al., 1991; Solomon,

1992). With adequate instruction, it is possible to have middle-school students understand that experimentation is guided by particular ideas and questions and that experiments are tests of ideas (Carey et al., 1989; Solomon et al., 1992). Whether it is possible for younger students to achieve this understanding needs further investigation.

Students of all ages may overlook the need to hold all but one variable constant, although elementary students already understand the notion of fair comparisons, a precursor to the idea of "controlled experiments" (Wollman, 1977a, 1977b; Wollman & Lawson, 1977). Another example of defects in students' skills comes with the interpretation of experimental data. When engaged in experimentation, students have difficulty interpreting covariation and noncovariation evidence (Kuhn, Amsel, & O'Loughlin, 1988). For example, students tend to make a causal inference based on a single concurrence of antecedent and outcome or have difficulty understanding the distinction between a variable having no effect and a variable having an opposite effect. Furthermore, students tend to look for or accept evidence that is consistent with their prior beliefs and either distort or fail to generate evidence that is inconsistent with these beliefs. These deficiencies tend to mitigate over time and with experience (Schauble, 1990).

Theory (explanation) and evidence. Students of all ages find it difficult to distinguish between a theory and the evidence for it, or between description of evidence and interpretation of evidence (Allen, Statkiewitz, & Donovan, 1983; Kuhn 1991, 1992; Roseberry, Warren, & Conant, 1992). Some research suggests students can start understanding the distinction between theory and evidence after adequate instruction, as early as middle school (Roseberry et al., 1992).

Nature of knowledge. Students' ideas about the nature of knowledge and how knowledge is justified develop through stages in which knowledge is initially perceived in terms of "right/wrong," then as a matter of "mere opinion," and finally as "informed" and supported with reasons (Kitchener, 1983; Perry, 1970). This research provides some guidance for sequencing the benchmarks about the nature of scientific knowledge. For example, it suggests that students may not understand before they abandon their beliefs about knowledge being either "right" or "wrong" that scientists can *legitimately* hold different explanations for the same set of observations. However, this research does not say when, how quickly, and with what experiences students can move through these stages given adequate instruction. Several studies show that a large proportion of today's high-school students are still at the first stage of this development (Kitchener, 1983; Kitchener & King, 1981). Further research is needed to specify what school graduates could understand, if from a young age they were taught that different people will describe or explain events differently and that opinions must have reasons and can be challenged on rational grounds.

1c THE SCIENTIFIC ENTERPRISE

When asked to describe their views about science in general, high-school students portray scientists as brilliant, dedicated, and essential to the world. However, when asked about science as a career, they respond with a negative image of scientific work and scientists. They see scientific work as dull and rarely rewarding, and scientists as bearded, balding, working alone in the laboratory, isolated and lonely (Mead & Metraux, 1957). This image of scientists has also been frequently documented among elementary- and middle-school students (Fort & Varney, 1989; Newton & Newton, 1992). Some research suggests that this image may represent students' knowledge of the public stereotype rather than their personal views and knowledge of science and scientists (Boylan, Hill, Wallace, & Wheeler, 1992).

Some students of all ages believe science mainly invents things or solves practical problems rather than exploring and understanding the world. Some high-school students believe that moral values and personal motives do not influence a scientist's contributions to the public debate about science and technology and that scientists are more capable than others to decide those issues (Aikenhead, 1987; Fleming 1986a, 1986b, 1987).

2 THE NATURE OF MATHEMATICS

Research related to students' beliefs about the nature of mathematics has started to receive increasing attention. For literature reviews of the available research see McLeod (1992) and Schoenfeld (1992). Studies of the National Assessment of Educational Progress have recently included items related to student beliefs about mathematics as a discipline (Brown et al., 1988; Carpenter et al., 1983; Dossey et al., 1988). In addition, research on mathematical problem solving has recently included investigations of the beliefs students hold about the nature of mathematics (Schoenfeld, 1985, 1989a, 1989b, 1992). These studies have examined students' perceptions of mathematics as rule-oriented versus process-oriented or as a static versus a dynamic discipline, students' beliefs about the nature of mathematical problem solving, and students' perceptions about the role of memorization in learning mathematics. Little emphasis has been given to students' understanding of mathematics as the study of patterns and relationships, or to the relationships between mathematics, science, and technology, or to the nature of mathematical inquiry as a modeling process.

2A PATTERNS AND RELATIONSHIPS

Preliminary research hints that students have difficulty making connections between mathematical expressions, sentences, and sequences that share common structural patterns. They focus instead upon incidental similarities or differences (Ericksen, 1991).

2B MATHEMATICS, SCIENCE, AND TECHNOLOGY

Middle-school and high-school students think that mathematics has practical, everyday uses and tend to think mathematics is more important for society than for them personally (Brown et al., 1988).

2C MATHEMATICAL INQUIRY

Typical student beliefs about mathematical inquiry include the following: There is only one correct way to solve any mathematics problem; mathematics problems have only one correct answer; mathematics is done by individuals in isolation; mathematical problems can be solved quickly or not at all; mathematical problems and their solutions do not have to make sense; and that formal proof is irrelevant to processes of discovery and invention (Schoenfeld, 1985, 1989a, 1989b). These beliefs limit students' mathematical behavior (Schoenfeld, 1985). Further research is needed to assess when and how students can understand that mathematical inquiry is a cycle in which ideas are represented abstractly, the abstractions are manipulated, and the results are tested against the original ideas. We must also learn at what age students can begin to represent something by a symbol or expression, and what standards students use to judge when solutions to mathematical problems are useful or adequate.

3 THE NATURE OF TECHNOLOGY

There is a very small body of research on student learning about what technology is and how it relates to science and society. Most of this research relies on samples of students outside the United States, and it assesses high-school students' knowledge about the role of science and technology, as well as their attitudes toward the decision making of scientists and engineers in issues of public concern.

3A TECHNOLOGY AND SCIENCE

Even in middle school, students typically do not distinguish between an engineering model of experimentation where the goal is to produce a desirable outcome and the scientific model of experimentation where the goal is to understand the relation between causes and effects (Carey et al., 1989; Schauble et al., 1991). Some research suggests that students can understand and use the engineering model before they can the scientific model—that is, that students inevitably will think about producing desirable outcomes before they are able to do the more analytic form of thinking involved in scientific inquiry (Schauble et al., 1991).

High-school students do not distinguish between the roles of science and technology unless explicitly asked to do so (Fleming, 1987). This is evidenced, for example, by students' view that science serves the public interest. More generally, some students believe science affects society in more positive ways than does technology. That is partly because students associate science with medical research but associate technology with pollution or weapons. Students appear to understand the impact of science on technology but they do not always appreciate the impact of technology on science (Fleming, 1987).

3B DESIGN AND SYSTEMS

Preliminary research gives some indication of two student perspectives on risk resulting from the failure of technological systems. In the first perspective, if the risk of failure involves the possibility of widespread harm, it is unacceptable; however, if the risk of failure is to oneself and voluntary, it is considered a part of life and hardly worthy of concern by others. In the second perspective, if the risk of failure involves harm to oneself and benefits to oneself, then it is of primary interest. Harm to others is simply ignored in this perspective (Fleming, 1986a, 1986b).

3C ISSUES IN TECHNOLOGY

Some high-school students believe scientists and engineers are more capable of making decisions about public issues related to science and technology than the general public. Students believe that scientists and engineers know all the facts and are not influenced by personal motives and interests (Fleming, 1987; Aikenhead 1987).

4 THE PHYSICAL SETTING

There is more research on student conceptions about The Physical Setting than in any other area. The Pfundt and Duit (1991) bibliography reveals that more than 70% of the published papers about students' conceptions in science were concerned with topics related to The Physical Setting benchmarks. Much research has focused on topics related to The Earth, Structure of Matter, Energy Transformations, and Motion. Topics related to The Universe and Forces of Nature have also received attention, but for the Processes That Shape the Earth, there is little research. Even in the frequently researched areas, relatively few studies report on long-term teaching interventions that try to improve students' ideas about the physical setting. The available literature on students' understanding of topics related to The Physical Setting has been reviewed in Driver, Guesne, & Tiberghien (1985). Conference proceedings on these topics include Driver & Millar (1985); Duit, Goldberg, & Niedderer (1992); Jung, Pfundt, & Rhoeneck (1981); Lijnse (1985); and Lijnse et al., (1990).

4A THE UNIVERSE

Research available on student understanding about The Universe focuses on their conceptions of the sun as a star and as the center of our planetary system. The ideas "the sun is a star" and "the earth orbits the sun" appear counter-intuitive to elementary-school students (Baxter, 1989; Vosniadou & Brewer, 1992) and are not likely to be believed or even understood in those grades (Vosniadou, 1991). Whether it is possible for elementary students to understand these concepts even with good teaching needs further investigation.

4B THE EARTH

Shape of the earth. Student ideas about the shape of the earth are closely related to their ideas about gravity and the direction of "down" (Nussbaum, 1985a; Vosniadou, 1991). Students cannot accept that gravity is center-directed if they do not know the earth is spherical. Nor can they believe in a spherical earth without some knowledge of gravity to account for why people on the "bottom" do not fall off. Students are likely to say many things that sound right even though their ideas may be very far off base. For example, they may say that the earth is spherical, but believe that people live on a flat place on top or inside of it—or believe that the round earth is "up there" like other planets, while people live down here (Sneider & Pulos, 1983; Vosniadou, 1991). Research suggests teaching the concepts of spherical earth, space, and gravity in close connection to each other (Vosniadou, 1991). Some research indicates that students can understand basic concepts of the shape of the earth and gravity by 5th grade if the students' ideas are directly discussed and corrected in the classroom (Nussbaum, 1985a).

Explanations of astronomical phenomena. Explanations of the day-night cycle, the phases of the moon, and the seasons are very challenging for students. To understand these phenomena, students should first master the idea of a spherical earth, itself a challenging task (Vosniadou, 1991). Similarly, students must understand the concept of "light reflection" and how the moon gets its light from the sun before they can understand the phases of the moon. Finally, students may not be able to understand explanations of any of these phenomena before they reasonably understand the relative size, motion, and distance of the sun, moon, and the earth (Sadler, 1987; Vosniadou, 1991).

Water cycle. Students' ideas about conservation of matter, phase changes, clouds, and rain are interrelated and contribute to understanding the water cycle. Students seem to transit a series of stages to understand evaporation. Before they understand that water is converted to an invisible form, they may initially believe that when water evaporates it ceases to exist, or that it changes location but remains a liquid, or that it is transformed into some other perceptible form (fog, steam, droplets, etc.) (Bar, 1989; Russell, Harlen, & Watt, 1989; Russell & Watt, 1990). With special instruction, some students in 5th grade can identify the air as the final location of evaporating water (Russel & Watt, 1990), but they must first accept air as a permanent substance (Bar, 1989). This appears to be a challenging concept for upper elementary students (Sere, 1985). Students can understand rainfall in terms of gravity in middle school but not the mechanism of condensation, which is not understood until early high school (Bar, 1989).

4C PROCESSES THAT SHAPE THE EARTH
Students of all ages may hold the view that the world was always as it is now, or that any changes that have occurred must have been sudden and comprehensive (Freyberg, 1985). The students in these studies did not, however, have any formal instruction on the topics investigated. Moreover, middle-school students taught by traditional means are not able to construct coherent explanations about the causes of volcanoes and earthquakes (Duschl, Smith, Kesidou, Gitomer, & Schauble, 1992).

4D STRUCTURE OF MATTER
Nature of matter. Elementary and middle-school students may think everything that exists is matter, including heat, light, and electricity (Stavy, 1991; Lee et al., 1993). Alternatively, they may believe that matter does not include liquids and gases or that they are weightless materials (Stavy, 1991; Mas, Perez, & Harris, 1987). With specially designed instruction, some middle-school students can learn the scientific notion of matter (Lee et al., 1993).

Middle-school and high-school students are deeply committed to a theory of continuous matter (Nussbaum, 1985b). Although some students may think that substances can be divided up into small particles, they do not recognize the particles as building blocks, but as formed of basically continuous substances under certain conditions (Pfundt, 1981).

Students at the end of elementary school and beginning of middle school may be at different points in their conceptualization of a "theory" of matter (Carey, 1991; Smith et al., 1985; Smith, Snir, & Grosslight, 1987). Although some 3rd graders may start seeing weight as a fundamental property of all matter, many students in 6th and 7th grade still appear to think of weight simply as "felt weight"—something whose weight they can't feel is considered to have no weight at all. Accordingly, some students believe that if one keeps dividing a piece of styrofoam, one would soon obtain a piece that weighed nothing (Carey, 1991).

Conservation of matter. Students cannot understand conservation of matter and weight if they do not understand what matter is, or accept weight as an intrinsic property of matter, or distinguish between weight and density (Lee et al., 1993; Stavy, 1990). By 5th grade, many students can understand qualitatively that matter is conserved in transforming from solid to liquid. They also start to understand that matter is quantitatively conserved in transforming from solid to liquid and qualitatively in transforming from solid or liquid to gas—if the gas is visible (Stavy, 1990). For chemical reactions, especially those that evolve or absorb gas, weight conservation is more difficult for students to grasp (Stavy, 1990).

Particles. Students of all ages show a wide range of beliefs about the nature and behavior of particles. They lack an appreciation of the very small size of particles; attribute macroscopic properties to particles; believe there must be something in the space between particles; have difficulty in appreciating the intrinsic motion of particles in solids, liquids and gases; and have problems in conceptualizing forces between particles (Children's Learning in Science, 1987). Despite these difficulties, there is some evidence that carefully designed instruction carried out over a long period of time may help middle-school students develop correct ideas about particles (Lee et al., 1993).

Chemical changes. Middle- and high-school student thinking about chemical change tends to be dominated by the obvious features of the change (Driver, 1985). For example, some students think that when something is burned in a closed container, it will weigh more because they see the smoke that was produced. Further, many students do not view chemical changes as interactions. They do not understand that substances can be formed by the recombination of atoms in the original substances. Rather, they see chemical change as the result of a separate change in the original substance, or changes, each one separate, in several original substances. For example, some students see the smoke formed when wood burns as

having been driven out of the wood by the flame (Andersson, 1990).

A clear picture has emerged of students' misunderstanding of the nature and behavior of matter. There is still a need, however, for detailed research on effective teaching strategies to correct this, especially to identify ways of leading students from a macroscopic to a microscopic understanding of matter. Although some likely precursors to a microscopic view have been suggested—for example, the notion of invisibly small constituents of substances (Millar, 1990)—they have not been formally evaluated.

4E ENERGY TRANSFORMATIONS

Heat and temperature. Even after some years of physics instruction, students do not distinguish well between heat and temperature when they explain thermal phenomena (Kesidou & Duit, 1993; Tiberghien, 1983; Wiser, 1988). Their belief that temperature is the measure of heat is particularly resistant to change. Long-term teaching interventions are required for upper middle-school students to start differentiating between heat and temperature (Linn & Songer, 1991).

Heat transfer. Middle-school students do not always explain the process of heating and cooling in terms of heat being transferred (Tiberghien, 1983; Tomasini & Balandi, 1987). Some students think that "cold" is being transferred from a colder to a warmer object, others that both "heat" and "cold" are transferred at the same time. Middle- and high-school students do not always explain heat-exchange phenomena as interactions. For example, students often think objects cool down or release heat spontaneously—that is, without being in contact with a cooler object (Kesidou, 1990; Wiser, 1986). Even after instruction, students don't always give up their naive notion that some substances (for example, flour, sugar, or air) cannot heat up (Tiberghien, 1985) or that metals get hot quickly because "they attract heat," "suck heat in," or "hold heat well" (Erickson, 1985). Middle-school students believe different materials in the same surroundings have different temperatures if they feel different (for example, metal feels colder than wood). As a result, they do not recognize the universal tendency to temperature equalization (Tomasini & Balandi, 1987). Few middle- and high-school students understand the molecular basis of heat transfer even after instruction (Wiser, 1986; Kesidou & Duit, 1993). Although specially designed instruction appears to give students a better understanding about heat transfer than traditional instruction, some difficulties often remain (Tiberghien, 1985; Lewis, 1991).

Energy conceptualization. Students' meanings for "energy" both before and after traditional instruction are considerably different from its scientific meaning (Solomon, 1983). In particular, students believe energy is associated only with humans or movement, is a fuel-like quantity which is used up, or is something that makes things happen and is expended in the process. Students rarely think energy is measurable and quantifiable (Solomon, 1985; Watts, 1983a). Although students typically hold these meanings for energy at all ages, upper elementary-school students tend to associate energy only with living things, in particular with growing, fitness, exercise, and food (Black & Solomon, 1983).

Energy forms and energy transformation. Middle- and high-school students tend to think that energy transformations involve only one form of energy at a time (Brook & Wells, 1988). Although they develop some skill in identifying different forms of energy, in most cases their descriptions of energy change focus only on forms that have perceivable effects (Brook & Driver, 1986). The transformation of motion to heat seems to be difficult for students to accept, especially in cases with no obvious temperature increase (Brook & Driver, 1986; Kesidou & Duit, 1993). Finally, it may not be clear to students that some forms of energy, such as light, sound, and chemical energy, can be used to make things happen (Carr & Kirkwood, 1988).

Energy conservation. The idea of energy conservation seems counter-intuitive to middle- and high-school students who hold on to the everyday use of the term energy, but teaching heat-dissipation ideas at the same time as energy-conservation ideas may help alleviate this difficulty (Solomon, 1983). Even after instruction, however, students do not seem to appreciate that energy conservation is a useful way to explain phenomena (Brook & Driver, 1984). Middle- and high-school students tend to use their intuitive conceptualizations of energy to interpret energy conservation ideas (Brook & Driver, 1986; Kesidou & Duit, 1993; Solomon, 1985). For example, some students interpret the idea that "energy is not created or destroyed" to mean that energy is stored up in the system and can even be released again in its original form (Solomon, 1985). Although teaching approaches that accommodate students' difficulties about energy appear to be more successful than traditional science instruction, the main deficiencies outlined above remain despite these approaches (Brook & Driver, 1986; Brook & Wells, 1988).

4F MOTION

Light. The majority of elementary students and some middle-school students who have not received any systematic instruction about light tend to identify light with its source (e.g., light is in the bulb) or its effects (e.g., patch of light). They do not have a notion of light as something that travels from one place to another. As a result, these students have difficulties explaining the direction and formation of shadows, and the reflection of light by objects. For example, some students simply note the similarity of shape between the object and the shadow or say that the object hides the light. Middle-school students often accept that mirrors reflect light but, at least in some situations, reject the idea that ordinary objects reflect light (Guesne, 1985; Ramadas & Driver, 1989). Many elementary- and middle-school students do not believe that their eyes receive light when they look at an object. Students' conceptions of vision vary from the notion that light fills space ("the room is full of light") and the eye "sees" without anything linking it to the object to the idea that light illuminates surfaces that we can see by the action of our eyes on them (Guesne, 1985). The conception that the eye sees without anything linking it to the object persists after traditional instruction in optics (Guesne, 1985); however, some 5th-graders can understand seeing as "detecting" reflected light after specially designed instruction (Anderson & Smith, 1983).

The concept of force. Students hold various meanings for the word "force." Typically, students think force is something that makes things happen or creates change. Their descriptions of force often include related words such as energy, momentum, pressure, power, and strength. Younger students associate the word "force" with living things (Watts, 1983b).

Students tend to think of force as a property of an object ("an object has force," or "force is within an object") rather than as a relation between objects (Dykstra, Boyle, & Monarch, 1992; Jung et al., 1981; Osborne, 1985). In addition, students tend to distinguish between active objects and objects that support or block or otherwise act passively. Students tend to call the active actions "force" but do not consider passive actions as "forces" (Gunstone & Watts, 1985). Teaching students to integrate the concept of passive support into the broader concept of force is a challenging task even at the high-school level (Minstrell, 1989).

Newton's laws of motion. Students believe constant speed needs some cause to sustain it. In addition, students believe that the amount of motion is proportional to the amount of force; that if a body is not moving, there is no force acting on it; and that if a body is moving there is a force acting on it in the direction of the motion (Gunstone & Watts, 1985). Students also believe that objects resist acceleration from the state of rest because of friction—that is, they confound inertia with friction (Jung et al., 1981; Brown & Clement, 1992). Students tend to hold onto these ideas even after instruction in high-school or college physics (McDermott, 1983). Specially designed instruction does help high-school students change their ideas (Brown & Clement, 1992; Minstrell, 1989; Dykstra et al., 1992).

Research has shown less success in changing middle-school students' ideas about force and motion (Champagne, Gunstone & Klopfer, 1985). Nevertheless, some research indicates that middle-school students can start understanding the effect of constant forces to speed up, slow down, or change the direction of motion of an object. This research also suggests it is possible to change middle-school students' belief that a force always acts in the direction of motion (White & Horwitz, 1987; White, 1990).

Students have difficulty appreciating that all interactions involve equal forces acting in opposite directions on the separate, interacting bodies. Instead they believe that "active" objects (like hands) can exert forces whereas "passive" objects (like tables) cannot (Gunstone & Watts, 1985). Alternatively, students may believe that the object with more of some obvious property will exert a greater force (Minstrell, 1992). Teaching high-school students to seek consistent explanations for the "at rest" condition of an object can lead them to appreciate that both "active" and "passive" objects exert forces (Minstrell, 1982). Showing them that apparently rigid or supporting objects actually deform might also help (Clement, 1987).

4G FORCES OF NATURE

The earth's gravity and gravitational forces in general form the bulk of research related to Forces of Nature. Elementary-school students typically do not understand gravity as a force. They see the phenomenon of a falling body as "natural" with no need for further explanation or they ascribe to it an internal effort of the object that is falling (Ogborn, 1985). If students do view weight as a force, they usually think it is the air that exerts this force (Ruggiero et al., 1985). Misconceptions about the causes of gravity persist after traditional high-school physics instruction (Brown & Clement, 1992) but can be overcome by specially designed instruction (Brown & Clement, 1992; Minstrell et al., 1992).

Students of all ages may hold misconceptions about the magnitude of the earth's gravitational force. Even after a physics course, many high-school students believe that gravity increases with height above the earth's surface (Gunstone & White, 1981) or are not sure whether the force of gravity would be greater on a lead ball than on a wooden ball of the same size (Brown & Clement, 1992). High-school students have also difficulty in conceptualizing gravitational forces as interactions. In particular, they have difficulty in understanding that the magnitudes of the gravitational forces that two objects of different mass exert on each other are equal. These difficulties persist even after specially designed instruction (Brown & Clement, 1992).

5 THE LIVING ENVIRONMENT

Several areas related to The Living Environment have received considerable research attention over recent years. These include student meanings of the terms animal, plant, and living; students' ideas about plant nutrition; and their understanding of genetics and natural selection. Little has been published about students' understanding of cells, or the dependence of organisms on one another and the environment, or the flow of energy through the living environment. Research has focused on what students understand about the living environment at isolated points in time or on how this understanding evolves naturally in students. Research on instructional interventions that improve students understanding is limited. Reviews of research can be found in Carey (1985), Good et al. (1993), and Mintzes et al. (1991).

5A DIVERSITY OF LIFE

Classification of organisms. Some research indicates that in 2nd grade there is a shift in children's understanding of organisms from representations based on perceptual and behavioral features to representations in which central principles of biological theory are most important. Children at this age can begin to understand that animals of the same species have similar internal parts and offspring (Keil, 1989).

When asked to group certain organisms, lower elementary-school students form groups of different status—for example, organisms that are able to fly and organisms that fight each other. Upper elementary-school students tend to use a number of mutually exclusive groups rather than a hierarchy of groups. Some groups are based on observable features; others on concepts. By middle school, students can group organisms hierarchically when asked to do so, whereas high-school students use hierarchical taxonomies without prompting (Leach, et al.).

Meaning of the words "animal" and "plant." Elementary-and middle-school students hold a much more restricted meaning than biologists for the word "animal" (Mintzes et al., 1991). For example, most students list only vertebrates as animals. Elementary- and middle-school students use such criteria as number of legs, body covering, and habitat to decide whether things are animals. High-school students frequently use attributes that are common to both plants and animals (e.g., reproduction and respiration) as criteria (Trowbridge & Mintzes, 1985). Because upper elementary-school students tend not to use hierarchical classification, they may have difficulty understanding that an organism can be classified as both a bird and an animal (Bell, 1981). Elementary- and middle-school students also hold a much more restricted meaning than biologists do for the word "plant." Students often do not recognize that trees, vegetables, and grass are all plants (Osborne & Freyberg, 1985).

Living and nonliving. Elementary- and middle-school students typically use criteria such as "movement," "breath," "reproduction," and "death" to decide whether things are alive. Thus, some believe fire, clouds, and the sun are alive, but others think plants and certain animals are nonliving. (Bell & Freyberg, 1985; Leach et al., 1992). High-school and college students also mainly use obvious criteria (e.g., "movement," "growth") to distinguish between "living" and "nonliving" and rarely mention structural criteria ("cells") or biochemical characteristics ("DNA") (Brumby, 1982; Leach et al., 1992).

5B HEREDITY

By the end of 2nd grade, students know that children resemble their parents and realize that reproduction underlies this resemblance. Students at this age can also begin to understand the difference between learned resemblance and inherited resemblance (Carey, 1985).

When asked to explain how physical traits are passed from parents to offspring, elementary-school, middle-school, and some high-school students express the following misconceptions: Some students believe that traits are inherited from only one of the parents (for example, the traits are inherited from the mother, because she gives birth or has most contact as children grow up; or the same-sex parent will be the determiner). Other students believe that certain characteristics are always inherited from the mother and others come from the father. Some students believe in a "blending of characteristics." It may not be until the end of 5th grade that some students can use arguments based on chance to predict the outcome of inherited characteristics from observing those characteristics in the parents (Deadman & Kelly, 1978; Kargbo, Hobbs, & Erickson, 1980; Clough & Wood-Robinson, 1985b).

Early middle-school students explain inheritance only in observable features, but upper middle-school and high-school students have some understanding that characteristics are determined by a particular genetic entity which carries information translatable by the cell. Students of all ages believe that some environmentally produced characteristics can be inherited, especially over several generations (Clough & Wood-Robinson, 1985b).

5C CELLS

Preliminary research indicates that it may be easier for students to understand that the cell is the basic unit of structure (which they can observe) than that the cell is the basic unit of function (which has to be inferred from experiments) (Dreyfus & Jungwirth, 1989). Research also shows that high-school students may hold various misconceptions about cells after traditional instruction (Dreyfus & Jungwirth, 1988).

5D INTERDEPENDENCE OF LIFE

Relationships between organisms. Lower elementary-school students can understand simple food links involving two organisms. Yet they often think of organisms as independent of each other but dependent on people to supply them with food and shelter. Upper elementary-school students may not believe food is a scarce resource in ecosystems, thinking that organisms can change their food at will according to the availability of particular sources (Leach et al., 1992). Students of all ages think that some populations of organisms are numerous in order to fulfill a demand for food by another population (Leach et al., 1992).

Habitat. Middle-school and high-school students may believe that organisms are able to effect changes in bodily structure to exploit particular habitats or that they respond to a changed environment by seeking a more favorable environment (Jungwirth, 1975; Clough & Wood-Robinson, 1985a). It has been suggested that the language about adaptation used by teachers or textbooks to make biology more accessible to students may cause or reinforce these beliefs (Jungwirth, 1975).

5E FLOW OF MATTER AND ENERGY

Food. Students of all ages tend to use the term "food" in ways that are consistent with the everyday meaning of the term, not the biological meaning. They see food as substances (water, air, minerals, etc.) that organisms take directly in from their environment (Anderson, Sheldon, & Dubay, 1990; Simpson & Arnold, 1985). In addition, some students of all ages think food is a requirement for growth, rather than a source of matter for growth. They have little knowledge about food being transformed and made part of a growing organism's body (Smith & Anderson, 1986; Leach et al., 1992).

Organisms as chemical systems. Middle-school and high-school students have difficulty thinking of the human body as a chemical system and have little knowledge about the elements composing the living body (Stavy, Eisen, & Yaakobi, 1987). In particular, middle-school students think organisms and materials in the environment are very different types of matter. For example, animals are made of bone, muscle, skin, etc.; plants are made of leaves, stems, and roots; and the nonliving environment is made of water, soil, and air. Students see these substances as fundamentally different and not transformable into each other (Smith & Anderson, 1986).

Plant and animal nutrition. Some students of all ages hold misconceptions about plant nutrition (Bell & Brook, 1984; Roth & Anderson, 1987; Anderson et al., 1990). They think plants get their food from the environment rather than manufacturing it internally, and that food for plants is taken in from the outside. These misconceptions are particularly resistant to change (Anderson et al., 1990). Even after traditional instruction, students have difficulty accepting that plants make food from water and air, and that this is their only source of food. Understanding that the food made by plants is very different from other nutrients such as water or minerals is a prerequisite for understanding the distinction between plants as producers and animals as consumers (Roth & Anderson, 1987; Anderson et al., 1990).

Some students of all ages have difficulty in identifying the sources of energy for plants and also for animals (Anderson et al., 1990). Students tend to confuse energy and other concepts such as food, force, and temperature. As a result, students may not appreciate the uniqueness and importance of energy conversion processes like respiration and photosynthesis (Anderson et al., 1990). Although specially designed instruction does help students correct their understanding about energy exchanges, some difficulties remain (Anderson et al., 1990). Careful coordination between The Physical Setting and The Living Environment benchmarks about conservation of matter and energy and the nature of energy may help alleviate these difficulties (Anderson et al., 1990).

Decay. Some middle-school students think dead organisms simply rot away. They do not realize that the matter from the dead organism is converted into other materials in the environment. Some middle-school students see decay as a gradual, inevitable consequence of time without need of decomposing agents (Smith & Anderson, 1986). Some high-school students believe that matter is conserved during decay, but do not know where it goes (Leach et al., 1992).

Matter cycling. Middle-school students seem to know that some kind of cyclical process takes place in ecosystems (Smith & Anderson, 1986). Some students see only chains of events and pay little attention to the matter involved in processes such as plant growth or animals eating plants. They think the processes involve creating and destroying matter rather than transforming it from one substance to another. Other students recognize one form of recycling through soil minerals but fail to incorporate water, oxygen, and carbon dioxide into matter cycles. Even after specially designed instruction, students cling to their

misinterpretations. Instruction that traces matter through the ecosystem as a basic pattern of thinking may help correct these difficulties (Smith & Anderson, 1986).

5F EVOLUTION OF LIFE

Natural selection. High-school and college students, even after some years of biology instruction, have difficulties understanding the notion of natural selection (Brumby, 1979; Bishop & Anderson, 1990). A major hindrance to understanding natural selection appears to be students' inability to integrate two distinct processes in evolution, the occurrence of new traits in a population and their effect on long-term survival (Bishop & Anderson, 1990). Many students believe that environmental conditions are responsible for changes in traits, or that organisms develop new traits because they need them to survive, or that they over-use or under-use certain bodily organs or abilities (Bishop & Anderson, 1990). By contrast, students have little understanding that chance alone produces new heritable characteristics by forming new combinations of existing genes or by mutations of genes (Brumby, 1979; Clough & Wood-Robinson, 1985b; Hallden, 1988). Some students believe that a mutation modifies an individual's own form during its life rather than only its germ cells and offspring (see almost any science-fiction movie). Students also have difficulties understanding that changing a population results from the survival of a few individuals that preferentially reproduce, not from the gradual change of all individuals in the population. Explanations about "insects or germs becoming more resistant" rather than "more insects or germs becoming resistant" may reinforce these misunderstandings (Brumby, 1979). Specially designed instruction can improve students' understanding of natural selection (Bishop & Anderson, 1990).

Adaptation. Middle-school and high-school students may have difficulties with the various uses of the word "adaptation" (Clough & Wood-Robinson, 1985a; Lucas, 1971; Brumby, 1979). In everyday usage, individuals adapt deliberately. But in the theory of natural selection, populations change or "adapt" over generations, inadvertently. Students of all ages often believe that adaptations result from some overall purpose or design, or they describe adaptation as a conscious process to fulfill some need or want. Elementary- and middle-school students also tend to confuse non-inherited adaptations acquired during an individual's lifetime with adaptive features that are inherited in a population (Kargbo et al., 1980).

Evolution and reasoning ability. Some research suggests that students' understanding of evolution is related to their understanding of the nature of science and their general reasoning abilities (Lawson & Thomson, 1988; Lawson & Worsnop, 1992; Scharmann & Harris, 1992). Findings indicate that poor reasoners tend to retain nonscientific beliefs such as "evolutionary change occurs as a result of need" because they fail to examine alternative hypotheses and their predicted consequences, and they fail to comprehend conflicting evidence. Thus, they are left with no alternative but to believe their initial intuitions or the misstatements they hear.

6   THE HUMAN ORGANISM

Several studies have investigated the spontaneous development of student conceptions about The Human Organism. A general pattern that emerged from this research is that until the age of seven, students have little knowledge about the human organism. By age nine or ten, there is a marked increase in their knowledge (Carey, 1985).

6B   HUMAN DEVELOPMENT

Fertilization. By the end of 5th grade, students know that babies result from the fusion of sperm and eggs. However, they often don't understand how the fusion brings new life. Before students have an early understanding of genetics, they may believe that the baby exists in the sperm but requires the egg for food and protection, or that the baby exists in the egg and requires the sperm as trigger to growth (Bernstein & Cowan, 1975; Goldman & Goldman, 1982).

Death. Lower elementary-school children understand that death is irreversible and inevitable (Lazar & Torney-Purta, 1991). They usually think death is caused by an external agent but do not connect death with what happens within the body as a result of these external events (Carey, 1985). Around 3rd or 4th grade, students understand that death means the cessation of bodily functions (Carey, 1985).

6C   BASIC FUNCTIONS

Internal organs. Lower elementary-school students may have little knowledge about internal bodily organs and think the contents of the body are what they have seen being put into or coming out of it (e.g., food, blood). Upper elementary students can list a large number of organs (Gellert, 1962); however, a sizeable proportion of adults has little knowledge of internal organs or their location (for example, few adults can draw the stomach and the liver in reasonable positions) (Blum, 1977).

Nervous system. By the end of 2nd grade, students know that thought is needed for different kinds of activities (e.g., motor acts) and as a result, know the brain is required for these activities (Carey, 1985). Fourth-graders know the brain helps the body parts but do not always realize that the body also helps the brain (Johnson & Wellman, 1982). Whether upper elementary-school students can achieve this understanding with adequate instruction needs further investigation. Upper elementary-school students attribute to nerves the functions of conducting messages, controlling activity, and stabilizing the body (Gellert, 1962), but even after traditional instruction about the brain and the nervous system, 5th-grade students appear not to understand yet the role of the brain in controlling involuntary behavior (Johnson & Wellman, 1982).

Circulatory system. Lower elementary-school students know about circulation and something of the blood's relation to breathing. Upper elementary-school students realize that the heart is a pump, but they are not aware that the blood returns to the heart (Carey, 1985). Students of all ages hold wrong ideas about the structure and function of blood, the structure and function of the heart, the circulatory pattern, the circulatory/respiratory relationships, and the closed system of circulation. Misconceptions concerning the circulatory pattern, the circulatory/respiratory relationships, and the closed system of circulation are difficult to change (Arnaudin & Mintzes 1985, 1986).

Digestive system. Lower elementary-school students know food is related to growing and being strong and healthy, but they are not aware of the physiological mechanisms. By 5th grade, students know that food undergoes a process of transformation in the body (Contento, 1981; Wellman & Johnson, 1982).

Respiratory system. Lower elementary-school students may not know what happens to air after it is inhaled. Upper elementary-school students associate the lungs' activities with breathing and may understand something about the exchange of gases in the lungs and that the air goes to all parts of the body (Carey, 1985).

6D LEARNING

In recent years, research on student views of the nature of the learning process has received increased attention. Several techniques for improving student knowledge of the learning process have been devised. Examples include encouraging students to construct concept maps (Novak & Gowin, 1984), to think about what they have learned and how they have learned it, and to document their learning in diaries. Students' ideas about learning appear to resist change (Baird & Mitchell, 1986), but long-term interventions can improve their knowledge of the learning process and their learning behavior (Baird, Fensham, Gunstone & White, 1989).

6E PHYSICAL HEALTH

Germs. Elementary-school students may have the following ideas about germs: Germs are microorganisms causing illness; germs enter the body through the mouth while eating and leave the body through the mouth; every illness is caused by germs; all diseases are caused by the same kind of germ; the process of infection is automatic; any infection in the body necessarily makes it ill; and when medicine is administered, healing takes place immediately (Nagy, 1953). (This admittedly dated study is still cited by many authors.)

Causes of illness. Lower elementary-school children may think that illness is the result of misbehavior and realize that they are ill only when they are told that they are by others or when their illness has a behavioral impact, such as having to stay in bed or to go to the doctor. Upper elementary-school children may believe that all illnesses are caused by germs and are contagious. As students grow older, their beliefs about causes of illness begin to include also the malfunctioning of internal organs and systems, poor health habits, and genetics. Upper elementary students can understand that a change in internal body state or the experience of symptoms is the consequence of illness (Hergenrather & Rabinowitz, 1991).

Health. Students of all ages focus on the physical dimensions of health and pay less attention to the mental and social dimensions. Students associate health primarily with food and fitness (Brumby et al., 1985; Moon et al., 1985). Middle-school and high-school students' wrong ideas about the causes of health and illness may derive from cultural knowledge (Rice, 1991; Prout, 1985).

Students of all ages tend to believe that many factors they consider important to their health and life span are also beyond their personal control (Brumby et al., 1985; Merkle & Treagust, 1987). After instruction, middle-school students appear to have accurate knowledge about nutrition and physical fitness, but they are often unable to explain their knowledge in scientific terms (Merkle & Treagust, 1987).

Nutrition. Lower elementary-school children know that there are different foods, that there are good foods and bad foods, and that there are different nutritional outcomes such as variations in size and health. In addition, they are aware of certain limits (drinking just water leads to death; eating only one thing—even one good food—is insufficient for good health). They still may believe, however, that food and water have equivalent nutritional consequences; height and weight are similarly influenced by amount of food eaten; and energy and strength result from exercise but not nutrition. These misconceptions tend to fade by the end of 5th grade (Wellman & Johnson, 1982).

7 HUMAN SOCIETY

A number of studies have examined the spontaneous development of students' conceptions and thinking in the social sciences. In these studies, student thinking is usually described by a series of levels or stages similar to those described by Piaget. Although such stages have been identified, little is known about how student developmental characteristics affect or are affected by formal instruction. As a result, it is difficult to draw conclusions from the research base about when and how students can learn this material. Also, the published research is spotty. For some topics, for example, those related to Political and Economic Systems, there is a small but growing literature base. Research on learning related to Cultural Effects on Behavior, Group Behavior, Social Change, Social Trade-offs, Social Conflict, and Global Interdependence is limited. Literature reviews can be found in Atwood (1986) and Shaver (1991).

7A CULTURAL EFFECTS ON BEHAVIOR

Although lower elementary-school children do not have the capacity to see social conventions from another point of view, they can learn about and enjoy many concrete manifestations of cultural diversity (Ramsey, 1986). Research also suggests that students under the age of ten may be more receptive than older students to learning about other people and more likely to develop a positive outlook toward people from other cultures and homelands (Stone, 1986).

Research into student thinking about people from the past indicates that students do not realize that values, beliefs, and attitudes may differ from culture to culture or that people from other cultures have different ideas because their situations are different. Before students can reason about different world views, they often have to abandon the belief that some human cultures are biologically subordinate (Shelmit, 1984).

Another complication is that students tend to impose contemporary values and ideas from their own culture upon other cultures (Shelmit, 1984).

7b GROUP BEHAVIOR

As children try to understand biological and social phenomena, they often overgeneralize information about racial and cultural differences. One must be cautious, however, not to assume that children are prejudiced or deliberately using stereotypes when they overgeneralize. They may simply be thinking typically for young children trying to make sense out of their limited experience with other groups (Ramsey, 1986). Research indicates that stereotypic attitudes begin to develop about 7th grade (Stone, 1986).

Research supports the view that lower elementary-school children are aware of some of the many ways in which rules vary. For example, children agree that some rules in their culture are more important than others and that some rules are more universally right than others. In addition, children are aware of the social function of different kinds of rules (Edwards, 1986). They may go through alternating developmental periods of "affirming" versus "negating" social conventions (Turiel, 1983). Only near the end of middle school and the beginning of high school do students start to accept the need for social conventions to facilitate social interactions in their groups (Mackey, 1991).

7c SOCIAL CHANGE

Some research has investigated student notions of laws. Findings indicate that elementary-school students mix legal and moral norms (Berti, 1988). In addition, students at that age often have an authoritarian view of laws. They believe that laws are automatically right and "are handed down from on high" (Moore et al., 1985).

7d SOCIAL TRADE-OFFS

Some research indicates that elementary-school children have difficulty with the concept of choice and the idea that whenever a choice is made, a cost is incurred (Schug & Birkey, 1985).

7e POLITICAL AND ECONOMIC SYSTEMS

Much of the research related to Political and Economic Systems relies on samples of students outside the United States. These studies need to be replicated with U.S. students from different social backgrounds because research suggests that students' environments may influence their economic ideas.

Scarcity. Lower elementary-school students have already formed a fairly accurate concept of scarcity. However, the majority of children at that age may still reason in a superficial and literal manner about economic problems concerning choice and trade-offs (Schug & Birkey, 1985).

Money circulation. Elementary-school students know that workers are paid by bosses for the work done. They also know that consumers pay shopkeepers and that shopkeepers pay middlemen and producers in exchange for goods. However, until 4th or 5th grade they do not connect these two areas of experience. As a result, they may believe that the boss must have access to sources external to the factory or store to pay workers. Or they may think that prices remain the same or even decrease in the passage from producer to consumer (Berti & Bombi, 1988).

Profit. The concept of profit, which is a prerequisite idea to understanding the notion of a market economy, emerges naturally in children between the upper elementary and the early middle-school grades (Jahoda, 1979, 1981). By 4th grade, students can usually understand the notion of profit, after specially designed instruction. Students' ability to compare expenses (total costs) and revenue is a prerequisite to this understanding (Berti, 1992).

Price. There are several student difficulties related to understanding the price mechanism in market economies. Even at the college level, students have difficulty understanding that price is not an inherent characteristic of goods but a function of demand and supply (Berti & Grivet, 1990; Marton, 1978). Students often fail to connect the different viewpoints of consumers and producers. For example, while believing that consumers buy less when prices increase, some students may also believe the reverse and that, if demand drops, producers can increase prices to earn the same amount of money as before (Berti & Grivet, 1990).

Source of goods and production. Lower elementary-school students do not have a notion of production based on the transformation of primary materials, for example, furniture from wood. In addition, students at this age have little knowledge about agricultural and industrial production. They believe that farmers themselves consume all the products from their crops and animals. It is only in 2nd grade that the majority of the children recognize the existence of a producer distinct from a shopkeeper (Berti & Bombi, 1988).

Political organization. Elementary-school children typically do not understand public institutions as institutions that provide collective services. For them, terms like "council," "state," or "government" do not specifically designate any particular body, nor do they use the terms in a sense that distinguishes them clearly from a private employer. The "state," the "council," or the "government" are perceived as important or wealthy persons who wield authority and pay people who work (Berti & Bombi, 1988). Whether young children can understand these concepts with adequate instruction needs further investigation. Research

indicates students acquire the intellectual capacity to construct a political order in hierarchical form at around 5th grade (Connell, 1971). From this age on, various authorities are no longer thought to have power over only a few persons close to them, but over whole populations through the enactment of laws and the control of power. By this age, students also know that political parties exist and their activities affect elections (Mackey, 1991).

7F SOCIAL CONFLICT

Studies on young children's recognition of conflict indicate that from age six on, children are able to recognize that a child's desires may conflict with those of his parents or friends (Berti, 1988; Damon, 1977), but it is not clear whether children at this age can also recognize conflict among adults. Upper elementary-school students may not recognize that making laws (to settle conflicts) is the job of the whole community as well as political leaders (Berti, 1988). Middle-school students may not recognize conflict involving social groups (Berti, 1988; Connell, 1971). Middle-school students do not recognize the role of debate, disagreement, and conflict in the operation of the democratic political system (Hess & Torney, 1968).

7G GLOBAL INTERDEPENDENCE

Some research suggests that middle-school and high-school students have an understanding of the global nature of trade, although they have only limited understanding about reciprocal benefits from trade (Schug & Lephardt, 1992).

8 THE DESIGNED WORLD

Although extensive research has focused on survey evaluations of technology programs or on student attitudes toward technology, there is only a small body of research on what students know and how they learn about concepts and systems in technology (Donnelly, 1992; Foster, 1992a, 1992b). For instance, the Assessment of Performance Unit in the United Kingdom has recently collected baseline data on 15-year-old students' performance in design and technology (Kimbell, Stables, Wheeler, Wosniak, & Kelly, 1991). In addition, preliminary research has investigated middle-school students' learning during technology tasks (Hennessy, McCormick, & Murphy, in press; McCormick, Hennessy, & Murphy, 1993). Results indicate that middle-school students have little understanding about the design process. They do not appear to understand what evaluation of design is or why it is important. Students also have trouble assessing and applying knowledge from other contexts while engaging in design and technology activities (McCormick et al., 1993).

The Designed World benchmarks are associated with knowledge and skills needed for other literacy goals. For example, they draw upon The Mathematical World for knowledge of shapes, estimation, measure, and the ability to use scale, and upon The Physical Setting for knowledge of materials and their properties, forces, and energy. As a result, the literature on student understanding of these topics provides some insight into when and how students may understand concepts of The Designed World. For example, research on student understanding of materials suggests that the tasks of classifying objects according to what they are made of and of comparing properties of materials can be challenging for early elementary-school children. In addition, elementary-school children may have limited knowledge or hold misconceptions about the origins and transformations of materials (Russell, Longden, & McGuigan, 1991).

Given the limited published research related to The Designed World benchmarks, recommendations on what students at different grade ranges should know about The Designed World have drawn primarily on existing good practice documented in The Technology Teacher and on recommendations from technology teachers.

9 THE MATHEMATICAL WORLD

Students' understanding of the Mathematical World has been extensively researched. The *Handbook of Research in Mathematics Teaching and Learning.* (Grouws, 1992) as well as the papers presented at the "research agenda conferences" on number concepts in the middle grades (Hiebert & Behr, 1988) and on the learning and teaching of algebra (Wagner & Kieran, 1989), reveal that there is a rich and growing research base related to Numbers, Symbolic Relationships, Shapes, and Uncertainty. There is still little research related to Reasoning, however. As in other domains, research has focused on what students understand about mathematical concepts at isolated points in time or on how this understanding evolves naturally in students. Research on instructional interventions that improve student understanding has received less attention.

9A NUMBERS

Whole numbers. During preschool and elementary-school years, children develop meanings for number words in which sequence, count, and cardinal meanings of number words become increasingly integrated (Fuson et. al., 1982; Fuson, 1988). Students' own meanings for number words determine to some extent their strategies for adding and subtracting and the complexity of problems they can solve. Elementary- and middle-school students may have limited ability with place value (Sowder, 1992a). Sowder reports that middle-school students are able to identify the place values of the digits that appear in a number, but they cannot use the knowledge confidently in context (for example, students have trouble determining how many boxes of 100 candy bars could be packed from 48,638 candy bars).

Rational numbers. Upper elementary- and middle-school students often do not understand that decimal fractions represent concrete objects that can be measured by units, tenths of units, hundredths of units, and so on (Hiebert, 1992). For example, students have trouble writing decimals for shaded parts of rectangular regions divided into 10 or 100 equal parts (Hiebert & Wearne, 1986). Other students have little understanding of the value represented by each of the digits of a decimal number or know the value of the number is the sum of the value of its digits. Students of all ages have problems choosing the largest or the smallest in a set of decimals with different numbers of digits to the right of the decimal points (Carpenter et al., 1981; Hiebert & Wearne, 1986; Resnick et al., 1989). Upper elementary-school students can establish rich meanings for decimal symbols and do a variety of decimal tasks well after specially designed instruction using base-10 blocks (Wearne & Hiebert, 1988, 1989).

Upper elementary- and middle-school students may exhibit limited understanding of the meaning of fractional number (Kieren, 1992). For example, many 7th-graders do not recognize that 5 1/4 is the same as 5 + 1/4 (Kouba et al., 1988). In addition, elementary-school students may have difficulties perceiving a fraction as a single quantity (Sowder, 1988), but rather see it as a pair of whole numbers. An intuitive basis for developing the concept of fractional number is provided by partitioning (Kieren, 1992) and by seeing fractions as multiples of basic units—for example, 3/4 is 1/4 and 1/4 and 1/4 rather than 3 of 4 parts (Behr et. al., 1983).

Estimation. Middle-school and even high-school students may have limited understanding about the nature and purpose of estimation. They often think it is inferior to exact computation and equate it with guessing (Sowder, 1992b), so that they do not believe estimation is useful (Sowder & Wheeler, 1989). Students who see estimation as a valuable tactic for obtaining information use estimation more frequently and successfully (Threadgill-Sowder, 1984).

Number symbols. There is very little research into student understanding of number symbols as arbitrary conventions. It does indicate that not until 11 years of age do most children consider that correct counting with nonstandard symbols is as adequate as correct counting with standard symbols (Saxe et. al, 1989).

9B SYMBOLIC RELATIONSHIPS

Research on Symbolic Relationships examines student understanding of the concepts of variable and algebraic equality, their ability to construct and interpret graphs, and their ability to solve algebraic equations. Several reviews on the literature base in this area are available (Herscovics, 1989; Kieran, 1989, 1992; Leinhardt et al., 1990).

Variables. Students have difficulty understanding how symbols are used in algebra (Kieran, 1992). They are often unaware of the arbitrariness of the letters chosen to represent variables in equations (Wagner, 1981). Middle-school and high-school students may regard the letters as shorthand for single objects, or as specific but unknown numbers, or as generalized numbers before they understand them as representations of variables (Kieran, 1992). These difficulties tend to persist even after instruction in algebra (Carpenter et al., 1981) and are evident even in college students (Clement, 1982). Long-term experience (3 years) in elementary computer programming has been shown to help middle-school students overcome these difficulties, although short-term experiences (less than 6 months) are less successful (Kieran, 1992; Sutherland, 1987).

Graphs. Students of all ages often interpret graphs of situations as literal pictures rather than as symbolic representations of the situations (Leinhardt, Zaslavsky, & Stein, 1990; McDermott, Rosenquist, & van Zee, 1987). Many students interpret distance/time graphs as the paths of actual journeys (Kerslake, 1981). In addition, students confound the slope of a graph with the maximum or the minimum value and do not know that the slope of a graph is a measure of rate (McDermott et al., 1987; Clement, 1989). When constructing graphs, middle-school and high-school students have difficulties with the notions of interval scale and coordinates even after traditional instruction in algebra (Kerslake, 1981; Leinhardt et al., 1990; Vergnaud & Errecalde, 1980; Wavering, 1985). For example, some students think it is legitimate to construct different scales for the positive and the negative parts of the axes. Alternatively, students think that the scales on the X and Y axes must be identical, even if that obscures the relationship. When

interpreting graphs, middle-school students do not understand the effect that a scale change would have on the appearance of the graph (Kerslake, 1981). Finally, students read graphs point-by-point and ignore their global features. This has been attributed to algebra lessons where students are given questions that they could easily answer from a table of ordered pairs. They are rarely asked questions about maximum and minimum values; intervals over which a function increases, decreases or levels off; or rates of change (Herscovics, 1989).

Students have difficulty translating between graphical and algebraic representations, especially moving from a graph into an equation (Leinhardt et al., 1990). Results from the second study of the National Assessment for Educational Progress showed, for instance, that given a line with indicated intercepts, only 5% of 17-year-olds could generate the equation (Carpenter et al., 1981).

Little is known about how graphic skills are learned and how graph production is related to graph interpretation. Microcomputer-based Laboratories (MBLs) are known to improve the development of students' abilities to interpret graphs. For instance, MBLs can help middle-school students learn that a graph is not a picture and overcome the height/slope confusion mentioned above (Mokros & Tinker, 1987).

Algebraic equations. Students of all ages often do not view the equality sign of equations as a symbol of the equivalence between the left and the right side of the equation, but rather interpret it as a sign to begin calculating (Kieran, 1992). For example, middle-school students may not accept statements like $3x + 4 = x + 8$ as legitimate because they think the right side should indicate the answer. Introducing the equal sign from the beginning as a symbol indicating "equivalence" between arithmetic equalities can ameliorate this difficulty (Kieran, 1981).

Beginning algebra students use various intuitive methods for solving algebraic equations (Kieran, 1992). Some of these methods may help their understanding of equations and equation solving. Students who are encouraged initially to use trial-and-error substitution develop a better notion of the equivalence of the two sides of the equation and are more successful in applying more formal methods later on (Kieran, 1988). By contrast, students who are taught to solve equations only by formal methods may not understand what they are doing. Students who are taught to use the method of "transposing" are found to only mechanically apply the change side/change sign rule (Kieran, 1988, 1989).

Students of all ages can often solve algebraic equations without a deeper understanding of what a solution is. For example, middle- and high-school students do not realize that an incorrect solution, when substituted into the equation, will yield different values for the two sides of the equation (Greeno, 1982; Kieran, 1984). More research is needed to identify how students can come to understand what a solution means and why anyone would want to find it.

9c SHAPES

Development of geometrical thinking. Students advance through levels of thought in geometry. Van Hiele has characterized them as visual, descriptive, abstract/relational, and formal deduction (Van Hiele, 1986; Clements & Battista, 1992). At the first level, students identify shapes and figures according to their concrete examples. For example, a student may say that a figure is a rectangle because it looks like a door. At the second level, students identify shapes according to their properties, and here a student might think of a rhombus as a figure with four equal sides. At the third level, students can identify relationships between classes of figures (e.g., a square is a rectangle) and can discover properties of classes of figures by simple

logical deduction. At the fourth level, students can produce a short sequence of statements to logically justify a conclusion and can understand that deduction is the method of establishing geometric truth.

Progress from one of Van Hiele's levels to the next is more dependent upon instruction than age. Given traditional instruction, middle-school students perform at levels one or two (Clements & Battista, 1992). Despite that, almost 40% of high-school graduates finish high-school geometry below level two (Burger & Shaughnessy, 1986; Clements & Battista, 1992; Suydam, 1985). Further research will help identify what levels of geometric thinking students can attain at different grades given effective instruction that takes account of their difficulties in learning geometry. Some evidence suggests it is possible for students to understand the abstract properties of geometric figures by 5th grade (Clements & Battista, 1989, 1990, 1992; Wirszup, 1976) and can understand the relations that connect the properties of shapes or make simple deductions by 8th or 9th grade (Clements & Battista, 1992).

Proof. Research on students' development of the ability to construct proofs reflects somewhat conflicting views (Clements & Battista, 1992). Piagetian research suggests that students can reason deductively from any assumptions once they reach the formal operational stage (roughly age 12 and beyond). Other research, however, suggests that the ability to construct proofs depends on the amount and organization of particular knowledge they have. For example, this research indicates that students are not likely to understand and construct geometric proofs before they can see the relationships between classes of figures (Senk, 1989). Still other research suggests that students may need to understand the nature of proof and how it differs from everyday argumentation before they are able to construct proofs (Clements & Battista, 1992). Clearly, further research is needed to identify how students can come to understand what it means to prove something in geometry and what such a proof entails.

9D UNCERTAINTY

Students' conceptions about uncertainty and students' probabilistic reasoning have been extensively researched, and there are several literature reviews on the topic (Garfield & Ahlgren, 1988; Hawkins & Kapadia, 1984; Shaughnessy, 1992). The research on summarizing data, which focuses on students' understanding of different measures of central tendency and dispersion, is less extensive.

Probability. Research presents somewhat contradictory results on elementary children's understanding of probability. Piagetian research says lower elementary children have no conception of probability (Piaget & Inhelder, 1975; Shayer & Adey, 1981), but other studies indicate that even lower elementary-school children have probabilistic intuitions upon which probability instruction can build. Falk et al. (1980) presented elementary-school students with two sets, each containing blue and yellow elements. Each time, one color was pointed out as the payoff color. The students had to choose the set from which they would draw at random a "payoff element" to be rewarded. From the age of six, children began to select the more probable set systematically. The ability to choose correctly precedes the ability to explain these choices.

Upper elementary students can give correct examples for certain, possible, and impossible events, but cannot calculate the probability of independent and dependent events even after instruction on the procedure (Fischbein & Gazit, 1984). That is partly because students at this age tend to create "part to part" rather than "part to whole" comparisons (e.g., 9 men and 11 women rather than 15% of men and 10% of women). By the end of 8th grade, students can use ratios to calculate probabilities in independent events, after adequate instruction (Fischbein & Gazit, 1984).

Upper elementary students begin to understand that there is an increase in regularity of a sample distribution with an increase in the sample size, but they can apply this idea only to relatively small numbers. It is postulated that to deal with large numbers, children must first cope with notions of ratio and proportion and that their failure to understand these notions creates "a law of small large numbers" (Bliss, 1978).

Extensive research points to several misconceptions about probabilistic reasoning that are similar at all age levels and are found even among experienced researchers (Kahneman, Slovic, & Tversky, 1982; Shaughnessy, 1992). One common misconception is the idea of representativeness, according to which an event is believed to be probable to the extent that it is "typical." For example, many people believe that after a run of heads in coin tossing, tails should be more likely to come up. Another common error is estimating the likelihood of events based on how easily instances of it can be brought to mind.

Summarizing data. The concept of the mean is quite difficult for students of all ages to understand even after several years of formal instruction. Several difficulties have been documented in the literature: Students of all ages can talk about the algorithm for computing the mean and relate it to limited contexts, but cannot use it meaningfully in problems (Mokros & Russell, 1992; Pollatsek, Lima, & Well, 1981); upper elementary- and middle-school students believe that the mean of a particular data set is not one precise numerical value but an approximation that can have one of several values (Mokros & Russell, 1992); some middle-school students cannot use the mean to compare two different-sized sets of data (Gal et al.,

1990); high-school students may believe the mean is the usual or typical value (Garfield & Ahlgren, 1988); students (or adults) may think that the sum of the data values below the mean is equivalent to the sum above the mean (rather than that the total of the deviations below the mean is equal to the total above) (Mokros & Russell, 1992).

Research suggests that a good notion of representativeness may be a prerequisite to grasping the definitions for measures of location like mean, median, or mode. Students can acquire notions of representativeness after they start seeing data sets as entities to be described and summarized rather than as "unconnected" individual values. This occurs typically around 4th grade (Mokros & Russell, 1992).

Research suggests students should be introduced first to location measures that connect with their emerging concept of the "middle," such as the median, and later in the middle-school grades, to the mean. Premature introduction of the algorithm for computing the mean divorced from a meaningful context may block students from understanding what averages are for (Mokros & Russell, 1992; Pollatsek et al., 1981).

10 HISTORICAL PERSPECTIVES

Research into student understanding of the history of science is limited. Much of the literature consists of descriptions of exemplary practices or untried prescriptions for effective teaching (see, for example, Herget, 1989; Hills, 1992; Matthews, 1991; Shortland & Warwick, 1989). Claims for the effectiveness of the methods recommended are seldom supported by systematic research into what learning and how much of it took place. Some evaluation of instructional materials for bringing the history of science into high-school science classes was undertaken during the sixties. These evaluations did not yield consistent conclusions about the effect these materials had on

teaching students about the nature of science, although they hint that historical materials can help change students' image of science so they come to see it as a more philosophical, historical, and humanitarian discipline than they had thought (Klopfer & Cooley, 1963; Welch, 1973; Welch & Walberg, 1968, 1972). Recent research in middle-school classrooms has shown that learning some history of science can lead students to a better understanding of the nature of science as well as the science itself (Solomon et al., 1992).

Research into the development of students' broader historical thinking (beyond just the history of science) reflects conflicting views on when history should be taught. On the one hand, some research indicates students are limited in their historical understanding before they reach Piaget's formal-operations stage (Hallam, 1970, 1979; Joyce et al., 1991). Elementary students, for example, have difficulties with time and related aspects like duration and succession (Downey & Levstik, 1988). These results have been used to argue that adolescence is the better time to begin history instruction (Joyce et al., 1991). On the other hand, recent studies indicate that young children know more history facts than has been thought and can think more maturely when they have good background knowledge (Downey & Levstik, 1991). Also, although some children have difficulty with some time concepts, young children can and do understand historical time in a variety of ways (Egan, 1982; Levstik & Pappas, 1987). They can see patterns and sequences in real events, though some of the patterns may be general and imprecise. These results have been used to argue for an earlier introduction to historical study (Downey & Levstik, 1991). Clearly, more research is needed to assess when and how historical understanding develops in young children and how it can be improved by instruction. Research is also needed to assess whether and how children's conceptions of time are connected to the development of historical understanding.

Some educators claim that simplified historical stories are appropriate content in the elementary school because they deal with basic emotions familiar even to young children (Egan, 1982). There is indeed evidence that historical narratives motivate historical interest (Levstik, 1986) and provide helpful contexts for historical learning (Levstik, 1988; Downey & Levstik, 1991). Specific research will help decide the value of using narratives to introduce young children to the study of history of science and technology.

Even high-school students have difficulties understanding the points of view of people in the past (Lee, 1984; Shelmit, 1984). In particular, students may think their predecessors were intellectually and morally inferior or may account for their thoughts and behavior with stereotypes before they understand that past values, beliefs, and attitudes were often different from those of today (Shelmit, 1984). Research suggests students may have similar difficulties in understanding the points of view of scientists in the past. Middle-school students show little regard for the thinking of scientists whose theories they know have been superseded (Solomon et al., 1992).

11 COMMON THEMES

Research related to Common Themes has focused on students' understanding of the notion of system, on the theoretical and tentative nature of models, and on the concept of conservation. Some research has found that student misconceptions about certain subjects can arise from their difficulty in recognizing natural phenomena as groups or systems of interacting objects.

11A SYSTEMS

The Science Curriculum Improvement Study (SCIS) curriculum led children to approach observation and analysis of natural phenomena by thinking of them as systems of interacting objects (Karplus & Thier, 1969). Research done in connection with SCIS indicates elementary students may believe that a system of objects must be doing something (interacting) in order to be a system or that a system that loses a part of itself is still the same system (Garigliano, 1975; Hill & Redden, 1985). Studies of student thinking show that, at all ages, they tend to interpret phenomena by noting the qualities of separate objects rather than by seeing the interactions between the parts of a system (Driver et al., 1985). Force, for instance, is considered as a property of bodies (forcefulness) rather than as an interaction between bodies. Similarly, students tend to think that whether a substance burns or not is being solely decided by the substance itself, whereas from a scientist's perspective, the process of burning involves the interaction of the burning substance and oxygen.

When students explain changes, they tend to postulate a cause that produces a chain of effects one after another (Driver et al., 1985). In considering a container being heated, students think of the process in directional terms with a source applying heat to the receptor. From a scientific point of view, of course, the situation is symmetrical, with two systems interacting, one gaining energy and the other losing it (Driver et al., 1985). Concentrating on the inputs and outputs of a system often requires a different, time-independent view, which students may not take to be an explanation. Students often do not seem to appreciate that the idea of energy conservation may help explain phenomena. Studies reporting students' difficulties with energy conservation suggest students should have opportunities to describe systems both as sequences of changes over time and as energy inputs and outputs (a systems approach) (Brook & Driver, 1984).

Student explanations of material change seldom include certain kinds of causes that are central to a scientific understanding of the world (Brosnan, 1990); for instance, that parts interact to produce wholes that have properties the parts do not. For children, wholes are like their parts. Brosnan (1990) summarizes all this by presenting two stereotypical views of the nature of change—the common-sense view and the scientific view (pp. 208-209):

Characteristics of a common-sense view of change:

Properties belong to *objects*.

The properties of an object *are the same* as those of the bits that make it up - not all of which may be visible at any one time.

There are *many* kinds of stuff.

Changes in macroscopic properties are the result of *equivalent changes* in the microscopic particles.

If properties change it is because the *bits that cause that property* have moved away, come into view, changed from, grown or disappeared. New properties can be caused by the arrival of new bits.

Characteristics of a scientific view of change:

Properties belong to *systems*.

The properties of an object are *different in kind* from those bits that make it up.

There are fundamentally only a *few* kinds of stuff.

Changes in macroscopic properties are the results of changes in *arrangements* of *unchanging* microscopic particles.

If properties appear or disappear it is because the arrangement of an *unchanging* set *of continuing* particles has altered—at a fundamental level substance is always conserved.

11b Models

There is important research into the use of interactive computer models to teach students certain scientific concepts (e.g., Smith et al., 1987; White, 1990). Most models being developed are qualitative for two reasons. Because the prior knowledge and models students bring to their science instruction are themselves usually qualitative, qualitative reasoning is closely connected to that prior knowledge. Moreover, problem-solving studies have shown that qualitative reasoning is not engaged if students move too quickly into memorizing and applying formal laws. There is still a need to examine student understanding and use of models in general and the characteristic knowledge and misunderstandings they hold about models.

Middle-school and high-school students typically think of models as physical copies of reality, not as conceptual representations (Grosslight et al., 1991). They lack the notion that the usefulness of a model can be tested by comparing its implications to actual observations. Students know models can be changed but changing a model for them means (typical of high-school students) adding new information or (typical of middle-school students) replacing a part that was made wrong.

Many high-school students think models help them understand nature but also believe that models do not duplicate reality. This is chiefly because they think that models have always changed and not because they are aware of the metaphorical status of scientific models (Aikenhead, 1987; Ryan & Aikenhead, 1992). These difficulties continue even for some undergraduate chemistry students (Ingham & Gilbert, 1991).

Students may not accept the explanatory role of models if the model shares only its abstract form with the phenomenon, but will usually accept the explanatory role of models if many of the material features are also the same (Brown & Clement, 1989). Middle-school students may have severe difficulties understanding the hydraulic analogue of an electric circuit and think the two circuits belong to entirely different areas of reality (Kircher, 1985).

Middle-school and high-school students may think everything they learn in science classes is factual and make no distinction between observation and theory (or model) (Brook et al., 1983). If this distinction is to be understood, it should be made explicit when models like the atomic/molecular model are introduced (Brook et al., 1983). Irrelevant aspects of the concrete model can distract students and should be pointed out.

11c Constancy and Change

Lower elementary-school students fail to conserve weight and volume of objects that change shape. When an object's appearance changes in several dimensions, they focus on only one. They cannot imagine a reversed or restored condition and focus mostly on the object's present appearance (Gega, 1986). The ability to conserve develops gradually. Students typically understand conservation of number between the ages of 6 and 7, of length and amount (solid and liquid) between 7 and 8, of area between 8 and 10, of weight between 9 and 11, and of displaced volume between 13 and 14. These ages will vary when different children are tested or the same children are tested in different contexts (Donaldson, 1978).

Many students cannot discern weight conservation in some tasks until they are 15 years old. The ability to conserve weight in a task involving transformation from liquid to gas or solid to gas may rise from 5% in 9-year-old children to about 70% in 14- to 15-year-old-children (Stavy, 1990). More complex changes, such as chemical reactions, especially those where gas is absorbed or released, are still more difficult to grasp as instances of weight conservation (Stavy, 1990).

Fourth-graders' representations of changes over time are "data-driven" in the sense that the particular data in the problem are the most important. This contrasts with "system-driven" representations in which the emphasis is on overall patterns. Unfortunately, students are typically introduced to system-driven representations while they still think it is a wrong or meaningless way to convey information (Tierney & Nemirovsky, 1991).

12 HABITS OF MIND

The research on Habits of Mind has focused on the development of students' Computation and Estimation skills and Critical-Response skills. There is a rich and growing literature on the development of basic arithmetic skills, on the development of estimation skills, and on students' ability to solve problems that involve proportional reasoning. The development of skills for interpreting evidence has also received considerable research attention.

12B COMPUTATION AND ESTIMATION

Operations with whole numbers. Research suggests using word problems as a basis for teaching addition and subtraction concepts, rather than teaching computational skills first and then applying them to solve problems (Carpenter & Moser, 1983). Students should be exposed to a large variety of addition and subtraction situations (compare, combine, equalize, change add to, and change taken from) and given opportunities to consider different meanings for the +, -, and = marks. For example, when 9 - 3 = 6 represents the situation "John has 3 cars. Andy has 9 cars. How many more cars does Andy have?" the minus sign means compare rather than take away (Fuson, 1992).

Research has identified a developmental progression of concepts and skills that students use for addition and subtraction (Fuson, 1988; Fuson, 1992). There is some evidence that instruction based on this progression can help (Romberg & Carpenter, 1986). For example, after a year-long instruction based on this progression, 2nd-graders could solve almost all add or subtract problems with sums up to 18 (Fuson & Willis, 1989).

Students make a variety of errors in multi-digit addition and subtraction calculations (Brown & Van Lehn, 1982). Given traditional instruction, a substantial number of 4th- and 5th-graders are not able to subtract some whole numbers successfully (Fuson, 1992). Student errors suggest students interpret and treat multi-digit numbers as single-digit numbers placed adjacent to each other, rather than using place-value meanings for the digits in different positions (Fuson, 1992). With specially designed instruction, 2nd-graders are able to understand place value and to add and subtract four-digit numbers more accurately and meaningfully than 3rd-graders receiving traditional instruction (Fuson, 1992). Research also suggests students interpret multiplication of whole numbers mainly as repeated addition. This interpretation is inadequate for many multiplication problems and can lead to restrictive intuitive notions such as "multiplication always makes larger" (Greer, 1992).

Operations with fractions and decimals. Elementary- and middle-school students make several errors when they operate on decimals and fractions (Benander & Clement, 1985; Kouba et al., 1988; Peck & Jencks, 1981; Wearne & Hiebert, 1988). For example, many middle-school students cannot add 4 + 0.3 correctly or 7 1/6 + 3 1/2 (Kouba et al., 1988; Wearne & Hiebert, 1988). These errors are due in part to the fact that students lack essential concepts about decimals and fractions and have memorized procedures that they apply incorrectly. Interventions to improve concept knowledge can lead to increased ability by 5th-graders to add and subtract decimals correctly (Wearne & Hiebert, 1988).

Students of all ages misunderstand multiplication and division (Bell et al., 1984; Graeber & Tirosh, 1988; Greer, 1992). Commonly held misconceptions include "multiplication always makes larger," "division always makes smaller," "the divisor must always be smaller than the dividend." Students may correctly select multiplication as the operation needed to calculate the cost of gasoline when the amount and unit cost are integers, then select division for the same problem when the amount and unit cost are decimal numbers (Bell et al., 1981). Numerous suggestions have been made to improve student concepts of multiplication (Greer, 1992), but further research is needed to determine how effective these suggestions will be in the classroom.

Converting between fractions and decimals. Lower middle-school students may have difficulties understanding the relationship between fractions and decimal numbers (Markovits & Sowder, 1991). They may think that fractions and decimals can occur together in a single expression, like 0.5 + 1/2, or they might believe that they must not change from one representation to the other (from 1/2 to 0.5 and back) within a given problem. Instruction that focuses on the meaning of fractions and decimals forms a basis on which to build a good understanding of the relationship between fractions and decimals. Instruction that merely shows how to translate between the two forms does not provide a conceptual base for understanding the relationship (Markovits & Sowder, 1991).

Number comparison. Lower elementary students do not have procedures to compare the size of whole numbers. By 4th grade, students generally have no difficulty comparing the sizes of whole numbers up to four digits (Sowder, 1988). Students are less successful when the number of digits is much larger or when more than two numbers are to be compared. This might be due to increased memory requirements of working with more or larger numbers (Sowder, 1988). Upper elementary- and middle-school students taught traditionally cannot successfully compare decimal numbers (Sowder, 1988, 1992a). Rather they overgeneralize the features of the whole number system to the decimal numbers (Resnick et al., 1989). They apply a "more digits make bigger" rule (according to which .1814 > .385). After specially designed instruction which develops good meanings for decimal symbols, many students are able to compare decimal numbers with understanding by 5th grade (Wearne & Hiebert, 1988). Upper elementary- and middle-school students taught traditionally cannot compare fractions successfully (Sowder, 1988). Students' difficulties here indicate they treat the numerator and the denominator separately. Specially designed instruction to teach meanings for fractions can help to improve ordering fractions by as early as the end of the 5th grade (Behr et al., 1984).

Calculators. The use of calculators in K-12 mathematics does not hinder the development of basic computation skills and frequently improves concept development and paper-and-pencil skills, both in basic operations and in problem solving (Hembree & Dessart, 1986; Kaput, 1992). The use of calculators in testing produces higher scores than paper-and-pencil efforts in problem solving as well as in basic operations (Hembree & Dessart, 1986).

Estimation skills. Good estimators use a variety of estimating tactics and switch easily between them. They have a good understanding of place value and the meaning of operations, and they are skilled in mental computation. Poor estimators rely on algorithms that are more likely to yield the exact answer. They lack an understanding of the notion and value of estimation and often describe it as "guessing" (Sowder, 1992b). Before 6th grade, students develop very few estimation skills from their natural experiences (Case & Sowder, 1990; Sowder, 1992b). As a result, some researchers caution that teaching estimation to young children may have as its single effect that they master specific procedures in a superficial manner (Sowder, 1992b).

Proportional reasoning. Early adolescents and also many adults have difficulty with proportional reasoning (Behr, 1987; Hart, 1988). Difficulty is influenced by the problem format, the particular numbers in the problem, the types of ratios used, and the problem situation (Heller et al., 1989; Karplus et al., 1983; Tournaire & Pulos, 1985; Vergnaud, 1988). Middle-school students can solve problems in proportions that involve simple numbers and simple wordings (Vergnaud, 1988), but troubles appear with more difficult numerical values or problem contexts. Problems using 2:1 ratios are easier than problems using n:1 ratios, and can be solved by elementary-school children (Shayer & Adey, 1981). Problems using n:1 ratios are easier than problems using other integer ratios (e.g., 6/2) which in turn are easier than problems using non-integer ratios (e.g., 6/4) (Tournaire & Pulos, 1985). Different ratio types (e.g., speed, exchange, mixture) appear to give more or less difficulty. For example, speed problems appear to be more difficult than exchange problems (Heller et al., 1989; Vergnaud, 1988). And these difficulties

compound one another. Unfamiliarity with the problem situation causes even more difficulty when it occurs with a difficult ratio type (Heller et al., 1989).

12c Manipulation and Observation

Upper elementary- and middle-school students who can use measuring instruments and procedures when asked to do so often do not use this ability while performing an investigation. Typically a student asked to undertake an investigation and given a set of equipment that includes measuring instruments will make a qualitative comparison even though she might be competent to use the instruments in a different context (Black, 1990). It appears students often know how to take measurements but not what or when.

12e Critical-Response Skills

Control of variables. Upper elementary-school students can reject a proposed experimental test where a factor whose effect is intuitively obvious is uncontrolled, at the level of "that's not fair" (Shayer & Adey, 1981). "Fairness" develops as an intuitive principle as early as 7 to 8 years of age and provides a sound basis for understanding experimental design. This intuition does not, however, develop spontaneously into a clear, generally applicable procedure for planning experiments (Wollman, 1977a, 1977b; Wollman & Lawson, 1977). Although young children have a sense of what it means to run a fair test, they frequently cannot identify all of the important variables, and they are more likely to control those variables that they believe will affect the result. Accordingly, student familiarity with the topic of the given experiment influences the likelihood that they will control variables (Linn & Swiney, 1981; Linn, et al., 1983). After specially designed instruction, students in 8th grade are able to call attention to inadequate data resulting from lack of controls (see for example Rowell & Dawson, 1984; Ross, 1988).

Theory and evidence. Middle-school students tend to invoke personal experiences as evidence to justify a particular hypothesis. They seem to think of evidence as selected from what is already known or from personal experience or second-hand sources, not as information produced by experiment (Roseberry et al., 1992). Most 6th-graders can judge whether evidence is related to a theory, although they do not always evaluate this evidence correctly (Kuhn et al., 1988). When asked to use evidence to judge a theory, students of all ages may make only theory-based responses with no reference made to the presented evidence. Sometimes this appears to be because the available evidence conflicts with the students' beliefs (Kuhn et al., 1988).

Interpretation of data. Students of all ages show a tendency to uncritically infer cause from correlation (Kuhn et al., 1988). Some students think even a single co-occurrence of antecedent and outcome is always sufficient to infer causality. Rarely do middle-school students realize the indeterminacy of single instances, although high-school students may readily realize it. Despite that, as covariant data accumulate, even high-school students will infer a causal relation based on correlations. Further, students of all ages will make a causal inference even when no variation occurs in one of the variables. For example, if students are told that light-colored balls are used successfully in a game, they seem willing to infer that the color of the balls will make some difference in the outcome without any evidence about dark-colored balls (Kuhn et al., 1988).

Faced with no correlation of antecedent and outcome, 6th-graders only rarely conclude that the variable has no effect on the outcome. Ninth-graders draw such conclusions more often. A basic problem appears to be understanding the distinction between a variable making no difference and a variable that is correlated with the outcome in the opposite way than the students initially conceived (Kuhn et al., 1988).

Inadequacies in arguments. Most high-school students will accept arguments based on inadequate sample size, accept causality from contiguous events, and accept conclusions based on statistically insignificant differences (Jungwirth & Dreyfus, 1990, 1992; Jungwirth, 1987). More students can recognize these inadequacies in arguments after prompting (for example, after being told that the conclusions drawn from the data were invalid and asked to state why) (Jungwirth & Dreyfus, 1992; Jungwirth, 1987).

REFERENCES

Aikenhead, G.S. (1987). High school graduates' beliefs about science-technology-society III. Characteristics and limitations of scientific knowledge. *Science Education, 71,* 459-487.

Allen, R.D., Statkiewitz, W.R., & Donovan, M. (1983). Student perceptions of evidence and interpretations. In J. Novak (Ed.), *Proceedings of the international seminar: Misconceptions in science and mathematics* (pp. 79-83). Ithaca, NY: Cornell University.

Anderson, C., Sheldon, T., & Dubay, J. (1990). The effects of instruction on college nonmajors' conceptions of respiration and photosynthesis. *Journal of Research in Science Teaching, 27,* 761-776.

Anderson, C., & Smith, E. (1983). *Children's conceptions of light and color: Understanding the concept of unseen rays.* East Lansing: Michigan State University, ERIC No. ED 270 318.

Andersson, B. (1990). Pupils' conceptions of matter and its transformations (age 12-16). In P. Lijnse, P. Licht, W. de Vos, & A.J. Waarlo (Eds.), *Relating macroscopic phenomena to microscopic particles* (pp. 12-35). Utrecht: CD-ß Press.

Arnaudin, M.W., & Mintzes, J.J. (1985). Students' alternative conceptions of the human circulatory system: A cross age study. *Science Education, 69,* 721-733.

Arnaudin, M.W., & Mintzes, J.J. (1986). The cardiovascular system: Children's conceptions and misconceptions. *Science and Children, 23*(5), 48-51.

Atwood, V. (Ed.). (1986). *Elementary school social studies: Research as a guide to practice.* Washington, DC: National Council for the Social Studies.

Baird, J., Fensham, P., Gunstone, R., & White, R. (1989). A study of the importance of reflection for improving teaching and learning. Paper presented at the annual meeting of the National Association for Research in Science Teaching, San Francisco, ERIC No. ED 307 151.

Baird, J.R., & Mitchell, J. (Eds.). (1986). *Improving the quality of teaching and learning: An Australian case study.* Melbourne, Australia: Monash University Press.

Bar, V. (1989). Children's views about the water cycle. *Science Education, 73,* 481-500.

Baxter, J. (1989). Children's understanding of familiar astronomical events. *International Journal of Science Education, 11,* 502-513.

Behr, M.J. (1987). Ratio and proportion: A synthesis of eight conference papers. In U. Bergson, N. Hescovits, & C. Kieran (Eds.), *Psychology and mathematics education* (Vol. II). Proceedings of the eleventh international conference, Montreal, Canada.

Behr, M., Lesh, R., Post, T., & Silver, E. (1983). Rational number concepts. In R. Lesh & M. Lindau (Eds.), *Acquisition of mathematical concepts and processes* (pp. 91-126). New York: Academic Press.

Behr, M., Wachsmuth, I., Post, T., & Lesh, R. (1984). Order and equivalence of rational numbers: A clinical teaching experiment. *Journal for Research in Mathematics Education, 15,* 323-341.

Bell, A., Fischbein, E., & Greer, B. (1984). Choice of operation in verbal arithmetic problems: The effects of number size, problem, structure, and context. *Educational Studies in Mathematics, 15,* 129-147.

Bell, A., Swan, M., & Taylor, G. (1981). Choice of operations in verbal problems with decimal numbers. *Educational Studies in Mathematics, 12,* 399-420.

Bell, B. (1981). When is an animal, not an animal? *Journal of Biological Education, 15,* 213-218.

Bell, B., & Brook, A. (1984). *Aspects of secondary students understanding of plant nutrition.* Leeds, UK: University of Leeds, Centre for Studies in Science and Mathematics Education.

Bell, B., & Freyberg, P. (1985). Language in the science classroom. In R. Osborne & P. Freyberg (Eds.), *Learning in Science* (pp. 29-40). Auckland, NZ: Heinemann.

Benander, L., & Clement, J. (1985). Catalogue of error patterns observed in courses in basic mathematics. Unpublished manuscript. (ERIC Reproduction Service No. ED 287 672).

Bernstein, A.C., & Cowan, P.A. (1975). Children's concepts of how people get babies. *Child Development*, 46, 77-91.

Berti, A. E. (1988). The development of political understanding in children between 6-15 years old. *Human Relations*, 41, 437-446.

Berti, A.E. (1992). Acquisition of the profit concept by third-grade children. *Contemporary Educational Psychology*, 17, 293-299.

Berti, A.E., & Bombi, A.S. (1988). *The child's construction of economics*. Cambridge: Cambridge University Press.

Berti, A.E., & Grivet, A. (1990). The development of economic reasoning in children from 8 to 13 years old: Price mechanism. *Contributi di Psicologia*, 3(III), 37-47.

Bishop, B., & Anderson, C. (1990). Student conceptions of natural selection and its role in evolution. *Journal of Research in Science Teaching*, 27, 415-427.

Black, P. (1990). Can pupils design their own experiments? In *Proceedings of the international conference on physics education through experiments* (pp. 281-299). Tianjin, China: ICPE.

Black, P., & Solomon, J. (1983). Life world and science world: Pupils' ideas about energy. In G. Marx (Ed.), *Entropy in the school: Proceedings of the 6th Danube seminar on physics education* (pp. 43-55). Budapest: Roland Eotvos Physical Society.

Bliss, J. (1978). Ideas of chance and probability in children and adolescents. *Physics Education*, 13, 408-413.

Blum, L.H. (1977). Health information via mass media: Study of the individual's concepts of the body and its parts. *Psychological Reports*, 40, 991-999.

Boylan, C., Hill, D., Wallace, A., & Wheeler, A. (1992). Beyond stereotypes. *Science Education*, 76, 465-476.

Brook, A., Briggs, H., & Bell, B. (1983). *Secondary students' ideas about particles*. Leeds, UK: The University of Leeds, Centre for Studies in Science and Mathematics Education.

Brook, A., & Driver, R. (1984). *Aspects of secondary students' understanding of energy: Summary report*. Leeds, UK: University of Leeds, Centre for Studies in Science and Mathematics Education.

Brook, A., & Driver, R. (1986). *The construction of meaning and conceptual change in the classroom: Case studies on energy*. Leeds, UK: University of Leeds, Centre for Studies in Science and Mathematics Education.

Brook, A., & Wells, P. (1988). Conserving the circus: An alternative approach to teaching and learning about energy. *Physics Education*, 23, 80-85.

Brosnan, T. (1990). Categorizing macro and micro explanations of material change. In P.L. Lijnse, P. Licht, W. de Vos, & A.J. Waarlo (Eds.), *Relating macroscopic phenomena to microscopic particles* (pp. 198-211). Utrecht, Holland: CD-ß Press.

Brown, C., Carpenter, T., Kouba, V., Lindquist, M., Silver, E., & Swafford, J. (May 1988). Secondary school results from the fourth NAEP mathematics assessment: Algebra, geometry, mathematical methods, and attitudes. *Mathematics Teacher*, 337-347, 397.

Brown, D., & Clement, J. (1989). Overcoming misconceptions via analogical reasoning: abstract transfer versus explanatory model construction. *Instructional Science*, 18, 237-261.

Brown, D., & Clement, J. (1992). Classroom teaching experiments in mechanics. In R. Duit, F. Goldberg, & H. Niedderer (Eds.), *Research in physics learning: Theoretical issues and empirical studies* (pp. 380-397). Kiel, Germany: Institute for Science Education at the University of Kiel.

Brown, J. & VanLehn, K. (1982). Towards a generative theory of "bugs." In T.P. Carpenter, J. Moser, & T. Romberg (Eds.), *Addition and subtraction: A cognitive perspective* (pp. 117-135). Hillsdale, NJ: Lawrence Erlbaum Associates.

Brumby, M. (1979). Problems in learning the concept of natural selection. *Journal of Biological Education*, 13, 119-122.

Brumby, M. (1982). Students' perceptions of the concept of life. *Science Education*, 66, 613-622.

Brumby, M. N., Garrard, J., & Auman, J. (1985). Students' perceptions of the concept of health. *European Journal of Science Education*, 7, 307-323.

Burger, W., & Shaughnessy, J. (1986). Characterizing the van Hiele levels of development in geometry. *Journal for Research in Mathematics Education*, 17, 31-48.

Carey, S. (1985). *Conceptual change in childhood.* Cambridge, MA: MIT Press.

Carey, S. (1991). Knowledge acquisition: Enrichment or conceptual change? In S. Carey & R. Gelman (Eds.), *The epigenesis of mind: Essays on biology and cognition* (pp. 257-291). Hillsdale, NJ: Lawrence Erlbaum Associates.

Carey, S., Evans, R., Honda, M., Jay, E., & Unger, C. (1989). An experiment is when you try it and see if it works: A study of grade 7 students' understanding of the construction of scientific knowledge. *International Journal of Science Education*, 11, 514-529.

Carpenter, T., Corbitt, M., Kepner, H., Lindquist, M., & Reys, R. (1981). Decimals: Results and implications from the second NAEP mathematics assessment. *Arithmetic Teacher*, 28(8), 34-37.

Carpenter, T., Lindquist, M., Matthews, W., & Silver, E. (1983). Results of the third NAEP mathematics assessment: Secondary school. *Mathematics Teacher*, 76, 652-659.

Carpenter, T., & Moser, J. (1983). The acquisition of addition and subtraction concepts. In R. Lesh & M. Landau (Eds.), *Acquisition of mathematics: Concepts and processes* (pp. 7-44). New York: Academic Press.

Carr, M., & Kirkwood, V. (1988). Teaching and learning about energy in New Zealand secondary school junior science classrooms. *Physics Education*, 23, 86-91.

Case, R., & Sowder, J. (1990). The development of computational estimation: A neo-Piagetian analysis. *Cognition and Instruction*, 7, 79-104.

Champagne, A., Gunstone, R., & Klopfer, L. (1985). Effecting changes in cognitive structures among physics students. In L. West & A. Pines (Eds.), *Cognitive structure and conceptual change* (pp. 61-90). Orlando, FL: Academic Press.

Children's Learning in Science (1987). *Approaches to teaching the particulate theory of matter.* Leeds, UK: University of Leeds, Centre for Studies in Science and Mathematics Education.

Clement, J. (1982). Algebra word problem solutions: Thought processes underlying a common misconception. *Journal for Research in Mathematics Education*, 13, 16-30.

Clement, J. (1987). Overcoming students' misconceptions in physics: The role of anchoring intuitions and analogical validity. In J. Novak (Ed.), Proceedings of the second international seminar misconceptions and educational strategies in science and mathematics (Vol. III, pp. 84-96). Ithaca, NY: Cornell University.

Clement, J. (1989).The concept of variation and mis-conceptions in Cartesian graphing. *Focus on Learning Problems in Mathematics*, 11(1-2), 77-87.

Clements, D., & Battista, M. (1989). Learning of geometric concepts in a Logo environment. *Journal for Research in Mathematics Education*, 20, 450 467.

Clements, D., & Battista, M. (1990). The effects of Logo on children's conceptualizations of angle and polygons. *Journal for Research in Mathematics Education*, 21, 356-371.

Clements, D., & Battista, M. (1992). Geometry and spacial reasoning. In D. Grouws (Ed.), *Handbook of research on mathematics teaching and learning* (pp. 420-464). New York: Macmillan Publishing Company.

Clough, E.E., & Wood-Robinson, C. (1985a). How secondary students interpret instances of biological adaptation. *Journal of Biological Education*, 19, 125-130.

Clough, E.E., & Wood-Robinson, C. (1985b). Children's understanding of inheritance. *Journal of Biological Education*, 19, 304-310.

Connell, R. (1971). *The child's construction of politics.* Calton, Australia: Melbourne University Press.

Contento, I. (1981). Children's thinking about food and eating: *A Piagetian-based study. Journal of Nutrition Education*, 13(1), 86-90.

Cooley, W., & Klopfer, L. (1961). *Test on understanding science, Form W.* Princeton: Educational Testing Service.

Damon, W. (1977). *The social world of the child.* San Francisco: Jossey-Bass.

Deadman, J., & Kelly, P. (1978). What do secondary school boys understand about evolution and heredity before they are taught the topics? *Journal of Biological Education*, 12, 7-15.

Dickinson, A.K., Lee, P.J., & Rogers, P.J. (Eds.). (1984). *Learning history* (pp. 39-84). London: Heinemann Educational Books.

Donaldson, M. (1978). *Children's minds.* New York: W. W. Norton & Company.

Donnelly, J.F. (1992). Technology in the school curriculum: A critical bibliography. *Studies in Science Education*, 20, 123-156.

Dossey, J., Mullis, I., Lindquist, M.M., & Chambers, D. (1988). *The mathematics report card: Trends and achievement based on the 1986 national assessment.* Princeton, NJ: Educational Testing Service.

Downey, M. & Levstik, L. (1988, September). Teaching and learning history: The research base. *Social Education*, 336-342.

Downey, M. & Levstik, L. (1991). Teaching and learning history. In J. Shaver (Ed.), *Handbook of research on social studies teaching and learning* (pp. 400-410). New York: Macmillan Publishing Company.

Dreyfus, A., & Jungwirth, E. (1988). The cell concept of 10th graders: Curricular expectations and reality. *International Journal of Science Education*, 10, 221-229.

Dreyfus, A., & Jungwirth, E. (1989). The pupil and the living cell: A taxonomy of dysfunctional ideas about an abstract idea. *Journal of Biological Education*, 23, 49-55.

Driver, R. (1985). Beyond appearances: The conservation of matter under physical and chemical transformations. In R. Driver, E. Guesne, & A. Tiberghien (Eds.), *Children's ideas in science* (pp. 145-169). Milton Keynes, UK: Open University Press.

Driver, R., Guesne, E., & Tiberghien, A. (1985). Some features of children's ideas and their implications for teaching. In R. Driver, E. Guesne, & A. Tiberghien (Eds.), *Children's ideas in science* (pp. 193-201). Milton Keynes, UK: Open University Press.

Driver, R., & Millar, R. (1985). *Energy matters.* Leeds, UK: University of Leeds, Centre for Studies in Science and Mathematics Education.

Duit, R., Goldberg, F., & Niedderer, H. (Eds.). (1992). *Research in physics learning: Theoretical issues and empirical studies.* Kiel, Germany: Institute for Science Education at the University of Kiel.

Duschl, R., Smith, M., Kesidou, S., Gitomer, D., & Schauble, L. (1992, April). Assessing student explanations for criteria to format conceptual change learning environments. Paper presented at the annual meeting of the American Educational Research Association, San Francisco, CA.

Dykstra, D., Boyle, C., & Monarch, I. (1992). Studying conceptual change in learning physics. *Science Education,* 76, 615-652.

Edwards, C. P. (1986). *Promoting social and moral development in young children.* New York: Teachers College Press.

Egan, K. (1982, March). Teaching history to young children. *Phi Delta Kappan,* 439-441

Ericksen, D. (1991). Students ability to recognize patterns. *School Science and Mathematics,* 91, 255-258.

Erickson, G. (1985). Heat and temperature: An overview of pupils' ideas. In R. Driver, E. Guesne, & A. Tiberghien (Eds.), *Children's ideas in science* (pp. 55-66). Milton Keynes, UK: Open University Press.

Falk, R., Falk, R., & Levin, I. (1980). A potential for learning probability in young children. *Journal for Research in Mathematics Education,* 11, 181-204.

Fischbein, E., & Gazit, A. (1984). Does the teaching of probability improve probabilistic intuitions? *Educational Studies in Mathematics,* 15, 1-24.

Fleming, R. (1986a). Adolescent reasoning in socio-scientific issues. Part I: Social cognition. *Journal of Research in Science Teaching,* 23, 677-687.

Fleming, R. (1986b). Adolescent reasoning in socio-scientific issues. Part II: Nonsocial cognition. *Journal of Research in Science Teaching,* 23, 688-698.

Fleming, R. (1987). High school graduates' beliefs about science-technology-society II. The interaction among science, technology, society. *Science Education,* 71, 163-186.

Fort, D., & Varney, H. (1989). How students see scientists: Mostly male, mostly white, and mostly benevolent. *Science and Children,* 26(8), 8-13.

Foster, T. (1992a). Technology education research: Looking to the future. *The Technology Teacher,* 52(1), 33-34.

Foster, T. (1992b). Topics and methods of recent graduate student research in industrial education and related fields. *Journal of Industrial Teacher Education,* 30, 59-72.

Freyberg, P. (1985). Implications across the curriculum. In R. Osborne & P. Freyberg (Eds.), *Learning in science* (pp. 125-135). Auckland, NZ: Heinemann.

Fuson, K. (1988). *Children's counting and concepts of number.* New York: Springer Verlag.

Fuson, K. (1992). Research on whole number addition and subtraction. In D. A. Grouws (Ed.), *Handbook of research on mathematics teaching and learning* (pp. 243-275). New York: Macmillan Publishing Company.

Fuson, K., Richards, J., & Briars, D. (1982). The acquisition and elaboration of the number word sequence. In C. Brainerd (Ed.), *Progress in cognitive development research Vol. 1: Children's logical and mathematical cognition* (pp. 33-92). New York: Springer Verlag.

Fuson, K., & Willis, G. (1989). Second graders' use of schematic drawings in solving addition and subtraction word problems. *Journal of Educational Psychology,* 81, 514-520.

Gal, I., Rothschild, K., & Wagner, D. (1990). Which group is better? The development of statistical reasoning in elementary school children. Paper presented at the American Educational Research Association, Boston, MA.

Garfield, J., & Ahlgren, A. (1988). Difficulties in learning probability and statistics: Implications for research. *Journal for Research in Mathematics Education*, 19, 44-63.

Garigliano, L. (1975). SCIS: Children's understanding of the systems concept. *School Science and Mathematics*, 75, 245-249.

Gega, P. (1986). *Science in elementary education.* New York: Macmillan Publishing Company.

Gellert, E. (1962). Children's conceptions of the content and functions of the human body. *Genetic Psychology Monographs*, 65, 293-305.

Goldman, R.J., & Goldman, J.D. (1982). How children perceive the origin of babies and the role of mothers and fathers in procreation: A cross-national study. *Child Development*, 53, 491-504.

Good, R., Trowbridge, J., Demastes, S., Wandersee, J., Hafner, M., & Cummins, C. (Eds.). (1993). Proceedings of the 1992 evolution education research conference. Baton Rouge, LA: Louisiana State University.

Graeber, A., & Tirosh, D. (1988). Multiplication and division involving decimals: Preservice elementary teachers' performance and beliefs. *Journal of Mathematical Behavior*, 7, 263-280.

Greeno, J. (1982, March). A cognitive learning analysis of algebra. Paper presented at the annual meeting of the American Educational Research Association, Boston, MA.

Greer, B. (1992). Multiplication and division as models of situations. In D. A. Grouws (Ed.), *Handbook of research on mathematics teaching and learning* (pp. 276-295). New York: Macmillan Publishing Company.

Grosslight, L., Unger, C., Jay, E., & Smith, C.L. (1991). Understanding models and their use in science: Conceptions of middle and high school students and experts. *Journal of Research in Science Teaching*, 28, 799-822.

Grouws, D. (Ed.) (1992). *Handbook of research on mathematics teaching and learning.* New York: Macmillan Publishing Company.

Guesne, E. (1985). Light. In R. Driver, E. Guesne, & A. Tiberghien (Eds.), *Children's ideas in science* (pp. 10-32). Milton Keynes, UK: Open University Press.

Gunstone, R., & Watts, M. (1985). Force and motion. In R. Driver, E. Guesne, & A. Tiberghien (Eds.), *Children's ideas in science* (pp. 85-104). Milton Keynes, UK: Open University Press.

Gunstone, R., & White, R. (1981). Understanding of gravity. *Science Education*, 65, 291-299.

Hallam, R.N. (1970). Piaget and thinking in history. In M. Ballard (Ed.), *New movements in the study and teaching of history.* Bloomington: Indiana University Press.

Hallam, R.N. (1979). Attempting to improve logical thinking in school history. *Research in Education*, 21, 1-24.

Hallden, O. (1988). The evolution of species: Pupils' perspectives and school perspectives. *International Journal of Science Education*, 10, 541-552.

Hart, K. (1988). Ratio and proportion. In J. Hiebert & M. Behr (Eds.), *Number concepts and operations in the middle grades* (pp. 198-219). Reston, VA: National Council of Teachers of Mathematics.

Hawkins, A., & Kapadia, R. (1984). Children's conceptions of probability: A psychological and pedagogical review. *Educational Studies in Mathematics*, 15, 349-377.

Heller, P., Ahlgren, A., Post, T., Behr, M., & Lesh, R. (1989). Proportional reasoning: The effect of two context variables, rate type, and problem setting. *Journal of Research in Science Teaching*, 26, 205-220.

Hembree, R., & Dessart, D. (1986). Effects of hand held calculators in precollege mathematics education: A meta-analysis. *Journal for Research in Mathematics Education*, 17, 83-89.

Hennessy, S., McCormick, R., & Murphy, P. (in press). The myth of general problem-solving capability: Design and technology as an example. *Curriculum Journal.*

Hergenrather, J., & Rabinowitz, M. (1991). Age-related differences in the organization of children's knowledge of illness. *Developmental Psychology*, 27, 952-959.

Herget, D. (Ed.). (1989). The history and philosophy of science in science teaching. Proceedings of the first international conference (Vols. 1-2). Tallahassee, FL: Florida State University.

Herscovics, N. (1989). Cognitive obstacles encountered in the learning of algebra. In S. Wagner & C. Kieran (Eds.), *Research issues in the learning and teaching of algebra* (pp. 60-86). Reston, VA: National Council of Teachers of Mathematics.

Hess, R., & Torney, J. (1968). *The development of political attitudes in children.* Garden City, NY: Doubleday & Company.

Hiebert, J. (1992). Mathematical, cognitive, and instructional analyses of decimal fractions. In G. Leinhardt, R. Putnam, & R. Hattrup (Eds.), *Analysis of arithmetic for mathematics teaching* (pp. 283-322). Hillsdale, NJ: Lawrence Erlbaum Associates.

Hiebert, J., & Behr, M. (Eds.). (1988). *Number concepts and operations in the middle grades.* Reston, VA: National Council of Teachers of Mathematics.

Hiebert, J., & Wearne, D. (1986). Procedures over concepts: The acquisition of decimal number knowledge. In J. Hiebert (Ed.), *Conceptual and procedural knowledge: The case of mathematics* (pp. 199-223). Hillsdale, NJ: Lawrence Erlbaum Associates.

Hill, D., & Redden, M. (1985). An investigation of the system concept. *School Science and Mathematics*, 85, 233-239.

Hills, S. (Ed.). (1992). *The history and philosophy of science in science education.* Proceedings of the second international conference (Vols. 1-2). Kingston, Ontario: Queen's University.

Ingham, A.M., & Gilbert, J.K, (1991). The use of analogue models by students of chemistry at higher education level. *International Journal of Science Education*, 13, 193-202.

Jahoda, G. (1979). The construction of economic reality by some Glaswegian children. *European Journal of Social Psychology*, 19, 115-127.

Jahoda, G. (1981). The development of thinking about economic institutions: the bank. *Cahiers de Psychologie Cognitive*, 1(1), 55-73.

Johnson, C., & Wellman, H. (1982). Children's developing conceptions of the mind and brain. *Child Development*, 53(1), 222-234.

Joyce, W., Little, T., & Wronski, S. (1991). Scope and sequence, goals, and objectives: Effects on social studies. In J. Shaver (Ed.), *Handbook of research on social studies teaching and learning* (pp. 321-331). New York: Macmillan Publishing Company.

Jung, W., Pfundt, H., & Rhoeneck, C. von. (Eds.). (1981). Proceedings of the international workshop on "problems concerning students' representations of physics and chemistry knowledge." Ludwigsburg: Paedagogische Hochschule.

Jung, W., Wiesner, H., & Engelhard, P. (1981). *Vorstellungen von Schuelern ueber Begriffe der Newtonschen Mechanik.* Bad Salzdetfurth: Didaktischer Dienst Franzbecker.

Jungwirth, E. (1987). Avoidance of logical fallacies: A neglected aspect of science education and science-teacher education. *Research in Science and Technological Education*, 5, 43-58.

Jungwirth, E. (1975). Preconceived adaptation and inverted evolution (a case study of distorted concept formation in high school biology) *Australian Science Teacher Journal*, 21, 95-100.

Jungwirth, E., & Dreyfus, A. (1990). Identification and acceptance of a posteriori causal assertions invalidated by faulty enquiry methodology: An international study of curricular expectations and reality. In D. Herget (Ed.), *More history and philosophy of science in science teaching* (pp. 202-211). Tallahassee, FL: Florida State University.

Jungwirth, E., & Dreyfus, A. (1992). After this, therefore because of this: One way of jumping to conclusions. *Journal of Biological Education*, 26, 139-142.

Kahneman, D., Slovic, P., & Tversky, A. (1982). *Judgment under certainty: Heuristics and biases*. Cambridge: Cambridge University Press.

Kaput, J. (1992). Technology and mathematics education. In D. A. Grouws (Ed.), *Handbook of research on mathematics teaching and learning* (pp. 515-556). New York: Macmillan Publishing Company.

Kargbo, D., Hobbs, E., & Erickson, G. (1980). Children's beliefs about inherited characteristics. *Journal of Biological Education*, 14, 137-146.

Karplus, R., Pulos, S., & Stage, E. (1983). Proportional reasoning of early adolescents. In R. Lesh & M. Landau (Eds.), *Acquisition of mathematics concepts and processes*. New York: Academic Press.

Karplus, R. & Thier, H. (1969). *A new look at elementary school science; science curriculum improvement study*. Chicago: Rand McNally.

Keil, F. (1989). *Concepts, kinds, and cognitive development*. Cambridge, MA: MIT Press.

Kerslake, D. (1981). Graphs. In K. M. Hart (Ed.), *Children's understanding of mathematics:* 11-16 (pp. 120-136). London: John Murray.

Kesidou, S. (1990). *Schuelervorstellungen zur Irreversibilitaet*. Kiel: Institute for Science Education at the University of Kiel.

Kesidou, S., & Duit, R. (1993). Students' conceptions of the second law of thermodynamics: An interpretive study. *Journal of Research on Science Teaching*, 30, 85-106.

Kieran, C. (1981). Concepts associated with the equality symbol. *Educational Studies in Mathematics*, 12, 317-326.

Kieran, C. (1984). A comparison between novice and more-expert algebra students on tasks dealing with the equivalence of equations. In J. Moser (Ed.), Proceedings of the sixth annual meeting of PME-NA (pp. 83-91). Madison: University of Wisconsin.

Kieran, C. (1988). Two different approaches among algebra learners. In A. F. Coxford (Ed.), *The ideas of algebra, K-12* (1988 Yearbook, pp. 91-96). Reston, VA: National Council of Teachers of Mathematics.

Kieran, C. (1989). The early learning of algebra: A structural perspective. In S. Wagner & C. Kieran (Eds.), *Research issues in the learning and teaching of algebra* (pp. 33-56). Reston, VA: National Council of Teachers of Mathematics.

Kieran, C. (1992). The learning and teaching of school algebra. In D. Grouws (Ed.), *Handbook of research on mathematics teaching and learning* (pp. 390-419). New York: Macmillan Publishing Company.

Kieren, T. (1992). Rational and fractional numbers as mathematical and personal knowledge: Implications for curriculum and instruction. In G. Leinhardt, R. Putnam, & R. Hattrup (Eds.), *Analysis of arithmetic for mathematics teaching* (pp. 323-372). Hillsdale, NJ: Lawrence Erlbaum Associates.

Kimbell, R., Stables, K., Wheeler, T., Wosniak, A., & Kelly, V. (1991). *The assessment of performance in design and technology*. London, UK: School Examinations and Assessment Council.

Kircher, E. (1985). Analogies for the electric circuit? In R. Duit, W. Jung, and C. von Rhoeneck (Eds.), *Aspects of understanding electricity* (pp. 299-310). Kiel, Germany: Institute for Science Education at the University of Kiel.

Kitchener, K. (1983, Fall). Educational goals and reflective thinking. *The Educational Forum*, 75-95.

Kitchener, K., & King, P. (1981). Reflective judgment: Concepts of justification and their relationship to age and education. *Journal of Applied Developmental Psychology*, 2, 89-116.

Klopfer, L., & Cooley, W. (1963). Effectiveness of the history of science cases for high schools in the development of student understanding of science and scientists. *Journal of Research in Science Teaching*, 1, 35-47.

Kouba, V., Brown, C., Carpenter, T., Lindquist, M., Silver, E., & Swafford, J. (1988). Results of the fourth NAEP assessment of mathematics: Numbers, operations, and word problems. *Arithmetic Teacher*, 35(8), 14-19.

Kuhn, D. (1991). *The skills of argument*. Cambridge: Cambridge University Press.

Kuhn, D. (1992). Thinking as argument. *Harvard Educational Review*, 62, 155-178.

Kuhn, D., Amsel, E., & O'Loughlin, M. (1988). *The development of scientific thinking skills*. San Diego, CA: Academic Press.

Lawson, A., & Thomson, L. (1988). Formal reasoning ability and misconceptions concerning genetics and natural selection. *Journal of Research in Science Teaching*, 25, 733-746.

Lawson, A., & Worsnop, W. (1992). Learning about evolution and rejecting a belief about natural creation: Effects of reasoning skill, prior knowledge, prior beliefs and religious commitment. *Journal of Research in Science Teaching*, 29, 143-166.

Lazar, A., & Torney-Purta, J. (1991). The development of the subconcepts of death in young children: A short-term longitudinal study. *Child Development*, 62, 1321-1333.

Leach, J., Driver, R., Scott, P., & Wood-Robinson, C. (1992). *Progression in understanding of ecological concepts by pupils aged 5 to 16*. Leeds, UK: The University of Leeds, Centre for Studies in Science and Mathematics Education.

Lederman, N. (1992). Students' and teachers' conceptions of the nature of science: A review of the research. *Journal of Research in Science Teaching*, 29, 331-359.

Lederman, N., & O'Malley, M. (1990). Students' perceptions of the tentativeness in science: Development, use, and sources of change. *Science Education*, 74, 225-239.

Lee, O., Eichinger, D.C., Anderson, C.W., Berkheimer, G.D., & Blakeslee, T.S. (1993). Changing middle school students' conceptions of matter and molecules. *Journal of Research in Science Teaching*, 30, 249-270.

Lee, P.J. (1984). Historical imagination. In A.K. Dickinson, P.J. Lee, & P.J Rogers (Eds.), *Learning history* (pp. 85-117). London: Heinemann Educational Books.

Leinhardt, G., Zaslavsky, O., & Stein, M. (1990). Functions, graphs, and graphing: Tasks, learning, and teaching. *Review of Educational Research*, 60, 1-64.

Levstik, L. (1986). The relationship between historical response and narrative in a sixth-grade classroom. *Theory and Research in Social Education*, 16(1), 1-19.

Levstik, L. (1988). Historical narrative and the young reader. *Theory into Practice*, XXVIII(2), 114-119.

Levstik, L. & Pappas, C. (1987). Exploring the development of historical understanding. *Journal of Research and Development in Education*, 21(1), 1-15.

Lewis, E.L. (1991, April). The development of understanding in elementary thermodynamics. Paper presented at the annual meeting of the American Educational Research Association, Chicago, IL, ERIC No. ED 344 744.

Lijnse, P. (Ed.). (1985). *The many faces of teaching and learning mechanics: Conference on physics education.* Utrecht: GIREP/SVO/UNESCO.

Lijnse, P., Licht, P., de Vos, W., & Waarlo A. (Eds.). (1990). Relating macroscopic phenomena to microscopic particles. Utrecht, Holland: CD-ßPress.

Linn, M., Clement, C., & Pulos, S. (1983). Is it formal if it's not physics? The influence of content on formal reasoning. *Journal of Research in Science Teaching, 20,* 755-776.

Linn, M., & Songer, N.B. (1991). Teaching thermo-dynamics to middle school students: What are appropriate cognitive demands? *Journal of Research in Science Teaching, 28,* 885-918.

Linn, M. & Swiney, J. (1981). Individual differences in formal thought: Role of cognitions and aptitudes. *Journal of Educational Psychology, 73,* 274-286.

Lucas, A. (1971). The teaching of adaptation. *Journal of Biological Education, 5,* 86-90.

Mackey, L. (1971). Development of understanding about the nature of science. *Journal of Research in Science Teaching, 8,* 57-66.

Mackey, J. (1991). Adolescents' social, cognitive, and moral development and secondary school social studies. In J. Shaver (Ed.), *Handbook of research on social studies teaching and learning.* New York: Macmillan Publishing Company.

Markovits, Z., & Sowder, J. (1991). Students' understanding of the relationship between fractions and decimals. *Focus on Learning Problems in Mathematics, 13*(1), 3-11.

Marton, F. (1978). Phenomenography: Describing conceptions of the world around us. *Instructional Science, 10,* 177-200.

Mas, C.J., Perez, J.H., & Harris, H. (1987). Parallels between adolescents' conceptions of gases and the history of chemistry. *Journal of Chemical Education, 64,* 616-618.

Matthews, M. (Ed.). (1991). *History, philosophy, and science teaching. Selected readings.* Toronto: OISE Press.

McCormick, R., Hennessy, S. & Murphy, P. (1993, April). Problem-solving processes in technology education. Paper presented at the 55th annual conference of the ITEA, Charlotte, NC.

McDermott, L. (1983). Critical review of research in the domain of mechanics. *Proceedings of the first international workshop research on physics education* (pp. 139-182). Paris: Editions du CNRS.

McDermott, L., Rosenquist, M., & van Zee, E. (1987). Student difficulties in connecting graphs and physics: Example from kinematics. *American Journal of Physics, 55,* 503-513.

McLeod, D. (1992). Affect in mathematics education: A reconceptualization. In D. Grouws (Ed.), *Handbook of research on mathematics teaching and learning* (pp. 575-596). New York: Macmillan Publishing Company.

Mead, M., & Metraux, R. (1957). Image of the scientist among high-school students: A pilot study. *Science, 26,* 384-390.

Merkle, D.G., & Treagust, D.F. (1987). Secondary school students' locus of control and conceptual knowledge relating to health and fitness. In J. Novak (Ed.), *Proceedings of the second international seminar misconceptions and educational strategies in science and mathematics* (Vol. II, pp. 325-335). Ithaca, NY: Cornell University.

Millar, R. (1990). Making sense: What use are particles to children? In P. Lijnse, P. Licht, W. de Vos, & A.J. Waarlo (Eds.), *Relating macroscopic phenomena to microscopic particles* (pp. 283-293). Utrecht: CD-ß Press.

Minstrell, J. (1982). Explaining the "at rest" condition of an object. *The Physics Teacher, 20,* 10-14.

Minstrell, J. (1989). Teaching science for understanding. In L. Resnick & L. Klopfer (Eds.), *Toward the thinking curriculum: Current cognitive research* (pp. 129-149). Alexandria, VA: Association for Supervision and Curriculum Development.

Minstrell, J. (1992). Facets of students' knowledge and relevant instruction. In R. Duit, F. Goldberg, & H. Niedderer (Eds.), *Research in physics learning: Theoretical issues and empirical studies* (pp. 110-128). Kiel, Germany: Institute for Science Education at the University of Kiel.

Minstrell, J., Stimpson, V., & Hunt, E. (1992, April). Instructional design and tools to assist teachers in addressing students' understanding and reasoning. Paper presented at the annual meeting of the American Educational Research Association, San Francisco.

Mintzes, J., Trowbridge, J., Arnaudin, M., & Wandersee, J. (1991). Children's biology: Studies on conceptual development in the life sciences. In S. Glynn, R. Yeany, & B. Britton (Eds.), *The psychology of learning science* (pp. 179-202). Hillsdale, NJ: Lawrence Erlbaum Associates.

Mokros, J., & Russell, S. (1992). Children's concepts of average and representativeness. *Working Paper* 4-92. Cambridge, MA: TERC.

Mokros, J., & Tinker, R. (1987). The impact of microcomputer-based labs on children's ability to interpret graphs. *Journal of Research in Science Teaching*, 24, 369-383.

Moon, A., Wetton, N., & Williams, D. (1985). Perceptions of young children concerning health. In P.J. Kelly & J.L. Lewis (Ed.), *Education and health* (pp. 27-34). Oxford: Pergamon Press.

Moore, S.W., Lare, J., & Wagner, K. (1985). *The child's political world: A longitudinal perspective*. New York: Praeger Publishers.

Nagy, M.H. (1953). The representations of "germs" by children. *Journal of Genetic Psychology*, 83, 227-240.

Newton, D., & Newton, L. (1992). Young children's perceptions of science and the scientist. *International Journal of Science Education*, 14, 331-348.

Novak, J., & Gowin, B. (1984). Learning how to learn. Cambridge: Cambridge University Press.

Nussbaum, J. (1985a). The earth as a cosmic body. In R. Driver, E. Guesne, & A. Tiberghien (Eds.), *Children's ideas in science* (pp. 170-192). Milton Keynes, UK: Open University Press.

Nussbaum, J. (1985b). The particulate nature of matter in the gaseous phase. In R. Driver, E. Guesne, & A. Tiberghien (Eds.), *Children's ideas in science* (pp. 124-144). Milton Keynes, UK: Open University Press.

Ogborn, J. (1985). Understanding students' understandings: An example from dynamics. *European Journal of Science Education*, 7, 141-150.

Osborne, R. (1985). Building on children's intuitive ideas. In R. Osborne & P. Freyberg (Eds.), *Learning in Science* (pp. 41-50). Auckland, NZ: Heinemann.

Osborne, R., & Freyberg, P. (1985). Children's science. In R. Osborne & P. Freyberg (Eds.), *Learning in Science* (pp. 5-14). Auckland, NZ: Heinemann.

Peck, D., & Jencks, S. (1981). Conceptual issues in the teaching and learning of fractions. *Journal for Research in Mathematics Education*, 12, 339-348.

Perry, W. G., Jr. (1970). *Forms of intellectual and ethical development in the college years*. Fort Worth, TX: HBJ College Publishers.

Pfundt, H. (1981). The atom– the final link in the division process or the first building block? *Chemica Didactica*, 7, 75-94.

Pfundt, H., & Duit, R. (1991). *Bibliography. Students' alternative frameworks and science education* (3rd Ed.). Kiel, Germany: Institute for Science Education at the University of Kiel.

Piaget, J., & Inhelder, B. (1975). *The origin of the idea of chance in children*. London: Routledge & Kegan Paul.

Pollatsek, A., Lima, S., & Well, A. (1981). Concept of computation: Students' understanding of the mean. *Educational Studies in Mathematics*, 12, 191-204.

Prout, A. (1985). Science, health, and everyday knowledge: A case study about the common cold. *European Journal of Science Education*, 7, 399-406.

Ramadas, J., & Driver, R. (1989). *Aspects of secondary students' ideas about light*. Leeds, UK: University of Leeds, Centre for Studies in Science and Mathematics Education.

Ramsey, P. (1986). Racial and cultural categories. In C.P. Edwards (Ed.), *Promoting social and moral development in young children* (pp. 78-101). Columbia University: Teachers College Press.

Resnick, L., Nesher, P., Leonard, F., Magone, M. Omanson, S., & Peled, I. (1989). Conceptual bases of arithmetic errors: The case of decimal fractions. *Journal for Research in Mathematics Education*, 20, 8-27.

Rice, P. (1991). Concepts of health and illness in Thai children. *International Journal of Science Education*, 13, 115-127.

Romberg, T., & Carpenter, T. (1986). Research on teaching and learning mathematics: Two disciplines of scientific inquiry. In M. Wittrock (Ed.), *Handbook of research on teaching* (pp. 850-873). New York: Macmillan Publishing Company.

Roseberry, A., Warren, B., & Conant, F. (1992). Appropriating scientific discourse: Findings from language minority classrooms (Working paper 1-92). Cambridge, MA: TERC.

Ross, J.A. (1988). Controlling variables: A meta-analysis of training studies. *Review of Educational Research*, 58, 405-437.

Roth, K., & Anderson, C. (1987). The power plant: Teacher's guide to photosynthesis. Occasional Paper no. 112. Institute for Research on Teaching. East Lansing: Michigan State University, ERIC No. ED 288 699.

Rowell, J., & Dawson, C. (1984). Controlling variables: Testing a programme for teaching a general solution strategy. *Research in Science and Technological Education*, 2(1), 37-46.

Ruggiero, S., Cartelli, A., Dupre, F., & Vicentini-Missoni, M. (1985). Weight, gravity and air pressure: Mental representations by Italian middle-school students. *European Journal of Science Education*, 7, 181-194.

Russell, T., Harlen, W., & Watt, D. (1989). Childrens' ideas about evaporation. *International Journal of Science Education*, 11, 566-576.

Russell, T., Longden, K., & McGuigan (1991). *Materials*. Primary Space Project Research Report. Liverpool, UK: Liverpool University Press.

Russell, T., & Watt, D. (1990). *Evaporation and condensation*. SPACE Project Research Report. Liverpool, UK: Liverpool University Press.

Ryan, A. & Aikenhead, G. (1992). Students' preconceptions about the epistemology of science. *Science Education*, 76, 559-580.

Sadler, P. (1987). Misconceptions in astronomy. In J. Novak (Ed.), *Proceedings of the second international seminar misconceptions and educational strategies in science and mathematics* (Vol. III, pp. 422-425). Ithaca, NY: Cornell University.

Saxe, G., Becker, J., & Sadeghpour, M. (1989). Developmental differences in children's understanding of number word conventions. *Journal for Research in Mathematics Education*, 20, 468-488.

Scharmann, L., & Harris, W. (1992). Teaching evolution: Understanding and applying the nature of science. *Journal of Research in Science Teaching*, 29, 375-388.

Schauble, L. (1990). Belief revision in children: The role of prior knowledge and strategies for generating evidence. *Journal of Experimental Child Psychology*, 49, 31-57.

Schauble, L., Klopfer, L.E., & Raghavan, K. (1991). Students' transition from an engineering model to a science model of experimentation. *Journal of Research in Science Teaching*, 28, 859-882.

Schoenfeld, A. (1985). *Mathematical problem solving*. New York: Academic Press.

Schoenfeld, A. (1989a). Explorations of students' mathematical beliefs and behavior. *Journal for Research in Mathematics Education*, 20, 338-355.

Schoenfeld, A. (1989b). Problem solving in context(s). In R. Charles & E. Silver (Eds.), *The teaching and assessing of mathematical problem solving* (pp. 82-92). Reston, VA: National Council of Teachers of Mathematics.

Schoenfeld, A. (1992). Learning to think mathematically: Problem solving, metacognition, and sense making in mathematics. In D. Grouws (Ed.), *Handbook of research on mathematics teaching and learning* (pp. 334-370). New York: Macmillan Publishing Company.

Schug, M., & Birkey, C. (1985). The development of children's economic reasoning. *Theory and Research in Social Education*, XIII(1), 31-42.

Schug, M., & Lephardt, N. (1992, September). Development in children's thinking about international trade. *The Social Studies*, 207-211.

Senk, S. (1989). Van Hiele levels and achievement in writing geometry proofs. *Journal for Research in Mathematics Education*, 20, 309-321.

Sere, M. (1985). The gaseous state. In R. Driver, E. Guesne, & A. Tiberghien (Eds.), *Children's ideas in science* (pp. 105-123). Milton Keynes, UK: Open University Press.

Shaughnessy, J. M. (1992). Research in probability and statistics: reflections and directions. In D. Grouws (Ed.), *Handbook of research on mathematics teaching and learning* (pp. 465-494). New York: Macmillan Publishing Company.

Shaver, J. (Ed.). (1991). *Handbook of research on social studies teaching and learning*. New York: Macmillan Publishing Company.

Shayer, M., & Adey, P. (1981). *Towards a science of science teaching*. London: Heinemann.

Shelmit, D. (1984). Beauty and the philosopher: Empathy in history and classroom. In A.K. Dickinson, P.J. Lee, & P.J. Rogers (Eds.), *Learning history* (pp. 39-84). London: Heinemann Educational Books.

Shortland, M., & Warwick, A. (Eds.). (1989). *Teaching the history of science*. Oxford: Basil, Blackwell.

Simpson, M., & Arnold, B. (1985). The inappropriate use of subsumers in biology learning. *European Journal of Science Education*, 4, 173-182.

Smith, C., Carey, S., & Wiser, M. (1985). On differentiation: A case study of development of the concepts of size, weight, and density. *Cognition*, 21, 177-237.

Smith, C., Snir, J., & Grosslight, L. (1987). Teaching for conceptual change using a computer modeling approach: The case of weight/density differentiation (Technical Report). Cambridge, MA: Harvard University, Educational Technology Center, ERIC No. ED 291 598.

Smith, E., & Anderson, C. (1986, April). Alternative conceptions of matter cycling in ecosystems. Paper presented at the annual meeting of the National Association for Research in Science Teaching, San Francisco, CA.

Sneider, C., & Pulos, S. (1983). Children's cosmographies: Understanding the earth's shape and gravity. *Science Education*, 67, 205-221.

Solomon, J. (1983). Learning about energy: How pupils think in two domains. *European Journal of Science Education*, 5, 49-59.

Solomon, J. (1985). Teaching the conservation of energy. *Physics Education*, 20, 165-170.

Solomon, J. (1991). Teaching about the nature of science in the British National Curriculum. *Science Education*, 75, 95-103.

Solomon, J. (1992). Images of physics: How students are influenced by social aspects of science. In R. Duit, F. Goldberg, & H. Niedderer (Eds.), *Research in physics learning: Theoretical issues and empirical studies* (pp. 141-154). Kiel, Germany: Institute for Science Education at the University of Kiel.

Solomon, J., Duveen, J., Scot, L., & McCarthy, S. (1992). Teaching about the nature of science through history: Action research in the classroom. *Journal of Research in Science Teaching*, 29, 409-421.

Sowder, J. (1988). Mental computation and number comparison: Their roles in the development of number sense and computational estimation. In J. Hiebert & M. Behr (Eds.), *Number concepts and operations in the middle grades* (pp. 182-197). Reston, VA: National Council of Teachers of Mathematics.

Sowder, J. (1992a). Making sense of numbers in school mathematics. In G. Leinhardt, R. Putnam, & R. Hattrup (Eds.), *Analysis of arithmetic for mathematics teaching* (pp.1-51). Hillsdale, NJ: Lawrence Erlbaum Associates.

Sowder, J. (1992b). Estimation and number sense. In D. Grouws (Ed.), *Handbook of Research on Mathematics Teaching and Learning* (pp. 371-389). New York: Macmillan Publishing Company.

Sowder, J. & Wheeler, M. (1989). The development of concepts and strategies used in computational estimation. *Journal for Research in Mathematics Education*, 20, 130-146.

Stavy, R. (1990). Children's conceptions of changes in the state of matter: From liquid (or solid) to gas. *Journal of Research in Science Teaching*, 27, 247-266.

Stavy, R. (1991). Children's ideas about matter. *School Science and Mathematics*, 91, 240-244.

Stavy, R., Eisen, Y., & Yaakobi, D. (1987). How students aged 13-15 understand photosynthesis. *International Journal of Science Education*, 9, 105-115.

Stone, L.C. (1986). International and multicultural education. In V. Atwood (Ed.), *Elementary Social Studies: Research as a guide to practice* (pp. 34-54). Washington, DC: National Council for the Social Studies.

Sutherland, R. (1987). A study of the use and understanding of algebra related concepts within a Logo environment. In J. Bergeron, N. Herscovics, & C. Kieran (Eds.), *Proceedings of the tenth international conference for the psychology of mathematics education* (Vol. I, pp. 241-247). Montreal: University of Montreal.

Suydam, M. (1985). The shape of instruction in geometry: Some highlights from research. *Mathematics Teacher*, 78, 481-486.

Threadgill-Sowder, J. (1984). Computational estimation procedures of young children. *Journal of Educational Research*, 77, 332-336.

Tiberghien, A. (1985). Heat and temperature: The development of ideas with teaching. In R. Driver, E. Guesne & A. Tiberghien (Eds.), *Children's ideas in science* (pp. 66-84). Milton Keynes, UK: Open University Press.

Tiberghien, A. (1983). Critical review of the research aimed at elucidating the sense that notions of temperature and heat have for students aged 10 to 16 years. In *Proceedings of the first international workshop research on physics education* (pp. 73-90). Paris: Editions du CNRS.

Tierney, C., & Nemirovsky, R. (1991, Fall). Children's spontaneous representations of changing situations. *Hands on!*, 7-10.

Tomasini, G., & Balandi, P. (1987). Teaching strategies and children's science: An experiment on teaching "hot and cold." In J. Novak (Ed.), *Proceedings of the second international seminar "misconceptions and educational strategies in science and mathematics"* (Vol. II, pp. 158-171). Ithaca, NY: Cornell University.

Tournaire, F., & Pulos, S. (1985). Proportional reasoning: A review of the literature. *Educational Studies in Mathematics*, 16, 181-204.

Trowbridge, J., & Mintzes, J. (1985). Students' alternative conceptions of animals and animal classification. *School Science and Mathematics*, 85, 304-316.

Turiel, E. (1983). *The development of social knowledge*. Cambridge: Cambridge University Press.

Van Hiele, P. (1986). *Structure and insight*. Orlando, FL: Academic Press.

Vergnaud, G. (1988). Multiplicative structures. In J. Hiebert & M. Behr (Eds.), *Number concepts and operations in the middle grades* (pp. 141-161). Reston, VA: National Council of Teachers of Mathematics.

Vergnaud, G., & Errecalde, P. (1980). Some steps in the understanding and the use of scales and axis by 10-13 year old students. In R. Karplus (Ed.), *Proceedings of the fourth international conference for the psychology of mathematics education* (pp. 285-291).(ERIC Reproduction Service No. ED 250 186).

Vosniadou, S. (1991). Designing curricula for conceptual restructuring; lessons from the study of knowledge acquisition in astronomy. *Journal of Curriculum Studies*, 23, 219-237.

Vosniadou, S., & Brewer, W. (1992). Mental models of the earth: A study of conceptual change in childhood. *Cognitive Psychology*, 24, 535-585.

Wagner, S. (1981). Conservation of equation and fun-ction under transformations of variable. *Journal for Research in Mathematics Education*, 12, 107-118.

Wagner, S., & Kieran, C. (Eds.) (1989). *Research issues in the learning and teaching of algebra*. Reston, VA: National Council of Teachers of Mathematics.

Waterman, M. (1983). Alternative conceptions of the tentative nature of scientific knowledge. In J. Novak (Ed.), *Proceedings of the international seminar misconceptions in science and mathematics* (pp. 282-291). Ithaca, NY: Cornell University.

Watts, M. (1983a). Some alternative views of energy. *Physics Education*, 18, 213-217.

Watts, M. (1983b). A study of school children's alternative frameworks of the concept of force. *European Journal of Science Education*, 5, 217-230.

Wavering, M. (1985, April). The logical reasoning necessary to make line graphs. Paper presented at the annual meeting of the National Association for Research in Science Teaching, French Lick Springs, Indiana.

Wearne, D., & Hiebert, J. (1988). Constructing and using meaning for mathematical symbols: The case of decimal fractions. In J. Hiebert & M. Behr (Eds.), *Number concepts and operations in the middle grades* (pp. 220-235) Reston, VA: National Council of Teachers of Mathematics.

Wearne, D., & Hiebert, J. (1989). Cognitive changes during conceptually based instruction on decimal fractions. *Journal of Educational Psychology*, 81, 507-513, ERIC No. ED 254 409.

Welch, W. (1973). Review of the research and evaluation program of Harvard Project Physics. *Journal of Research in Science Teaching*, 10, 365-378.

Welch, W., & Pella, M. (1967). The development of an instrument for inventorying knowledge of the processes of science. *Journal of Research in Science Teaching*, 5, 64-68.

Welch, W., & Walberg, H. (1972). A national experiment in curriculum evaluation. *American Educational Research Journal*, 9, 373-383.

Welch, W., & Walberg, H. (1968). A design for curriculum evaluation. *Science Education*, 52, 10-16.

Wellman, H.M., & Johnson, C. (1982). Children's understanding of food and its functions: A preliminary study of the development of concepts of nutrition. *Journal of Applied Developmental Psychology*, 3, 135-148.

White, B. (1990). Reconceptualizing science and engineering education. Unpublished manuscript. Cambridge, MA: BBN Laboratories.

White, B., & Horwitz, P. (1987). Thinker tools: Enabling children to understand physical laws. BBN Laboratories Report. Cambridge, MA: BBN Laboratories.

Wirszup, I. (1976). Breakthroughs in the psychology of learning and teaching geometry. In J. Martin & D. Bradbard (Eds.), *Space and geometry. Papers from a research workshop* (pp. 75-97). Athens, GA: University of Georgia, Georgia Center for the Study of Learning and Teaching Mathematics. (ERIC Reproduction Service No. ED 132 033).

Wiser, M. (1986). The differentiation of heat and temperature: An evaluation of the effect of microcomputer teaching on students' misconceptions. Technical report. Cambridge, MA: Harvard Graduate School of Education.

Wiser, M. (1988). The differentiation of heat and temperature: History of science and novice-expert shift. In S. Strauss (Ed.), *Ontogeny, phylogeny, and historical development*. Norwood, NJ: Ablex Publishing Corporation.

Wollman, W. (1977a). Controlling variables: Assessing levels of understanding. *Science Education*, 61, 371-383.

Wollman, W. (1977b). Controlling variables: A neo-Piagetian developmental sequence. *Science Education*, 61, 385-391.

Wollman, W., & Lawson, A. (1977). Teaching the procedure of controlled experimentation: A Piagetian approach. *Science Education*, 61, 57-70.

MAURITS CORNELIS ESCHER
Liberation, 1955.

Chapter 16 BEYOND BENCHMARKS

A central premise of the Project 2061 strategy is that significant, lasting reform in education will happen only when people charged with operating the schools become part of the creative process. The simple implementation of reforms proposed by others will not work. To help states and school districts with this enormous responsibility, Project 2061 has begun an extended effort to perform some tasks that states or districts find difficult and to foster collaborative action with them. To this end, the Project will

- press for nationwide acceptance of the goals and philosophies of *Science for All Americans (SFAA)*,

- form alliances with educators who are ready to work in concert to help schools accomplish reforms, and

- develop field-tested reform tools to guide educators in writing and implementing curricula that will ensure science-literate graduates.

Teacher participation in curriculum reform varies widely. In some schools and school districts, teachers expect to have a major part in actually designing whatever curriculum will be introduced; in other schools, they prefer to adopt a ready-made curriculum pretty much as it is, concentrating on implementation rather than design. Perhaps the teachers in most schools are somewhere in between. They like to start with a curriculum framework that is well worked out but incomplete, so that they can modify it to fit their circumstances and preferences. The Project 2061 response to this has been to start developing an array of tools that educators can use to design curricula and to guide reform.

SFAA and *Benchmarks for Science Literacy* will soon be joined by *Designs for Science Literacy*,

Blueprints for Reform,

Resources for Science Literacy, and the

Project 2061 Curriculum-Design & Resource System.

Each of these is briefly described below. The figures on the following pages indicate the relationships among them.

Reform Tools

Project 2061 is not in the business of producing a national curriculum. Rather, with the assistance of its school-district teams, it is developing a set of tools to help schools and districts assemble their own curricula.

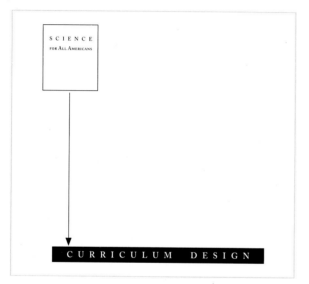

SFAA, the first such tool, defines and recommends a coherent set of learning goals for high-school graduates. *SFAA* also lays out the principles of effective learning and teaching that underlie all of the curriculum design tools being developed by Project 2061. Since *SFAA's* publication in 1989, educators have been using these goals and principles to stimulate thought and discussion about classroom practices such as finding out what students already think about major topics; giving students enough evidence and time to change their incorrect ideas; increasing the use of team approaches that increase active participation by every student; shifting classwork toward ideas and thinking

and away from vocabulary and predetermined answers; making sure that females, minorities, and the disabled are fully engaged in all class activities in science, mathematics, and technology; and expecting and rewarding clear, accurate reports, both written and oral, of student thinking and activity.

Benchmarks for Science Literacy elaborates the *SFAA* recommendations by suggesting the progress students should make along the way. By describing precursor understandings (at 2nd, 5th, and 8th grades) of 12th-grade understandings, *Benchmarks* can help educators decide what content to include (or exclude) in a core curriculum, what order to teach things in, and why. At the same time that others will be using *Benchmarks* to inform local curriculum design, Project 2061 will be using it as a guide in developing other resources.

To move from goals to the design or redesign of an entire K-12 curriculum requires more than adding or subtracting some instructional units here and there or

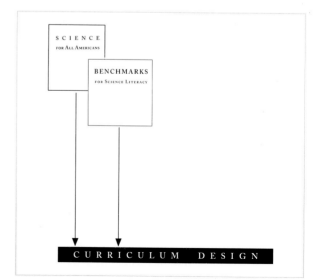

changing materials, tests, and techniques. Piecemeal changes are unlikely to lead to significant and lasting curriculum reform. What is needed is a coordinated K-12 plan that guides the curriculum-building process.

Designs for Science Literacy will be a handbook for educators wishing to take a systematic, goal-oriented approach to curriculum design. It will include the following:

- The principles of learning and teaching on which a Project 2061 curriculum should be based. This will extend and develop more fully the arguments made in Chapter 13 of *SFAA*.

- Principles of curriculum design for science literacy. This will spell out and give a rationale for various content and pedagogical conditions that need to be met in any K-12 curriculum that targets *Benchmarks* and *SFAA*. A curriculum should, for example, achieve a balance among activities that emphasize inquiry, design, explanation, and issues, and it should provide all students with learning experiences in the form of didactic instruction, seminars, projects (individual and group), independent study, and peer teaching.

- A process for configuring K-12 curricula that meet Project 2061 standards and take into account local and state policies, resources, and preferences. At the heart of the process is the notion that, guided by design principles and with models to refer to, educators can assemble a Project 2061 curriculum from components keyed to *Benchmarks*.

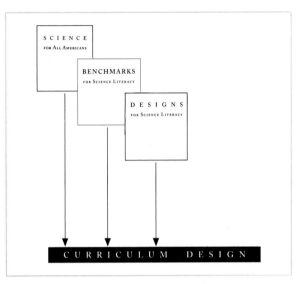

- Descriptions of curriculum blocks and models. Project 2061 does not intend to develop curriculum components but rather to set out, in *Designs for Science Literacy*, the requirements for the different kinds of components, collectively referred to as curriculum blocks, from which educators can assemble a curriculum. Samples of various kinds of curriculum blocks suitable for a Project 2061 curriculum will be included along with sketches of several alternative K-12 curriculum models to illustrate the range of possibilities that exist.

If lasting reform is to occur, changes are needed throughout the education system. Curriculum reform, however inspired, cannot do the job alone. Hence, Project 2061 has convened expert groups to prepare, in collaboration with the six Project 2061 teams, a dozen concept papers on aspects of the system that must

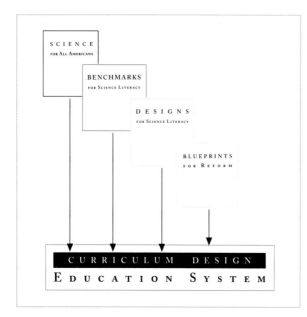

change to accommodate the curriculum reforms being proposed by the Project. The findings of these papers, individually and in relation to one another, will be presented in *Blueprints for Reform*. The *Blueprint* topics follow:

- **Teacher Education** will identify the changes needed in pre-service and in-service teacher education to produce teachers with the knowledge and skills needed to implement curricula based on Project 2061 goals and principles.

- **Materials and Technology** will identify what new resources are needed, what mechanisms to identify and access them will be effective, and what kinds of policies must be adopted to support the development and use of such resources.

- **Assessment** will specify what immediate and future assessment needs are demanded by Project 2061 curriculum-design principles—from in-class assessment during instruction, to program evaluation by schools, to monitoring education progress at state and national levels.

- **Curriculum Connections** will identify important linkages among the natural and social sciences, mathematics, and technology, and also between them and the arts and humanities, and will suggest how such linkages can be fostered in the curriculum.

- **School Organization** will suggest what alternatives for school organization will best enable Project 2061 curricula to work. This paper will discuss such issues as grade structure, teacher collaboration, control of curriculum materials and assessment, how time and space in school might be organized, and the school as a learning community.

- **Parents and Community** will specify what will be needed for parents and the community to understand Project 2061 reform recommendations and what kinds of commitment and effort from them are needed.

- **Business and Industry** will examine such issues as preparing students to enter an increasingly technological workplace and marketplace, the role of science literacy in U.S. competitiveness, appropriate partnerships between business and education, and resources and leadership that local business can bring to science instruction.

- **Higher Education** will address such issues as changes in admission requirements needed to accommodate changes in high-school course structure and assessment methods, and how undergraduate education should build on *SFAA*—especially for college students who may become teachers.

- **Policy** will examine the entire policy picture, including how policy has inhibited past reform initiatives, challenges posed by the current education system for the implementation of Project 2061 reform, changes that may be needed in laws and regulations that govern schools, and how modifications of current policy might be achieved.

- **Finance** will consider the implications of Project 2061 reform recommendations (including those in the other *Blueprint* papers) for the allocation of money and other resources. In an era of severely constrained resources, it is especially important to examine the financial base for education and the potential availability of resources for changes needed to implement reform, including possibilities for changing schools without incurring greater costs. Distinctions will be drawn between the cost of reform and the annual cost of operating the schools.

- **Equity** will recommend education equity policies to ensure that science literacy is attainable by *all* students. It will also contribute to other *Blueprint* papers and serve as a check for them.

- **Research** will discuss the research questions that arise in other *Blueprint* papers, *Benchmarks*, and *Designs*, as well as in initial attempts to implement Project 2061 reform. In addition, this paper will consider what mechanisms can permanently link research with practice.

Project 2061 curricula will need to draw upon a diverse array of learning resources—physical, print, video, audio, and computer. The traditional textbook will not do. A search is currently under way to identify learning materials with the greatest relevance for Project 2061. Excellent materials have been and continue to be developed by funded projects, science and technology museums, commercial organizations, and individual teachers. The best of these—including children's stories, novels, plays, poetry, reference

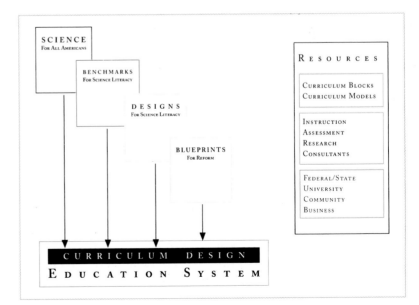

works, and nonfiction trade books—will be entered into a computer database and cross-referenced to *Benchmarks* and one another. The computerized database will be updated as new contributions to science literacy become available. References to appropriate human and institutional resources will also be added as they are identified. Educators will be able to access the database for designing curriculum.

Introducing new learning materials into traditional subjects and courses can help transform the curriculum. However, as noted earlier in the description of *Designs for Science Literacy,* Project 2061 will make a more comprehensive approach possible. This alternative approach to K-12 curriculum design calls for educators to select and configure a set of curriculum blocks that meet Project 2061 standards (from a large pool of such curriculum blocks).

Curriculum blocks are coherent units of instruction that can be assembled to create K-12 curricula. The description of any completed curriculum block will include an overview of the main topics and experiences; a list of benchmarks targeted in the block; suggestions for how to assess the learning; a fairly detailed description of the enabling activities and required resources; and the estimated time, space, and money required. A complete K-12 collection of blocks will advance all the learning goals in *SFAA* and their supporting benchmarks.

Project 2061 cannot hope to produce the necessary pool of hundreds of diverse curriculum blocks needed for building alternative K-12 curricula. Rather, with the help of its school-district teams, it plans to provide some examples that will help curriculum developers all over the nation produce the kinds of curriculum blocks that are needed. As high-quality blocks appear, they will be added to the resource database.

In the process of developing these examples and trying them out in classrooms, Project 2061 hopes to learn how much and what kind of information must be included in block descriptions for them to be helpful to various potential users. Some users will likely take rough sketches of blocks and design their own materials to fit, other users will use sketches plus various existing materials to make testable blocks, and still others will select blocks tested by others.

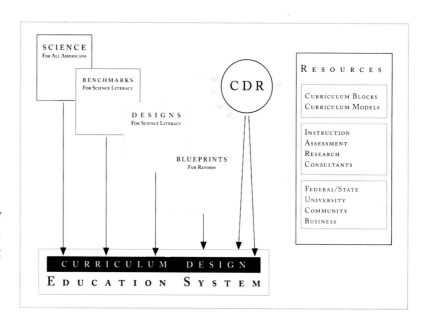

Another new kind of Project 2061 resource for guiding curriculum design will be curriculum models. To illustrate that it is possible to design 13 years of schooling to achieve literacy in science, mathematics, and technology for all Americans, Project 2061 is working on some alternative curriculum models. Each model is a K-12 configuration of blocks that accounts for all of the benchmarks. Descriptions of the models will be included in the resource database. Curriculum models are not "model curricula" in the sense often used in education and are not intended for exact replication. Their purpose is to suggest possibilities for planning curriculum and instruction and to show the degree to which models can vary and still provide a common core of instruction aimed at science literacy.

If Project 2061's vision of reform is realized, curriculum design in the future will be quite different from today. Schools will be able to plan K-12 curricula so that teaching and learning are coordinated over long spans of time and across subjects. Groups of teachers will, as indicated above, be able to draw from a much greater variety of resources. Project 2061's interactive, computer-based, multimedia Curriculum-Design & Resource System (CDR) will make these resources available to local curriculum designers, probably on CD-ROM and possibly on-line. It will provide a versatile, user-friendly system to access Project 2061 *Benchmarks, Blueprints,* and other Project 2061 tools and resources. Beyond its use in curriculum design, the system will be able to serve as a management tool for teachers. As students progress toward science literacy, teachers could use the CDR system to monitor their progress in terms of benchmarks and to suggest learning experiences for new concepts.

Toward Science Literacy

What is the Project 2061 vision for schooling in the 21st century?

First, an identifiable *common core* of learning in science, mathematics, and technology will focus on science literacy as its main goal and be closely allied with a common core of learning in the arts and humanities. Instructional units will be justified by referring to grade-level learning goals derived from expectations for what adults should retain. A comparison of this new core curriculum with the traditional curriculum of today would show far fewer topics than before, so that students can concentrate on learning well a basic set of ideas and skills that will lead to science literacy—and optimally promote further learning. Equally important, the curriculum will provide ample opportunity for students to go beyond the core in response to their individual interests, talent, and plans for the future.

Second, all students in a Project 2061 school will have wide-ranging learning experiences. Instructional units will employ a variety of instructional methods and learning materials. Students will have many opportunities for hands-on activities and, equally important, for the reflective thinking that enables them to make sense of their experiences—including connecting ideas among science, mathematics, and technology, and between them and the arts and humanities. Students' activities and reflections will engage them in using their knowledge in ways characteristic of literate adults—to explain everyday phenomena, to solve practical problems, to inform decisions about issues, and thereby to learn more and have more personal satisfaction.

Third, teachers will have primary responsibility for planning and implementing curriculum within their individual systems. K-12 teams will have planned for K-12 continuity of experiences, and cross-discipline groups will have planned for how students will encounter connections within the curriculum. Designers will not have to develop a curriculum from scratch but will be able to select from a wide variety of instructional units that foster *SFAA* goals while matching the needs of the local community. Furthermore, the teams that formulated the curriculum will monitor and coordinate its operation.

Finally, the school environment will support the science literacy goals and the curriculum designed to achieve them. The scheduling of time and personnel will be suited to the demands of the learning experience. Schools will welcome information and participation from outsiders who can contribute to the specified learning goals. Students and teachers will be able to leave school grounds to participate in activities in the community or to learn science in the field rather than in the school. Resources will be readily available for teachers and administrators to learn about new research findings and their implications for practice and to engage in similar study themselves. And everyone, including parents, policy makers, and teachers will understand that reform is a continuing process requiring time and consistent effort. ■

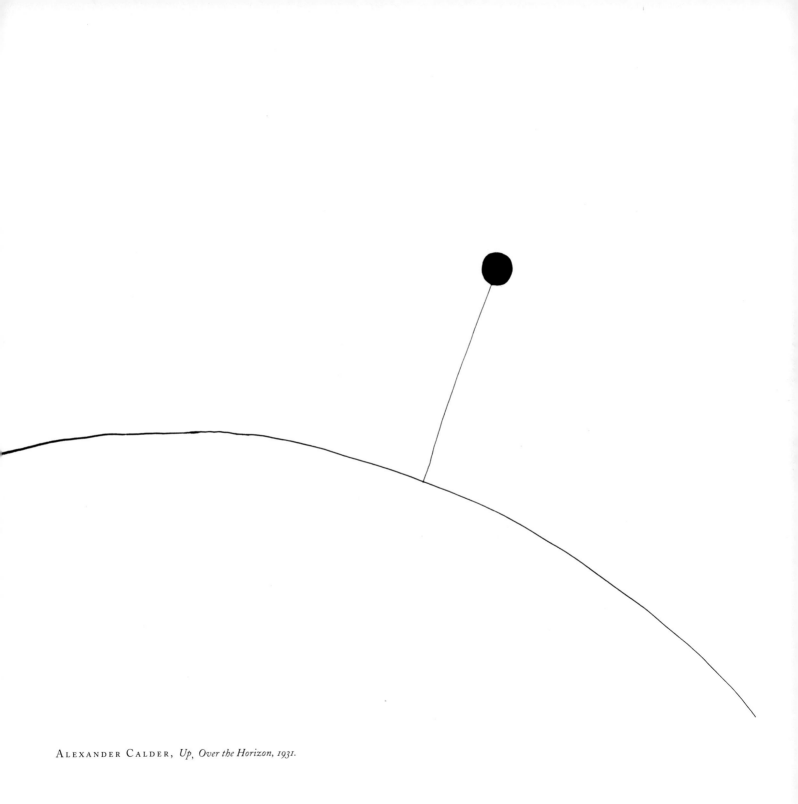

ALEXANDER CALDER, *Up, Over the Horizon, 1931.*

PARTICIPANTS

Benchmarks for Science Literacy is the result of the efforts of hundreds of educators and scientists. Over a period of four years, these individuals contributed to the project in different ways and at different times, but always with a high level of thoughtfulness, professional integrity, and commitment to a future in which all children can achieve science literacy. This statement, however, should not be taken to imply that the individual participants endorse all of the recommendations in *Benchmarks for Science Literacy*.

Members of the National Council on Science and Technology Education, our advisory and oversight group, have offered wise and patient counsel. Their support throughout Project 2061 has sustained us in the vision we share with many others—of a citizenry that is science literate. Members of the National Council are listed at the beginning of this report.

PROJECT 2061 SCHOOL-DISTRICT TEAMS have worked tirelessly for four years on how to move from *SFAA* to curriculum designs. Team leaders and many other team members have made exceptional contributions.

OTHER CONTRIBUTORS include colleagues whose generous help ranged from a day's discussion to many weeks of creative work, from clarifying current ideas in their fields to generating and wrestling with benchmark decisions and drafts.

In addition to the participants listed here, over 1300 individuals participated in the review of draft *Benchmarks for Science Literacy*. They came from Project 2061 school-district teams, Statewide Systemic Initiative and state framework groups, educational associations, R&D groups, AAAS Sections, National Council members, scientific societies, state academies of science, and volunteer groups and individuals throughout the United States and in five foreign countries. Although they are too numerous to list here, we could not have produced this report without them and will seek their help again as we look to future revisions of *Benchmarks for Science Literacy*.

Joseph Cabibbo *McFarland High School, MF*

Ruby Cabibbo *McFarland High School, MF*

Peggy S. Carnahan *Northside Independent School District, SA* *

Michael M. Casey *Montgomery Junior High School, SD*

Christina A. Castillo-Comer *Longfellow Middle School, San Antonio Independent School District, SA*

Rose Mary Castro *Fox Technical High School, San Antonio Independent School District, SA*

Dee Ann Chenn *Darnall Elementary School, SD*

Lonnie K. Chin *Spring Valley Elementary School, SF*

Herb Chu *International Studies Academy, SF*

Mary Clark *Elbert County Comprehensive High School, GA*

Steve Clark *Elbert County Comprehensive High School, GA*

Anthony I. Clerico *Office of Curriculum Support, PH* *

Edward Colación *Horace Mann Middle School, SF*

Ray Conser *Grant Elementary School, SD*

Linda A. De Leon *San Antonio Independent School District, SA*

Robert Dean *Madison High School, SD*

William Decker *Oglethorpe County High School, GA*

Alvin Dickens *Union Point Elementary School, GA*

Albert Dillard *Challenger Junior High School, SD*

Boris J. Dimbach *High School for the Creative and Performing Arts, PH*

Debra Leigh Dixon *Beaverdam Elementary School, GA*

George W. Dougherty *Oglethorpe County Middle School, GA*

Sandee W. Drake *Elbert County Middle School, GA*

Joan M. Drennan-Taylor *San Antonio Independent School District, SA*

Zola M. Dunn *Kennedy High School, Edgewood Independent School District, SA*

Garland A. Earnest *Elbert County Comprehensive High School, GA*

Rhonda Eaton *Greensboro Primary School, GA*

Gail W. Engel *Greene-Taliaferro Comprehensive High School, GA*

John E. Evans, Jr. *Fitzsimmons Middle School, PH* *

Patricia Everett *University City High School, PH*

Michael Eversoll *Indian Mound Middle School, MF*

Danine Ezell *Bell Junior High School, SD*

Richard Fabian *Computer Education Office, SD*

Bernard Farges *Mathematics and Science Resource Center, SF* *

Myrtle Faucette *Knox Elementary School, SD*

Jeanne M. Faust *McFarland High School, MF*

Phyllis A. Fees *Mark Twain Middle School, San Antonio Independent School District, SA*

Rochelle A. Fellman *Reynolds Elementary School, PH*

David Fenner *Serra High School, SD*

Aubrey M. Finch *Oglethorpe County High School, GA*

Susan A. Forthman *Hoelscher Elementary School, Edgewood Independent School District, SA*

Ruth Fortney *Indian Mound Middle School, MF*

Andrea S. Foster *Stinson Middle School, Northside Independent School District, SA*

Joseph Foster *Leeds Middle School, PH*

Jonathan Frank *J. Eugene McAteer High School, SF*

Martin Friedman *University City High School, PH*

Kimiko Fukuda *Wilson Middle School, SD*

George Fuller *McFarland High School, MF*

Philip D. Gay *Math/Science/Educational Technology Office, SD* *

Ethel M. Gilliam *Floyd T. Corry Elementary School, GA*

Irene Goetze *Winder Barrow Middle School, GA*

Rebecca E. Gonzalez-Luna *Carvajal Elementary School, San Antonio Independent School, SA*

Marian Jacqui Grabel *Hickman Elementary School, SD*

Douglas P. Griffin *Oglethorpe County High School, GA*

Ina Gustafson *Northside Independent School District, SA*

Erla T. Hackett *A.P. Giannini Middle School, SF*

Richard W. Halsey *Serra High School, SD*

Paul J. Hampel *Washington High School, PH*

Delayne Harmon *Darnall Elementary School, SD*

Robert Harmon *University City High School, PH* *

Roger W. Harris *Taft High School, Northside Independent School District, SA*

Deborah Hatchell-Carter *Holmes Elementary School, SD*

Marc Heuer *Conrad Elvehjem Elementary School, MF*

Marlene Hilkowitz *Carver High School of Engineering and Science, PH* *

Natalie A. Hiller *Simon Gratz High School, PH*

Steve Hirabayashi *International Studies Academy, SF*

Deborah K. Hodges *Northside Independent School District, SA*

Ambra Hook *Morrison Elementary School, PH*

Paul Horton *John Jay High School, Northside Independent School District, SA*

Curtis Howard *Office of Curriculum Support, PH*

Virginia L. Hutto *Stinson Middle School, Northside Independent School District, SA*

Alex Insaurralde *Mission High School, SF*

Richard Ivik *Indian Mound Middle School, MF*

Gail Janes *Doyle Elementary School, SD*

James Jennings *Crawford High School, SD*

Joan H. Jordan *Beaver Dam Elementary School, GA*

Sherrill Joseph *Chollas Elementary School, SD*

Patricia Kalhagen *McFarland High School, MF*

Rhoda Kanevsky *Powell Elementary School, PH*

Jim Knerl *International Studies Academy, SF* *

Linda Kolman *Southeast Regional Office, PH*

Joan E. Kunzler *Roosevelt Junior High School, SD*

Deborah Larson *McFarland High School, MF* *

James Lawless *McFarland High School, MF*

Pak Kee Lee *Masterman Lab Demonstration School, PH*

Martha Jean Lee *Oglethorpe County Middle School, GA*

Mamie L. Lew *Brackenridge High School, San Antonio Independent School District, SA*

Elliot H. Lewis *Edison High School, PH*

Janet Lewis *Oglethorpe County High School, GA*

Lynn Lipetzky *Darnall Elementary School, SD*

Mary C. Loal *Huey Elementary School, PH*

Monica Lochner *Conrad Elvehjem Elementary School, MF*

Joan K. Loomis *Math/Science/Educational Technology Office, SD*

Douglas Main *Greene-Taliaferro Comprehensive School, GA*

Evelyn Mancilla *Sheridan Elementary School, SF*

Bates Mandel *Carver High School of Engineering and Science, PH*

Brenda J. Martin *Jerabek Elementary School, SD*

Thomas A. Martz *Greene-Taliaferro Comprehensive School, GA*

Richard Matthews *Elbert County Middle School, GA*

Sue Matthews *Elbert County Middle School, GA*

Victoria Matthews *Elbert County School District, GA*

George Mavroulis *Indian Mound Middle School, MF*

Katharine McCosky *McFarland High School, MF*

Bruce McGirr *Holmes Elementary School, SD*

Joan McGirr *Holmes Elementary School, SD*

Esperanza McGuigan *San Diego High School, SD*

Kathleen McKinley *Stetson Middle School, PH*

James McSherry *Carver High School of Engineering and Science, PH*

Pilar Mejía *Cesar Chavez Elementary School, SF*

Marilyn Melstein *Rush Middle School, PH*

Miles Michaels *Wilson Middle School, SD*

Linda Milicia *Fell Elementary School, PH*

Carolyn Minor *Steel Elementary School, PH*

Thomas Mooney *McFarland Elementary School, MF*

Todd Morano *De Portola Middle School, SD*

Candido Munumer *Elbert County Middle School, GA* *

Lynne Munumer *Elbert County Middle School, GA*

Edward I. Muse *Randolph Skills Center, PH*

Bonnie Nessmith *Greensboro Primary School, GA*

Cam Van Nguyen *George Washington High School, SF*

Mary Nielsen *McFarland High School, MF*

Linda C. Nott *Wrenn Junior High School, Edgewood Independent School District, SA*

Margaret Nye *Horace Mann Middle School, SF*

Molly O'Malley *Treasure Island Elementary School, SF*

Darrel J. Ochoa *Edgewood Independent School District, SA*

Gary Oden *Math/Science/Educational Technology Office, SD**

Eugene Olson *McFarland High School, MF*

Iris Othrow *McFarland Elementary School, MF*

Sabrina Page *Greensboro Primary School, GA*

Beth Phillips *Elbert County Middle School, GA*

Leacy A. Piper *Anson Jones Middle School, Northside Independent School District, SA*

David A. Pope *Hoover High School, SD*

Larry Prager *Potrero Hill Middle School, SF*

Bruce A. Rachild *Conwell Middle Magnet School, PH*

Celia T. Rainwater *Clark High School, Northside Independent School District, SA*

Colleen W. Reeve *Leal Middle School, Harlandale Independent School District, SA*

Raymond L. Reeve *Morrill Elementary School, Harlandale Independent School District, SA*

Rita I. Rice *Central East Regional Office, PH*

Ollie H. Ritchey *Knox Elementary School, SD*

Pat Rompala *Darnall Elementary School, SD*

Mark Rosen *Carver High School of Engineering and Science, PH*

Patsy Rossman *Conrad Elvehjem Elementary School, MF*

Gwen Rountree *Rhodes Middle School, PH*

Ruth Rubinstein *McCall Elementary School, PH*

Marlene Ruskin *Office of Curriculum Support, PH*

Maria Santos *Department of Instructional Improvement and Professional Development, SF*

Patricia Saybolt *Mastbaum Area Vocational and Technical School, PH*

Lana Scott *Indian Mound Middle School, MF*

Thomas L. Scott *Simon Gratz High School, PH*

Vicki Scott *Elbert County Middle School, GA*

Jayne Seawright *Oglethorpe County Elementary School, GA*

Carmelo Sgarlato *Phillip and Sala Burton High School, SF*

Rose Shambourger *Sulzberger Middle School, PH*

Robin Sharp *San Francisco Community School, SF*

Lynn Shoff *Roosevelt Junior High School, SD*

William C. Shumake *University City High School, PH*

Lynette Smith *Office of Curriculum Support, PH*

Carol Smythe *Moffet Elementary School, PH*

Donald C. Snyder *South Philadelphia High School, PH*

Catherine Somers *De Portola Middle School, SD*

Jacqueline Stang *McFarland High School, MF*

Garth Story *Challenger Junior High School, SD*

Jana Swanson *Torrey Pines Elementary School, SD*

Dacotah Swett *Lowell Alternative High School, SF*

Mary Ellen Taylor *Conrad Elvehjem Elementary School, MF*

Ernie Thieding *Indian Mound Middle School, MF*

Janet Thomas *Elbert County Middle School, GA*

Clara F. Tolbert *Central High School, PH**

Ramon Torres *Clairemont High School, SD*

Susan Traganza *Doyle Elementary School, SD*

Terry C. Tripp *Elbert County Middle School, GA*

Norma Trost *Education Public Information Office, SD*

Janis M. Tschoepe *Clark High School, Northside Independent School District, SA*

Janet Tuomi *Jose Ortega Elementary School, SF*

Susie Turner *Woodrow Wilson High School, SF*

William Urban *McFarland High School, MF*

OTHER CONTRIBUTORS*

Marjorie E. Valentine *John Jay High School, Northside Independent School District, SA*

Sara R. Valenzuela *Taft High School, Northside Independent School District, SA*

Carolyn Vega *Computer Education Office, SD*

Fred C. Vincent *Edison High School, PH*

John Volk *West Philadelphia High School, PH*

Guillermina von Borstel *Gompers Secondary School, SD*

Susan J. Wachowiak *Old Town Program School, SD*

Judy Walker *Doyle Elementary School, SD*

Eleanor Walls *Fell Elementary School, PH*

Katherine Hill Walsh *Oak Park Elementary School, SD*

Mary H. Washington *Levering Science Magnet School, PH*

Gerald F. Weaver *University City High School, PH*

Anne Marie Webb *Greensboro Primary School, GA*

Louis Webb *School for Business and Commerce, SF*

Thomas J. Weldon *Burbank High School, San Antonio Independent School District, SA*

Ellen White-Volk *Stonewall Child Development Center, Harlandale, Independent School, SA*

Scott Wichmann *McFarland High School, MF*

Randall T. Wilbanks *Clark High School, Northside Independent School District, SA*

George Williams *Taft High School, Northside Independent School District, SA*

Jane Williamson *Stafford Elementary School, Edgewood Independent School District, SA*

Carol Withrow *Elbert County Comprehensive High School, GA*

David C. Withrow *Greene-Taliaferro Comprehensive High School, GA*

Edward Worrell *Dr. Charles Drew Elementary School, SF*

Hilda A. Zamora *Collier Elementary School, Edgewood Independent School District, SA*

Tim Aaronson *Lawrence Hall of Science, University of California, Berkeley*

Lewis Allen *Department of Educational Leadership, University of Georgia*

Shannon Almquist *Curriculum Development Associates, Washington, DC*

Charles Anderson *Institute for Research on Teaching, Michigan State University*

Frank Barnes *Department of Electrical and Computer Engineering, University of Colorado, Boulder*

Deanna Banks Beane *National Urban Coalition, Washington DC*

Jerry A. Bell *Directorate for Education and Human Resources, American Association for the Advancement of Science*

Dale Boatright *American Federation of Teachers, Washington, DC*

John C. Brandt *Department of Astrophysical, Planetary and Atmospheric Sciences, University of Colorado, Boulder*

Theodore Britton *College of Education, University of Florida*

Mary Jo McGee Brown *Educational Psychology Department, University of Georgia*

Stephen G. Brush *Institute for Physical Science and Technology, University of Maryland*

Reid A. Bryson *Departments of Geography and Meteorology, University of Wisconsin, Madison*

Gail Burrill *Department of Mathematics, University of Wisconsin, Madison*

Rodger W. Bybee *Biological Sciences Curriculum Study, Colorado Springs, CO*

Cynthia Carey *Environmental, Population and Organismic Biology, University of Colorado, Boulder*

Thomas R. Cech *Department of Chemistry, University of Colorado, Boulder*

Phillip R. Certain *Department of Chemistry, University of Wisconsin, Madison*

Donald L. Chambers *School of Education, University of Wisconsin, Madison*

Audrey B. Champagne *Directorate for Education and Human Resources, American Association for the Advancement of Science*

Richard Clark *Minnesota Department of Education*

Kathy Comfort *California State Department of Education*

John A. Conkling *Department of Chemistry, Washington College*

Catherine Corley *Baltimore County Public Schools*

James F. Crow *Department of Zoology, University of Wisconsin, Madison*

Donald W. Crowe *Department of Mathematics, University of Wisconsin, Madison*

June Danaher *Division of Instruction, Maryland State Department of Education*

Denice D. Denton *Department of Electrical & Computer Engineering, University of Wisconsin, Madison*

Mark Dubin *Department of Biology, University of Colorado, Boulder*

Sara Duff *Baltimore City Public Schools*

Jack Easley *College of Education, University of Illinois, Urbana-Champaign*

David J. Egan *Department of Plant Pathology, University of Wisconsin, Madison*

Paul Elliott *Wordsworth Communication, Alexandria, VA*

Arthur B. Ellis *Department of Chemistry, University of Wisconsin, Madison*

Arthur K. Ellis *School of Education, Seattle Pacific University*

Larry Esposito *Astrophysical, Planetary, and Atmospheric Sciences, University of Colorado, Boulder*

Batsheva Eylon *Science Teaching Department, Weizmann Institute of Science*

Stuart Feldman *Computer Systems Research, Bellcore*

Allan D. Franklin *Department of History of Science, University of Colorado, Boulder*

R. Igor Gamow *Department of Chemical Engineering, University of Colorado, Boulder*

Tracy Gath *Directorate for Education and Human Resources, American Association for the Advancement of Science*

Michael Grant *Environmental, Population and Organismic Biology, University of Colorado, Boulder*

Albert Gregory *Boston, MA*

John M. Greenler *Department of Botany, University of Wisconsin, Madison*

William Harton *Education Systems, IBM Corporation*

Michael Hartoonian *Wisconsin Department of Public Instruction*

Henry W. Heikkinen *Mathematics and Science Teaching Center and Department of Chemistry, University of Northern Colorado*

Patricia M. Heller *Department of Curriculum and Instruction, University of Minnesota, Minneapolis*

Leon Henkin *Department of Mathematics, University of California, Berkeley*

Ingo Hentschel *Education Systems, IBM Corporation*

Robert E. Hollon *Department of Curriculum and Instruction, University of Wisconsin, Madison*

Allen Hunter *Department of Sociology, University of Wisconsin, Madison*

Roberta Jaffee *Life Lab Science Program, Santa Cruz, CA*

David Jenness *Division of Research, Evaluation, and Dissemination, National Science Foundation*

Edward I. Johnson *Montgomery Blair Magnet Program, Silver Spring, MD*

Noel Johnson *St. Francis High School, Minnesota*

Roger T. Johnson *Department of Curriculum and Instruction, University of Minnesota, Minneapolis*

Alice B. Kehoe *Department of Anthropology, Marquette University*

David Kennedy *State Department of Education, Washington*

Judith T. Kildow *Department of Ocean Policy/Ocean Engineering, Massachusetts Institute of Technology*

Clifford Konold *Scientific Reasoning Research Institute, University of Massachusetts*

Steven E. Kornguth *Department of Neurology, University of Wisconsin, Madison*

Melvin Kranzberg *School of History, Technology and Society, Georgia Institute of Technology*

Gerald Kulm *Department of Curriculum and Instruction, Texas A&M University*

Nancy Landis *Biological Science Curriculum Study, Colorado Springs, CO*

James Landwehr *Statistical Models and Methods Research, AT & T Bell Laboratories*

Margaret Laughlin *Education Department, University of Wisconsin, Green Bay*

Leroy Lee *Wisconsin Academy of Sciences, Arts, and Letters*

Gregory A. Letterman *Quince Orchard High School, Montgomery County, MD*

Thomas T. Liao *Department of Technology and Society, State University of New York, Stony Brook*

Susan Loucks-Horsley *National Center for Improving Science Education*

Peter McIntyre *Department of Physics, Texas A&M University*

John J. Magnuson *Limnology Center and Department of Zoology, University of Wisconsin, Madison*

Shirley M. Malcom *Directorate for Education and Human Resources, American Association for the Advancement of Science*

Virginia Malone *Psychological Corporation, San Antonio, TX*

Cora E. Marrett *Departments of Afro-American Studies and Sociology, University of Wisconsin, Madison*

James F. Mason *Mathematics and Computer Education, San Diego State University*

Miriam J. Masullo *Computer Sciences Department, IBM Corporation*

Marsha L. Matyas *Directorate for Education and Human Resources, American Association for the Advancement of Science*

Marian P. Meyer *Department of Zoology, University of Wisconsin, Madison*

James Minstrell *Mercer Island High School, Washington*

Rob Newberry *Department of Physics, Texas A&M University*

Fred M. Newmann *Center for Educational Research and Department of Curriculum and Instruction, University of Wisconsin, Madison*

Gilbert S. Omenn *Department of Medicine and Environmental Health, University of Washington*

Maria de Los Angeles Ortiz *Ana G. Mendez Educational Foundation, Puerto Rico*

Meredith E. Ostrom *Department of Geology and Geophysics, University of Wisconsin, Madison*

Kathleen O'Sullivan *School of Education, San Francisco State University*

Michael Padilla *College of Education, University of Georgia*

Walter Parker *Department of Curriculum and Instruction, University of Washington, Seattle*

Robert Pois *Department of History, University of Colorado, Boulder*

Richard A. Pollack *Emerging Technologies Inc., St. Paul, MN*

David M. Prescott *Molecular, Cellular and Developmental Biology, University of Colorado, Boulder*

Philip J. Regal *Department of Zoology, University of Minnesota, Minneapolis*

Thomas A. Romberg *Department of Curriculum and Instruction, University of Wisconsin, Madison*

Kathleen Roth *College of Education, Michigan State University*

Mary Budd Rowe *College of Education, Stanford University*

Thomas P. Sachse *California Department of Education*

Alan Sandler *American Institute of Architects, Washington, DC*

Mario Salvadori *School of Architecture and Engineering, Columbia University*

Mark Schug *Department of Curriculum and Instruction, University of Wisconsin, Madison*

Sally Schuler *National Science Resources Center, Washington DC*

Ethel L. Schultz *Salem, MA*

Linda Scott *Department of Education, St. Cloud State University*

Steve Seidel *Project Zero, Harvard University*

Robert Siegfried *Department of History of Science, University of Wisconsin, Madison*

Deborah Smith *Office of the Director of Education, University of Delaware*

Cary Sneider *Lawrence Hall of Science, University of California, Berkeley*

Randall Souviney *Teacher Education Program, University of California, San Diego*

Barbara Spector *College of Education, University of South Florida*

Elizabeth K. Stage *California Science Project, University of California*

Fred M. Stein *Philadelphia Renaissance in Science and Mathematics*

Karen L. Steudel *Department of Zoology, University of Wisconsin, Madison*

David Sugg *Target 90, San Antonio, TX*

Anne Taylor *Department of Architecture and Planning, University of New Mexico*

John Taylor *Department of Physics, University of Colorado, Boulder*

Karen Testoni *Quantitative Literacy Project, Perry Hall, MD*

Jerry Thompson *International Association of Machinists and Aerospace Workers, Washington, DC*

Robert F. Tinker *Technical Education Research Center, Cambridge, MA*

Ronald Todd *TIES Magazine, Drexel University*

James Varnadore *San Diego, CA*

Decker F. Walker *School of Education, Stanford University*

John Watkins *The NETWORK, Inc., Andover, MA*

Jeannette Wedel *Development Office, American Association for the Advancement of Science*

Richard Weitz *Department of Physics, Texas A&M University*

Kenneth D. Welty *Technology Teacher Education Program, University of Wisconsin, Stout*

Mariamne H. Whatley *Department of Curriculum and Instruction, University of Wisconsin, Madison*

Grayson Wheatley *Department of Curriculum and Instruction, Florida State University*

Paul H. Williams *Department of Plant Pathology, University of Wisconsin, Madison*

Elaine Wizda *Division of Instruction, Maryland State Department of Education*

Karen Worth *Educational Development Center, Newton, MA*

Dave Youngs *Graduate Mathematics/Science Department, Fresno Pacific College*

* *Affiliation is at the time of contribution.*

Index

CREDITS

This book was set in a version of a type designed by William Caslon (1692-1766).
Caslon type has a special place in American history as it was used
in the first printing of the Declaration of Independence.

❧

Designed by
ALBERT GREGORY

Printed by
R. R. DONNELLEY & SONS

Since its founding in 1848, the American Association for the Advancement of Science has continually worked to advance science. From its early, specific aims concerned with communication and cooperation among scientists, the Association's goals now encompass the broader purposes of ". . . furthering the work of scientists, facilitating cooperation among them, fostering scientific freedom and responsibility, improving the effectiveness of science in the promotion of human welfare, advancing education in science, and increasing the public understanding and appreciation of the importance of the methods of science in human progress."

Beginning with 461 founding members in 1848, AAAS now enrolls over 140,000 scientists, engineers, science educators, policymakers, and others interested in science and technology who live and work in the United States and in many other countries throughout the world. In addition, AAAS is the world's largest federation of scientific and engineering societies, with 292 organizations that cooperate with the Association on a variety of projects, including Annual Meeting symposia, fellowships, international programs, annual analysis of the federal research and development budget, equal opportunity activities, and science education.

AAAS has been continuously engaged in elementary and secondary science education since 1955. Starting with *Science—A Process Approach* in the Sputnik era, the Association has developed a large array of publications and projects to advance science literacy in the nation. Some of the current publications are:

Science Books & Films, published nine times a year, provides critical reviews of the scientific accuracy and quality of presentation of print, audiovisual, and electronic resources intended for use in science, mathematics, and technology education.

Sourcebook for Science, Mathematics, & Technology Education, published annually, provides the names and addresses of leaders in associations, scientific academies, museums, educational research centers, funded projects, and state and federal agencies.

Science Education News, published eight times a year, carries information on current reform initiatives in science, mathematics, and technology education, and *2061 TODAY,* published twice yearly, follows the progress of Project 2061.

This Year in School Science is a monograph that is published each year in conjunction with the **Forum on School Science,** a national meeting of educators, scientists, and policy leaders convened by AAAS to explore together a topic such as technology, science literacy, or teacher preparation.

Books published by AAAS include, among others, *Best Books for Children, Science Assessment in the Service of Reform, Assessing Higher-Order Thinking in Mathematics, When Science Meets the Public,* and *The Liberal Art of Science.*

An Invitation

We would like you to help us revise *Benchmarks for Science Literacy*.

Are the Benchmarks clearly stated? Are they correctly placed by grade level? Are there some that should be eliminated? Are some additional ones needed? What would it take to make the commentary more helpful? Did we miss some important research? How can *Benchmarks* be made more convenient to use?

The more specific your suggestions, the better. If you believe certain Benchmarks should be moved, say which ones and why. Rewrite Benchmarks that are unsatisfactory as they stand but should be saved. Where passages in the commentary are not clear or well put, indicate the wording that you would like to see. Cite articles that we should know about. And so forth. But if your concerns are more general or philosophical, let us know that, too.

We do not want to lose what is best about this report while trying to improve it. We would, therefore, like to know if there are some parts or aspects of *Benchmarks for Science Literacy* that you believe should not be changed under any circumstances.

Project 2061 is happy to receive your suggestions for improving *Benchmarks* in any form that is convenient for you—hand-written notes, letters, or photocopies of marked-up pages. If you are responding as an individual, please indicate in what capacity, such as elementary-school teacher, middle-school principal, curriculum specialist, scientist, parent, student, etc. If you are responding on behalf of a group, we would like to know the group's composition and purpose.

Your response should be sent to:

Project 2061 BENCHMARKS
AAAS
1333 H Street, N.W.
Washington, D.C. 20005